Meromorphic Functions over Non-Archimedean Fields

Mathematics and Its Applications

Managing Editor:

M. HAZEWINKEL

Centre for Mathematics and Computer Science, Amsterdam, The Netherlands

Volume 522

Meromorphic Functions over Non-Archimedean Fields

by

Pei-Chu Hu

Shandong University,
Shandong, P.R. China

and

Chung-Chun Yang

University of Science and Technology,
Clearwater Bay, Hong Kong

KLUWER ACADEMIC PUBLISHERS
DORDRECHT / BOSTON / LONDON

A C.I.P. Catalogue record for this book is available from the Library of Congress.

ISBN 978-90-481-5546-0

Published by Kluwer Academic Publishers,
P.O. Box 17, 3300 AA Dordrecht, The Netherlands.

Sold and distributed in North, Central and South America
by Kluwer Academic Publishers,
101 Philip Drive, Norwell, MA 02061, U.S.A.

In all other countries, sold and distributed
by Kluwer Academic Publishers,
P.O. Box 322, 3300 AH Dordrecht, The Netherlands.

Printed on acid-free paper

Contents

Preface

Nevanlinna theory (or value distribution theory) in complex analysis is so beautiful that one would naturally be interested in determining how such a theory would look in the non-Archimedean analysis and Diophantine approximations. There are two "main theorems" and defect relations that occupy a central place in Nevanlinna theory. They generate a lot of applications in studying uniqueness of meromorphic functions, global solutions of differential equations, dynamics, and so on. In this book, we will introduce non-Archimedean analogues of Nevanlinna theory and its applications.

In value distribution theory, the main problem is that given a holomorphic curve f : $\mathbb{C} \longrightarrow M$ into a projective variety M of dimension n and a family $\mathscr{A}$ of hypersurfaces on M, under a proper condition of non-degeneracy on f, find the defect relation. If $\mathscr{A}$ is a family of hyperplanes on $M = \mathbb{P}^n$ in general position and if the smallest dimension of linear subspaces containing the image $f(\mathbb{C})$ is k, Cartan conjectured that the bound of defect relation is $2n - k + 1$. Generally, if $\mathscr{A}$ is a family of admissible or normal crossings hypersurfaces, there are respectively Shiffman's conjecture and Griffiths-Lang's conjecture. Here we list the process of this problem:

A. Complex analysis:

(i) Constant targets: R. Nevanlinna[98] for $n = k = 1$; H. Cartan [20] for $n = k > 1$; E. I. Nochka [99], [100],[101] for $n > k \geq 1$; Shiffman's conjecture partially solved by Hu-Yang [71]; Griffiths-Lang's conjecture (open).

(ii) Moving targets: N. Steinmetz [126] for $n = k = 1$; M. Ru and W. Stoll [112] for $n = k > 1$; M. Ru and W. Stoll [113] for $n > k \geq 1$; Shiffman's conjecture (open); Griffiths-Lang's conjecture (open).

B. Non-Archimedean analysis:

(i) Constant targets: H. K. Hà [44], H. K. Hà and V. Q. My [50], and A. Boutabaa [11] for $n = k = 1$; H. K. Hà and V. T. Mai [49], W. Cherry and Zh. Ye [26] for $n = k > 1$; P. C. Hu and C. C. Yang for $n > k \geq 1$; Shiffman's conjecture solved by Hu-Yang (see chapter 6); Griffiths-Lang's conjecture (open).

(ii) Moving targets: P. C. Hu and C. C. Yang [57] for $n = k \geq 1$; P. C. Hu and C. C. Yang for $n > k \geq 1$; other (open).

C. Diophantine analysis:

(i) Constant targets: Roth's theorem (cf. [133]) for $n = k = 1$; Schmidt's subspace theorem (cf. [133]) for $n = k \geq 1$; M. Ru and P. M. Wong [115] for $n > k \geq 1$; other (open).

(ii) Moving targets: P. Vojta [134] for $n = k = 1$; M. Ru and P. Vojta [114] for $n \geq k \geq 1$; other (open).

In chapter 2, 6, 7 and appendix A, we will discuss the main problem mentioned before for the cases in non-Archimedean Analysis and Diophantine Analysis.

In [62], we have unified Wiman-Valiron theory and Nevanlinna theory for meromorphic functions over non-Archimedean fields. Value distribution of differential polynomials also is an important topic in Nevanlinna theory. Also in [61], we have proved a general second main theorem dealing with differential polynomials which is an unification of the second main theorems of Nevanlinna, Milloux, Xiong and Hu. We obtain a non-Archimedean analogue of the result (see Theorem 2.47). The facts are reviewed in chapter 2 after given some basic notations and facts of non-Archimedean fields and meromorphic functions in chapter 1.

Gross and Yang [42] first introduced unique range sets for meromorphic functions (abbreviated URSM) on $\mathbb{C}$, and given a unique range set of entire functions (abbreviated URSE) on $\mathbb{C}$ with infinitely many elements. Recently, URSE and also URSM with finitely many elements have been found by Yi ([140], [141]), Li and Yang ([86], [87]), Mues and Reinders [96], and Frank and Reinders [33]. The non-Archimedean analogue of the uniqueness theory of meromorphic (or entire) functions and URSM or URSE are surveyed in chapter 3.

There are a lot of results on meromorphic solutions, in particular, on Malmquist type theorems, of algebraic differential equations. For example, see [37], [57], [65]-[67], [80], [81], [88], [108], [129], [131], [135] and [144]. In chapter 4, we will prove non-Archimedean analogues related to Malmquist type theorems, and show that some non-Archimedean algebraic differential equations have no admissible transcendental meromorphic solutions. In chapter 5, we will give Fatou-Julia type theory for non-Archimedean entire (or rational) functions by proving the Riemann-Hurwitz relation, theorems on fixed points, Montel's theorems, and so on.

This is a unique book for introducing non-Archimedean value distribution theory in which we will systematically discuss the theory and its applications mentioned above. Each chapter is self-contained and this book is appended with a comprehensive and up-dated bibliography. It is hoped that this book will provide some new research directions that follow unified approaches as well as challenging problems in studying non-Archimedean value distribution theory. Here we mention that some important physical, biological and social models can be constructed on the base of non-Archimedean analysis, but it is not in the aims of our book (cf. [74], [75]).

We wish to thank Hong Kong's Research Grants Council (R. G. C), China's Natural Science Foundation and Outstanding Youth Fund of Shandong University for their financial supports, which enabled us to do research jointly for past years and complete the writing of the book at the Hong Kong University of Science & Technology—a scenic campus with excellent library and computer facilities. Last but not the least, we are indebted to our wives, Jin and Chwang-Chia, for their patience and support throughout this project.

Pei-Chu Hu & Chung-Chun Yang
Hong Kong, 1999

Chapter 1

Basic facts in p-adic analysis

In this chapter, we will introduce some notations, terminologies and basic facts used in this book.

1.1 p-adic numbers

We will denote the fields of complex, real, and rational numbers by $\mathbb{C}$, $\mathbb{R}$, and $\mathbb{Q}$, respectively, and let $\mathbb{Z}$ be the ring of integers. If κ is a subset of $\mathbb{R}$, we write

$$\kappa_+ = \{x \in \kappa \mid x \geq 0\}, \quad \kappa^+ = \{x \in \kappa \mid x > 0\}.$$

For $a, b \in \kappa$ with $a \leq b$, we also write

$$\kappa[a, b] = \{x \in \kappa \mid a \leq x \leq b\}.$$

Let κ be a field and denote the multiplicative group $\kappa - \{0\}$ by κ_*.

Definition 1.1. *An absolute value on a field κ is a function*

$$|\cdot| : \kappa \longrightarrow \mathbb{R}_+ = [0, +\infty)$$

that satisfies the following conditions:
   1) $|x| = 0$ iff $x = 0$;
   2) $|xy| = |x||y|$ for all $x, y \in \kappa$;
   3) $|x + y| \leq |x| + |y|$ for all $x, y \in \kappa$,
and is said to be non-Archimedean if it satisfies the additional condition:
   4) $|x + y| \leq \max\{|x|, |y|\}$ for all $x, y \in \kappa$;
otherwise, it is called Archimedean.

Let κ be a field with an absolute value $|\cdot|$. The absolute value $|\cdot|$ is said to be *trivial* if

$$|x| = \begin{cases} 1 & : \quad x \in \kappa_* \\ 0 & : \quad x = 0, \end{cases}$$

and *dense* if the set

$$|\kappa| = \{|x| \mid x \in \kappa\}$$

1

is dense in $\mathbb{R}_+$. If $|\cdot|$ is non-Archimedean, in fact we have

$$|x + y| = \max\{|x|, |y|\},$$

if $|x| \neq |y|$. Generally, an absolute value $|\cdot|$ on κ induces a *distance function d* defined by

$$d(x, y) = |x - y|,$$

for any two elements $x, y \in \kappa$, and hence induces a topology on κ. For a positive real number r and a point $x \in \kappa$, denote the open and closed balls of radius r centered at x, respectively, by

$$\kappa(x; r) = \{y \in \kappa \mid d(x, y) < r\}, \quad \kappa[x; r] = \{y \in \kappa \mid d(x, y) \leq r\},$$

and denote the circle by

$$\kappa\langle x; r\rangle = \{y \in \kappa \mid d(x, y) = r\} = \kappa[x; r] - \kappa(x; r).$$

By using the distance, then $|\cdot|$ is non-Archimedean iff the induced metric satisfies

$$d(x, y) \leq \max\{d(x, z), d(z, y)\}$$

for any $x, y, z \in \kappa$. For this case, we know that

$$y \in \kappa[x; r] \Longrightarrow \kappa[y; r] = \kappa[x; r];$$

$$y \in \kappa\langle x; r\rangle \Longrightarrow \kappa(y; r) \subset \kappa\langle x; r\rangle.$$

Thus $\kappa[x; r]$, $\kappa(x; r)$ and $\kappa\langle x; r\rangle$ all are open and closed. Such sets usually are called *clopen*.

For a general metric space M, instead of the standard triangle inequality

$$d(x, y) \leq d(x, z) + d(z, y), \quad x, y, z \in M,$$

if the strong triangle inequality

$$d(x, y) \leq \max\{d(x, z), d(z, y)\}, \quad x, y, z \in M,$$

is valid, the metric d is called an *ultrametric*, and M is called an *ultrametric space*. In an ultrametric space M, two balls U and V have only two possibilities: (i) balls are ordered by inclusion, that is, $U \subset V$ or $V \subset U$; (ii) balls are disjoint. By using this fact, it is easy to show that any nonempty open subset D of M has a partition into balls.

Two absolute values on κ are said to be *equivalent* if they induce the same topology on κ. We have the following more accessible criteria (cf. [39]):

Lemma 1.2. *Let $|\cdot|_1$ and $|\cdot|_2$ be absolute values on a field κ. The following statements are equivalent:*
  1) $|\cdot|_1$ and $|\cdot|_2$ are equivalent absolute values;
  2) $|x|_1 < 1$ iff $|x|_2 < 1$ for any $x \in \kappa$;
  3) there exists a positive real number α such that for each $x \in \kappa$, one has $|x|_1 = |x|_2^\alpha$.

For a real constant $p > 1$, a function $v_p : \kappa \longrightarrow \mathbb{R} \cup \{+\infty\}$ is defined by

$$v_p(x) = \begin{cases} -\log_p |x| & : \quad x \in \kappa_* \\ +\infty & : \quad x = 0 \, , \end{cases}$$

where $\log_p$ is the real logarithm function of base p and is named the *(additive) valuation associated to the absolute value* $|\cdot|$. Setting $p = e^\alpha$ for some $\alpha > 0$, we have $v_e = \alpha v_p$. By Lemma 1.2, one has

Lemma 1.3. *Let κ be a field with two absolute values whose associated valuations are v and w, respectively. They are equivalent iff there exists $\alpha > 0$ such that $v = \alpha w$.*

A non-Archimedean absolute value can be characterized as follows:

Lemma 1.4. *An absolute value on a field κ is non-Archimedean iff its associated valuation v satisfies the following conditions:*
1) $v(x) = +\infty$ iff $x = 0$;
2) $v(xy) = v(x) + v(y)$ for all $x, y \in \kappa$;
3) $v(x + y) \geq \min\{v(x), v(y)\}$ for all $x, y \in \kappa$.

The image of κ_* by the valuation v_p associated with the absolute value $|\cdot|$ is a subgroup of the additive group $(\mathbb{R}, +)$ called the *valuation group of κ*. The valuation of κ is said to be *discrete* (resp., *dense*) if its valuation group is a discrete (resp., dense) subgroup of $\mathbb{R}$.

Lemma 1.5. *If κ is algebraically closed, then the valuation v_p associated with a non-trivial absolute value $|\cdot|$ is dense and hence the absolute value $|\cdot|$ also is dense.*

Proof. Since $|\cdot|$ is non-trivial, then there is an element $a \in \kappa_*$ with $|a| \neq 0, 1$. We may assume $0 < |a| < 1$. Since κ is algebraically closed, then the following equation

$$z^n - a^q \ (n \in \mathbb{Z}^+, \ q \in \mathbb{Z})$$

always has solutions. Thus, we have

$$v_p(z) = \frac{q}{n} v_p(a) \in v_p(\kappa_*).$$

Obviously, the set

$$\left\{ \frac{q}{n} v_p(a) \mid n \in \mathbb{Z}^+, \ q \in \mathbb{Z} \right\}$$

is dense in $\mathbb{R}$. Thus, the valuation v_p is dense. $\square$

There is another characterization for non-Archimedean absolute values. We begin by noting that for any field κ there is a mapping $\mathbb{Z} \longrightarrow \kappa$ defined by

$$n \mapsto \begin{cases} \underbrace{1 + 1 + \cdots + 1}_{n} & : \quad n > 0 \\ 0 & : \quad n = 0 \\ -\underbrace{(1 + 1 + \cdots + 1)}_{-n} & : \quad n < 0 \, , \end{cases}$$

where 1 is the unit of κ. We will identify $\mathbb{Z}$ with its image under the mapping in κ. Then an absolute value $|\cdot|$ on κ is Archimedean iff

$$\sup_{n \in \mathbb{Z}} |n| = +\infty,$$

and is non-Archimedean iff

$$\sup_{n \in \mathbb{Z}} |n| = 1.$$

If $|\cdot|$ is non-Archimedean, the subset

$$\mathcal{O}_\kappa = \kappa[0; 1] = \{x \in \kappa \mid |x| \leq 1\}$$

is a subring of κ that is called the *valuation ring* of $|\cdot|$. Its subset $\kappa(0; 1)$ is an ideal of $\mathcal{O}_\kappa$, which is called the *valuation ideal* of $|\cdot|$. Furthermore, $\kappa(0; 1)$ is a maximal ideal in $\mathcal{O}_\kappa$, and every element of the complement $\kappa\langle 0; 1\rangle$ is invertible in $\mathcal{O}_\kappa$. Rings that contain a unique maximal ideal whose complement consists of invertible elements are called *local rings*. Thus, $\mathcal{O}_\kappa$ is a local ring of κ. The field

$$\mathbb{F}(\kappa) = \mathcal{O}_\kappa / \kappa(0; 1)$$

is called the *residue field* of κ. The characteristic of $\mathbb{F}(\kappa)$ is named the *residue characteristic* of κ.

Let $p \in \mathbb{Z}$ be a prime number. Any integer $a \in \mathbb{Z} - \{0\}$ can be expressed uniquely as

$$a = p^v a', \quad p \nmid a', \quad a' \in \mathbb{Z} - \{0\},$$

where v is uniquely determined by p and a, which is just the multiplicity of p as a divisor of a, and is denoted by $v_p(a)$. Thus we obtain a function

$$v_p : \mathbb{Z} - \{0\} \longrightarrow \mathbb{Z}_+ = \mathbb{Z} \cap [0, +\infty),$$

which can be extended to $\mathbb{Q}$ as follows: if $x = a/b \in \mathbb{Q}$, set

$$v_p(x) = \begin{cases} v_p(a) - v_p(b) & : \ x \neq 0 \\ +\infty & : \ x = 0 . \end{cases}$$

Obviously, the function v_p on $\mathbb{Q}$ satisfies the condition in Lemma 1.4 and is called the *p-adic valuation* on $\mathbb{Q}$. It has the following property:

$$x = p^{v_p(x)} \cdot \frac{a'}{b'}, \quad p \nmid a'b', \quad x \in \mathbb{Q}.$$

Thus for any $x \in \mathbb{Q}$, one obtains the p-adic absolute value of x defined by

$$|x|_p = \begin{cases} p^{-v_p(x)} & : \ x \neq 0 \\ 0 & : \ x = 0 . \end{cases}$$

The function $|\cdot|_p$ is a non-Archimedean absolute value on $\mathbb{Q}$, called the *p-adic absolute value*.

For a number field κ, an embedding $\phi : \kappa \longrightarrow \mathbb{C}$ causes the absolute value on $\mathbb{C}$ to induce an absolute value on κ; such absolute values are called *infinite* and are written as $|\cdot|_\infty$, which obviously are Archimedean.

Theorem 1.6 (Ostrowski). *Every non-trivial absolute value on* $\mathbb{Q}$ *is equivalent to one of the absolute values* $|\cdot|_p$, *where either* p *is a prime number or* $p = \infty$.

Based on the theorem, (non-Archimedean or Archimedean) absolute values on number fields are also sometimes called (*finite* or *infinite*) *places* or *primes*. Thus for any $x \in \mathbb{Q}_*$, the *product formula*

$$\prod_{p \leq \infty} |x|_p = 1$$

holds, where $p \leq \infty$ means that we take the product over all of the primes of $\mathbb{Q}$ including the "prime at infinity".

According to the standard theory in p-adic analysis, the completion of $\mathbb{Q}$ relative to the topology induced by $|\cdot|_p$ is a field that is denoted by $\mathbb{Q}_p$, and the absolute value $|\cdot|_p$ on $\mathbb{Q}$ extends to a non-Archimedean absolute value on $\mathbb{Q}_p$, which also is denoted by $|\cdot|_p$, such that

i) there exists an inclusion $\mathbb{Q} \hookrightarrow \mathbb{Q}_p$, and the absolute value induced by $|\cdot|_p$ on $\mathbb{Q}$ via the inclusion is the p-adic absolute value, we will from now on identify $\mathbb{Q}$ with its image under the inclusion in $\mathbb{Q}_p$;

ii) $\mathbb{Q}$ is dense in $\mathbb{Q}_p$; and

iii) $\mathbb{Q}_p$ is complete.

The field $\mathbb{Q}_p$ satisfying (i), (ii) and (iii) is unique up to unique isomorphism preserving the absolute values and is called the field of *p-adic numbers*, which also has the following property:

iv) for each $x \in \mathbb{Q}_{p*}$, there exists an integer $v_p(x)$ such that $|x|_p = p^{-v_p(x)}$, i.e., the p-adic valuation v_p on $\mathbb{Q}$ extends to $\mathbb{Q}_p$. In other words, the set of values of $\mathbb{Q}$ and $\mathbb{Q}_p$ under $|\cdot|_p$ is the same, which is equal to the set $\{p^n \mid n \in \mathbb{Z}\} \cup \{0\}$.

By (iv), it is easy to prove

$$\mathbb{Q}_p(x; r) = \mathbb{Q}_p\left[x; \frac{r}{p}\right], \quad x \in \mathbb{Q}_p, \ r \in \mathbb{R}^+.$$

Thus, the valuation ring

$$\mathcal{O}_{\mathbb{Q}_p} = \mathbb{Q}_p[0; 1] = \mathbb{Q}_p(0; p)$$

is both open and closed, which is also called the ring of *p-adic integers* and denoted by $\mathbb{Z}_p$. For any $n \in \mathbb{Z}^+$, the ring $\mathbb{Z}_p$ is covered by $\mathbb{Q}_p[k; p^{-n}] = k + p^n \mathbb{Z}_p$ ($k = 0, 1, ..., p^n - 1$). This means that $\mathbb{Z}_p$ is compact and, hence, $\mathbb{Q}_p$ is locally compact. Thus, one has

$$\mathbb{Z}_p/p^n \mathbb{Z}_p \cong \mathbb{Z}/p^n \mathbb{Z},$$

and the cosets of $p^n \mathbb{Z}_p$ in $\mathbb{Z}_p$ are also balls in the p-adic topology. The sets $\mathbb{Q}_p[0; p^{-n}] = p^n \mathbb{Z}_p$ ($n \in \mathbb{Z}$) form a fundamental system of neighborhoods of $0 \in \mathbb{Q}_p$ which covers $\mathbb{Q}_p$. The space $\mathbb{Q}_p$ is also totally disconnected, that is, the connected component of any point is the set consisting of only the point, but it is a Hausdorff topological space.

1.2   Field extensions

We will assume that κ is a field of characteristic zero so that it contains $\mathbb{Q}$.

Definition 1.7. *Let V be a vector space over κ and let $|\cdot|$ be an absolute value on κ. A norm on V is a function*

$$\|\cdot\| : V \longrightarrow \mathbb{R}_+$$

that satisfies the following conditions:
  1) $\|X\| = 0$ *if and only if* $X = 0$;
  2) $\|X + Y\| \leq \|X\| + \|Y\|$ *for all* $X, Y \in V$;
  3) $\|aX\| = |a| \cdot \|X\|$ *for all* $a \in \kappa$ *and all* $X \in V$.
A vector space V with a norm is called a normed vector space over κ.

Let V be a normed vector space. Then any norm $\|\cdot\|$ induces a metric d as follows

$$d(X, Y) = \|X - Y\|,$$

which makes V a topological space. We say two norms $\|\cdot\|_1$ and $\|\cdot\|_2$ on V are *equivalent* if they define the same topology on V. It is equivalent to that there are positive real numbers c_1 and c_2 such that

$$c_1 \|\cdot\|_1 \leq \|\cdot\|_2 \leq c_2 \|\cdot\|_1.$$

If V is a finite-dimensional and if κ is complete, then any two norms on V are equivalent, and V is complete with respect to the metric induced by any norm. Further, if κ is locally compact, then V also is locally compact.

Let K be a *field extension* of κ, that is, K is a field that contains κ as a subfield. Then the field K can always be regarded as a vector space over κ. The dimension $\dim_\kappa K$ of K as an κ-vector space is called the *degree* of K over κ. It will be denoted by

$$[K : \kappa] = \dim_\kappa K.$$

If $[K : \kappa] < \infty$, then K is said to be a *finite field extension* of κ. Take an element α in K. The field extension of κ, which is generated by α, will be denoted by $\kappa(\alpha)$, that is, $\kappa(\alpha)$ is the smallest field containing κ and α. We denote the ring generated by α over κ by $\kappa[\alpha]$. It consists of all elements of K that can be written as polynomials in α with coefficients in κ:

$$a_n \alpha^n + \cdots + a_1 \alpha + a_0, \quad a_i \in \kappa. \tag{1.1}$$

The field $\kappa(\alpha)$ is isomorphic to the field of fractions of $\kappa[\alpha]$. Its elements are ratios of elements of the form (1.1). The element α is said to be *algebraic over κ* if it is the root of some nonzero polynomial with coefficients in κ. Assume that α is algebraic over κ. The lowest degree irreducible monic polynomial with coefficients in κ such that $f(\alpha) = 0$ is called the *minimal polynomial for α over κ*. Thus, the first important property of the degree is that $[\kappa(\alpha) : \kappa]$ is the degree of the minimal polynomial for α over κ, which is also called the *degree of α over κ*. The second important property of the degree is that it is multiplicative in towers of fields, that is, if there are three fields $\kappa \subset F \subset K$, then

$$[K : \kappa] = [K : F][F : \kappa]. \tag{1.2}$$

One important case of a tower of field extensions is that K is a given extension of κ and α is an element of K. Then the field $\kappa(\alpha)$ is an intermediate field:

$$\kappa \subset \kappa(\alpha) \subset K.$$

Thus, one has

$$[K : \kappa] = [K : \kappa(\alpha)][\kappa(\alpha) : \kappa].$$

Note that $[\kappa(\alpha) : \kappa]$ is the degree of α over κ if α is algebraic, otherwise $[\kappa(\alpha) : \kappa] = \infty$. Then, if K is a finite field extension of κ, the element α of K is algebraic. A field extension K of κ is called an *algebraic extension*, or K is said to be *algebraic over* κ, if all its elements are algebraic. Thus, each finite field extension of κ is algebraic over κ. A field extension K of κ is said to be an *algebraic closure* of κ if K is algebraic over κ, and K is *algebraically closed*, that is, every polynomial $f(x) \in K[x]$ of positive degree has a root in K. By using Zorn's lemma, one can prove that every field κ has an algebraic closure, and that if K_1, K_2 are two algebraic closures of κ, there is an isomorphism $\sigma : K_1 \longrightarrow K_2$, which is the identity mapping on κ. Thus the algebraic closure is essentially unique.

An absolute value $|\cdot|_K$ of K is said to be an extension of the absolute value $|\cdot|$ of κ if

$$|x|_K = |x|, \quad x \in \kappa.$$

Obviously, if $|\cdot|_K$ is an extension of $|\cdot|$, then $|\cdot|_K$ is also a norm on K as a κ-vector space, and, further, if $|\cdot|$ is non-Archimedean, then $|\cdot|_K$ has to be non-Archimedean because this depends only on the absolute values of the elements of $\mathbb{Z}$, which are in κ.

Lemma 1.8. *If K is a finite field extension of κ, and if κ is complete, then there is at most one absolute value on K extending the absolute value of κ.*

Proof. Suppose $|\cdot|_1$ and $|\cdot|_2$ are two absolute values on K that extend the absolute value $|\cdot|$ of κ. For any $x \in K$, it is easy to show that $x^n \to 0$ with respect to the topology induced by $|\cdot|_1$ (resp., $|\cdot|_2$) if and only if $|x|_1 < 1$ (resp., $|x|_2 < 1$). Note that $|\cdot|_1$ and $|\cdot|_2$ are equivalent as norms on the κ-vector space K, i.e., they define the same topology. Therefore, we have convergence with respect to $|\cdot|_1$ if and only if we have convergence with respect to $|\cdot|_2$, or $|x|_1 < 1$ if and only if $|x|_2 < 1$. Hence $|\cdot|_1$ and $|\cdot|_2$ also are two equivalent absolute values. Thus, there exists a positive real number α such that for each $x \in K$, one has $|x|_1 = |x|_2^\alpha$, and hence $\alpha = 1$ since $|x|_1 = |x|_2 = |x|$ when $x \in \kappa$, i.e., the two absolute values are the same. $\qquad \square$

Let K be a finite field extension of κ of characteristic zero, which will be denoted by K/κ. Take $\alpha \in K$. Then α induces a κ-linear mapping

$$\mathbf{A}_\alpha : K \longrightarrow K$$

defined by $\mathbf{A}_\alpha x = \alpha x$. If $P_\alpha \in \kappa[x]$ is the minimal polynomial for α over κ, then one has

$$\det(\mathbf{A}_\alpha) = \left\{ (-1)^d P_\alpha(0) \right\}^{[K:\kappa(\alpha)]} = \left\{ (-1)^d P_\alpha(0) \right\}^{\frac{[K:\kappa]}{d}},$$

where $d = \deg(P_\alpha) = [\kappa(\alpha) : \kappa]$. We will denote the element of κ by $\mathbf{N}_{K/\kappa}(\alpha)$, and hence obtain a mapping

$$\mathbf{N}_{K/\kappa} : K \longrightarrow \kappa$$

called the *norm of K over κ*, which satisfies

$$\mathbf{N}_{K/\kappa}(\alpha\beta) = \mathbf{N}_{K/\kappa}(\alpha)\mathbf{N}_{K/\kappa}(\beta), \quad \alpha, \beta \in K.$$

If we have three fields $\kappa \subset F \subset K$, then (1.2) implies

$$\mathbf{N}_{F/\kappa} \circ \mathbf{N}_{K/F} = \mathbf{N}_{K/\kappa}.$$

Let $\mathbf{C}$ be any algebraically closed field containing κ. The field extension K/κ is said to be *normal* if all the injective homomorphisms $\sigma : K \longrightarrow \mathbf{C}$ which induce the identity on κ have the same image. We know that for any finite field extension F/κ, there exists a normal finite field extension of κ containing F. The smallest such is said to be *normal closure* of F/κ. Now let K/κ be normal and identify K with one of its images $\sigma(K)$. Then we can think of the mappings σ as automorphisms $\sigma : K \longrightarrow K$, which induces the identity on κ. It is known that the automorphisms of K form a finite group whose order is equal to $[K : \kappa]$, which is called the *Galois group* of the field extension and is denoted by $G_{K/\kappa}$. Then for $\alpha \in K$, one has

$$\mathbf{N}_{K/\kappa}(\alpha) = \prod_{\sigma \in G_{K/\kappa}} \sigma(\alpha).$$

Let κ be complete and let $|\cdot|_K$ be the unique absolute value on K extending the absolute value of κ. Note that each $\sigma \in G_{K/\kappa}$ induces an absolute value $|\cdot|_{K,\sigma}$ on K defined by

$$|x|_{K,\sigma} = |\sigma(x)|_K, \quad x \in K.$$

Since the restriction of σ on κ is the identity, then $|\cdot|_{K,\sigma}$ also is an extension of the absolute value of κ, and hence $|\cdot|_{K,\sigma} = |\cdot|_K$ for each $\sigma \in G_{K/\kappa}$. Therefore for any $x \in K$, one has

$$\left|\mathbf{N}_{K/\kappa}(x)\right| = \left|\prod_{\sigma \in G_{K/\kappa}} \sigma(x)\right| = |x|_K^{[K:\kappa]},$$

or, taking the root,

$$|x|_K = \left|\mathbf{N}_{K/\kappa}(x)\right|^{1/[K:\kappa]}. \tag{1.3}$$

Actually, if κ is complete with respect to a non-trivial non-Archimedean absolute value $|\cdot|$, the function $|\cdot|_K : K \longrightarrow \mathbb{R}_+$ defined by (1.3) is a non-Archimedean absolute value on K that extends the absolute value of κ (see [32]). In particular, one obtains the following result (also see [39]):

Theorem 1.9. *Let $K/\mathbb{Q}_p$ be a finite field extension. The function $|\cdot|_K : K \longrightarrow \mathbb{R}_+$ defined by*

$$|x|_K = \left|\mathbf{N}_{K/\mathbb{Q}_p}(x)\right|_p^{1/[K:\mathbb{Q}_p]}$$

is a non-Archimedean absolute value on K that extends the p-adic absolute value of $\mathbb{Q}_p$.

This unique absolute value on K that extends the p-adic absolute value on $\mathbb{Q}_p$ is called the *p-adic absolute value* on K, which also is denoted by $|\cdot|_p$. It makes K a locally compact topological field, and makes K complete. From the formula for the absolute value on K, we immediately see that for any $x \in K_*$,

$$|x|_p = p^{-v}, \quad v = \frac{1}{[K:\mathbb{Q}_p]} v_p(\mathbf{N}_{K/\mathbb{Q}_p}(x)) \in \frac{1}{[K:\mathbb{Q}_p]} \mathbb{Z}.$$

The unique rational number v is also denoted by $v_p(x)$. Setting $v_p(0) = +\infty$, one obtains a valuation v_p on K, which also is called the *p-adic valuation* on K.

Now we show that the p-adic absolute value on $\mathbb{Q}_p$ can be extended to an algebraic closure $\overline{\mathbb{Q}}_p$ of $\mathbb{Q}_p$. In fact, given any $x \in \overline{\mathbb{Q}}_p$, then x is in the finite extension $\mathbb{Q}_p(x)$, and hence one can define $|x|_p$ by using the unique extension of the p-adic absolute value to $\mathbb{Q}_p(x)$. Therefore, one obtains a function

$$|\cdot| : \overline{\mathbb{Q}}_p \longrightarrow \mathbb{R}_+$$

that extends the p-adic absolute value on $\mathbb{Q}_p$, and it is easy to prove that this function is an absolute value. The unique absolute value on $\overline{\mathbb{Q}}_p$ is also called the p-adic absolute value. However, $\overline{\mathbb{Q}}_p$ is not complete with respect to the p-adic absolute value.

The completion of $\overline{\mathbb{Q}}_p$ relative to the topology induced by $|\cdot|_p$ is a field that is denoted by $\mathbb{C}_p$, and the absolute value $|\cdot|_p$ on $\overline{\mathbb{Q}}_p$ extends to a non-Archimedean absolute value on $\mathbb{C}_p$, which is also denoted by $|\cdot|_p$, such that
i) there exists an inclusion $\overline{\mathbb{Q}}_p \hookrightarrow \mathbb{C}_p$, and the absolute value induced by $|\cdot|_p$ on $\overline{\mathbb{Q}}_p$ via the inclusion is the p-adic absolute value; we will from now on identify $\overline{\mathbb{Q}}_p$ with its image under the inclusion in $\mathbb{C}_p$;
ii) $\overline{\mathbb{Q}}_p$ is dense in $\mathbb{C}_p$; and
iii) $\mathbb{C}_p$ is complete.
The field $\mathbb{C}_p$ satisfying (i), (ii) and (iii) is unique up to the unique isomorphism preserving the absolute values and is of the following properties:
iv) for each $x \in \mathbb{C}_{p*}$, there exists a rational number $v_p(x)$ such that $|x|_p = p^{-v_p(x)}$, i.e., the p-adic valuation v_p on $\overline{\mathbb{Q}}_p$ extends to $\mathbb{C}_p$, and the image of $\mathbb{C}_{p*}$ under v_p is $\mathbb{Q}$; and
v) $\mathbb{C}_p$ is algebraically closed, but it is not locally compact.

1.3 Maximum term of power series

Let κ be an algebraically closed field of characteristic zero, complete for a non-trivial non-Archimedean absolute value $|\cdot|$.

Definition 1.10. *Let $U \subset \kappa$ be an open set. A function $f : U \longrightarrow \kappa$ is said to be continuous at $z_0 \in U$ if for every $\varepsilon > 0$ there exists a $\delta > 0$ such that for every $z \in \kappa(z_0; \delta)$, one has $|f(z) - f(z_0)| < \varepsilon$. The function f is said to be differentiable at z_0 if the limit*

$$f'(z_0) = \lim_{h \to 0} \frac{f(z_0 + h) - f(z_0)}{h}$$

exists, and is called differentiable in U if it is differentiable at each point in U.

It is easy to prove that differentiable functions are continuous. In this section, we will study the functions defined by power series. To do this, we recall some basic facts.

Lemma 1.11. *A sequence $\{a_n\}$ in κ is a Cauchy sequence, and therefore convergent, if and only if it satisfies*

$$\lim_{n \to \infty} |a_{n+1} - a_n| = 0.$$

Proof. For any $m = n + k > n$, note that

$$
\begin{aligned}
|a_m - a_n| &= |(a_{n+k} - a_{n+k-1}) + (a_{n+k-1} - a_{n+k-2}) + \cdots + (a_{n+1} - a_n)| \\
&\leq \max\{|a_{n+k} - a_{n+k-1}|, |a_{n+k-1} - a_{n+k-2}|, \cdots, |a_{n+1} - a_n|\}.
\end{aligned}
$$

The result follows at once. $\square$

Corollary 1.12. *An infinite series $\sum_{n=0}^{\infty} a_n$ with $a_n \in \kappa$ is convergent if and only if*

$$\lim_{n \to \infty} a_n = 0,$$

in which case we have

$$\left| \sum_{n=0}^{\infty} a_n \right| \leq \max_n |a_n| .$$

Proof. Define

$$S_n = \sum_{k=0}^{n} a_k.$$

We know that the infinite series $\sum_{n=0}^{\infty} a_n$ converges if and only if $\{S_n\}$ converges, if and only if $\{S_n\}$ is a Cauchy sequence, if and only if the sequence $\{S_n - S_{n-1} = a_n\}$ converges by using Lemma 1.11. The estimate for the sum is a straight extension of the non-Archimedean inequality. $\square$

Consider a power series

$$f(z) = \sum_{n=0}^{\infty} a_n z^n, \quad (a_n \in \kappa). \tag{1.4}$$

Then we can give a value $\sum_{n=0}^{\infty} a_n z^n$ to $f(z)$ whenever an $z \in \kappa$ for which $|a_n z^n| \to 0$ as $n \to \infty$. Define the *radius ρ of convergence* by

$$\frac{1}{\rho} = \limsup_{n \to \infty} |a_n|^{\frac{1}{n}}.$$

If $\rho = 0$ (resp., $\rho = +\infty$), then the series converges only at $z = 0$ (resp., κ). If $0 < \rho < +\infty$, the series converges if $|z| < \rho$ and diverges if $|z| > \rho$. It is easy to prove that the function f defined by the power series is continuous on $\kappa(0; \rho)$ and can be extended to $\kappa[0; \rho]$ if $|a_n|\rho^n \to 0$ as $n \to \infty$. Assume $0 < \rho \leq +\infty$ and take $r \in \mathbb{R}$ with $0 < r < \rho$. Define the *maximum term*:

$$\mu(r, f) = \max_{n \geq 0} |a_n| r^n$$

with the associated the *central index*:

$$\nu(r, f) = \max_{n \geq 0}\{n \mid |a_n|r^n = \mu(r, f)\}.$$

We also write

$$\mu(0, f) = \lim_{r \to 0+} \mu(r, f), \quad \nu(0, f) = \lim_{r \to 0+} \nu(r, f).$$

Obviously, if $z \in \kappa$ satisfying $|z| \leq r$, one has

$$|f(z)| \leq \max_{n \geq 0} |a_n||z|^n \leq \mu(r, f).$$

Lemma 1.13. *The central index $\nu(r, f)$ increases as $r \to \rho$ and satisfies the formula:*

$$\log \mu(r, f) = \log |a_{\nu(0,f)}| + \int_0^r \frac{\nu(t, f) - \nu(0, f)}{t} dt + \nu(0, f) \log r \quad (0 < r < \rho).$$

Proof. Take $0 < r_1 < r_2 < \rho$ and set $\nu_1 = \nu(r_1, f)$. We will prove

$$|a_n|r_2^n < |a_{\nu_1}|r_2^{\nu_1},$$

for $n < \nu_1$, which implies $\nu(r_2, f) \geq \nu_1$. That is obvious if $a_n = 0$. If $a_n \neq 0$, noting that

$$|a_n|r_1^n \leq |a_{\nu_1}|r_1^{\nu_1},$$

we have

$$\log |a_n| + n \log r_1 \leq \log |a_{\nu_1}| + \nu_1 \log r_1,$$

i.e.,

$$\frac{\log |a_n| - \log |a_{\nu_1}|}{\nu_1 - n} \leq \log r_1 < \log r_2,$$

which implies our claim.

W.l.o.g, we prove the formula for the case $|f(0)| = |a_0| = 1$. Assume $\nu(t, f) = \nu_{k-1}$ for $t \in [r_{k-1}, r_k)$ $(k = 1, 2, ...)$, where

$$0 = \nu_0 < \nu_1 < \cdots; \quad 0 = r_0 < r_1 < \cdots .$$

We can easily find the following relations:

$$|a_{\nu_k}|r_k^{\nu_k} = |a_{\nu_{k-1}}|r_k^{\nu_{k-1}}, \quad k = 1, 2, ...,$$

by continuity of the function $|a_{\nu_k}|r^{\nu_k}$ at r_k. Note that when $r_n \leq r < r_{n+1}$,

$$\mu(r, f) = |a_{\nu_n}|r^{\nu_n}.$$

We obtain

$$\begin{aligned}
\int_0^r \frac{\nu(t, f)}{t} dt &= \sum_{k=1}^n \int_{r_{k-1}}^{r_k} \frac{\nu(t, f)}{t} dt + \int_{r_n}^r \frac{\nu(t, f)}{t} dt \\
&= \sum_{k=2}^n \nu_{k-1} \log \frac{r_k}{r_{k-1}} + \nu_n \log \frac{r}{r_n} \\
&= \log\{\frac{|a_{\nu_n}|r^{\nu_n}}{|a_{\nu_n}|r_n^{\nu_n}} \prod_{k=2}^n \frac{|a_{\nu_{k-1}}|r_k^{\nu_{k-1}}}{|a_{\nu_{k-1}}|r_{k-1}^{\nu_{k-1}}}\} \\
&= \log\{|a_{\nu_n}|r^{\nu_n}\} = \log \mu(r, f),
\end{aligned}$$

which is just the formula. $\qquad\qquad\qquad\qquad\qquad\qquad\qquad\qquad\qquad$ $\square$

Denote the ring of power series $f(z) = \sum a_n z^n$ ($a_n \in \kappa$) that satisfies the condition $\lim |a_n| r^n = 0$ by $\mathcal{A}_r(\kappa)$ and let $\mathcal{A}_{(r}(\kappa)$ be the set of the power series in z whose radius of convergence is superior or equal to r. Obviously, $f \in \mathcal{A}_{(r}(\kappa)$ if and only if $f \in \bigcap_{s<r} \mathcal{A}_s(\kappa)$. We will abbreviate

$$\mathcal{A}(\kappa) = \mathcal{A}_{(\infty}(\kappa).$$

It is easy to prove that the identity theorem holds for elements of $\mathcal{A}(\kappa)$, i.e., two elements of $\mathcal{A}(\kappa)$ that agree, say, on a disc, must agree everywhere.

Theorem 1.14. *For $r > 0$, the function $\mu(r, \cdot) : \mathcal{A}_r(\kappa) \longrightarrow \mathbb{R}_+$ satisfies the following properties:*

1) $\mu(r, f) = 0$ if and only if $f \equiv 0$;
2) $\mu(r, f + g) \le \max\{\mu(r, f), \mu(r, g)\}$; and
3) $\mu(r, fg) = \mu(r, f)\mu(r, g)$.

Proof. The statements (1) and (2) follow at once from the properties of the non-Archimedean absolute value. To prove (3), write

$$f(z) = \sum_{n=0}^{\infty} a_n z^n, \quad g(z) = \sum_{n=0}^{\infty} b_n z^n.$$

Then we have

$$f(z)g(z) = \sum_{n=0}^{\infty} c_n z^n, \quad c_n = \sum_{i+j=n} a_i b_j.$$

Therefore

$$\begin{aligned}
\mu(r, fg) &= \max_n |c_n| r^n \le \max_n \{\max_{i+j=n} |a_i||b_j| r^n\} \\
&= \max_n \{\max_{i+j=n} (|a_i| r^i)(|b_j| r^j) \le (\max_i |a_i| r^i)(\max_j |b_j| r^j) \\
&= \mu(r, f)\mu(r, g).
\end{aligned}$$

To get the converse inequality, choose I and J such that

$$|a_i| r^i < \mu(r, f) \ (i < I), \quad |a_I| r^I = \mu(r, f),$$

$$|b_j| r^j < \mu(r, g) \ (j < J), \quad |b_J| r^J = \mu(r, g).$$

Then one obtains

$$|a_I b_J| = r^{-I-J} \mu(r, f)\mu(r, g),$$

$$|a_i b_j| < r^{-I-J} \mu(r, f)\mu(r, g),$$

where $i < I$ or $j < J$, but $i + j = I + J$. This means that in the sum c_{I+J} there is one largest term, and hence

$$|c_{I+J}| = |a_I b_J| = r^{-I-J} \mu(r, f)\mu(r, g),$$

which we can rewrite as

$$|c_{I+J}|r^{I+J} = \mu(r, f)\mu(r, g).$$

Thus, we obtain the converse inequality:

$$\mu(r, fg) = \max_n |c_n|r^n \geq |c_{I+J}|r^{I+J} = \mu(r, f)\mu(r, g).$$

$\square$

Theorem 1.15. *Consider the power series (1.4) with a non-zero radius ρ of convergence and take $z \in \kappa$. If $f(z)$ converges, then $f'(z)$ exists such that*

$$f'(z) = \sum_{n=1}^{\infty} na_n z^{n-1}. \tag{1.5}$$

Moreover, the radius of convergence of the series (1.5) is the same as that of f, and f' satisfies

$$\mu(r, f') \leq \frac{1}{r}\mu(r, f) \quad (0 < r < \rho).$$

Proof. Set

$$g(z) = \sum_{n=1}^{\infty} na_n z^{n-1}.$$

By Lemma 1.2 and Theorem 1.6, there exists a positive real number α such that for each positive integer n, the inequality

$$\frac{1}{n^\alpha} \leq |n| \leq 1$$

holds. As a consequence, we have

$$\lim_{n \to +\infty} |n|^{\frac{1}{n}} = 1.$$

Therefore, it is easily seen that

$$\limsup_{n \to +\infty} |a_n|^{\frac{1}{n}} = \limsup_{n \to +\infty} |na_n|^{\frac{1}{n-1}},$$

and the series (1.5) has same radius of convergence as f. Suppose that $f(z)$ converges, which is equivalent to saying that $a_n z^n \to 0$. If $z = 0$, then it is clear that $g(z)$ converges. If $z \neq 0$, notice that

$$|na_n z^{n-1}| \leq |a_n z^{n-1}| = \frac{1}{|z|}|a_n z^n| \to 0,$$

and again we see that $g(z)$ converges.

Note that either $f(z)$ converges in $\kappa[0; \rho]$ or in $\kappa(0; \rho)$. In the first case, set $R = \rho$. In the second case, choose R such that $|z| \leq R < \rho$. If $z \neq 0$, assume that $|h| < |z| \leq R$. Otherwise, $z = 0$ and we assume $|h| \leq R$. Thus $|z + h| \leq \max\{|z|, |h|\} \leq R$ and $f(z + h)$ converges such that

$$\frac{f(z + h) - f(z)}{h} = \sum_{n=1}^{\infty} \sum_{j=1}^{n} \binom{n}{j} a_n z^{n-j} h^{j-1}.$$

Since $R < \rho$, we have

$$\left| \binom{n}{j} a_n z^{n-j} h^{j-1} \right| \leq |a_n| R^{n-1} \to 0.$$

This shows that the series converges uniformly in h. Hence we can take the limit term-by-term, which gives

$$f'(z) = \lim_{h \to 0} \frac{f(z+h) - f(z)}{h} = \sum_{n=1}^{\infty} n a_n z^{n-1}.$$

Finally, if $0 < r < \rho$, then

$$\mu(r, f') = \max_n |n a_n| r^{n-1} \leq \frac{1}{r} \max_n |a_n| r^n = \frac{1}{r} \mu(r, f).$$

$\square$

Corollary 1.16. *An element $f \in \mathcal{A}_\rho(\kappa)$ has a derivative identically equal to 0 if and only if it is equal to a constant.*

1.4   Weierstrass preparation theorem

For the discussion in this section, please refer to [39]. Let κ be an algebraically closed field of characteristic zero, complete for a non-trivial non-Archimedean absolute value $|\cdot|$.

Lemma 1.17. *For $r > 0$, $\mathcal{A}_r(\kappa)$ is complete with respect to the norm $\mu(r, \cdot)$.*

Proof. Given a Cauchy sequence of power series

$$f_i(z) = a_{i0} + a_{i1} z + a_{i2} z^2 + \cdots, \quad i = 1, 2, \ldots$$

in $\mathcal{A}_r(\kappa)$, i.e., for each $\varepsilon > 0$ there exists an N such that

$$\mu(r, f_i - f_j) = \max_n |a_{in} - a_{jn}| r^n < \varepsilon, \quad i, j > N.$$

In other words, $\{a_{in}\}_{i \geq 1}$ is a Cauchy sequence for each n. Since κ is complete, then they are all convergent. So let

$$a_n = \lim_{i \to \infty} a_{in}, \quad n = 0, 1, 2, \ldots,$$

and define a series

$$f(z) = a_0 + a_1 z + a_2 z^2 + \cdots.$$

Since $|a_{in} - a_{jn}| r^n < \varepsilon$ for $i, j > N$ and for each n, letting $j \to \infty$, it follows that if $i > N$, one has $|a_{in} - a_n| r^n \leq \varepsilon$ for each n, which means that

$$\mu(r, f_i - f) = \max_n |a_{in} - a_n| r^n \leq \varepsilon, \quad i > N,$$

so that $f_i \to f$ with respect to the norm $\mu(r, \cdot)$.

To prove $f \in \mathcal{A}_r(\kappa)$, note that $|a_{in}| r^n \to 0$ as $n \to \infty$, since $f_i \in \mathcal{A}_r(\kappa)$. That is, for each i there exists an N_i such that $|a_{in}| r^n < \varepsilon$ for all $n > N_i$. Fix any $i > N$. Then one has

$$|a_n| r^n \leq |a_{in} - a_n| r^n + |a_{in}| r^n < 2\varepsilon, \quad n > N_i.$$

It follows that $|a_n| r^n \to 0$ as $n \to \infty$, and hence $f \in \mathcal{A}_r(\kappa)$. $\square$

Lemma 1.18. *The ring $\kappa[z]$ of polynomials is dense in $\mathcal{A}_r(\kappa)$.*

Proof. Let
$$f(z) = a_0 + a_1 z + a_2 z^2 + \cdots$$
be a power series in $\mathcal{A}_r(\kappa)$ and let
$$f_n = \sum_{k=0}^{n} a_k z^k.$$
Then
$$\mu(r, f - f_n) = \max_{k>n} |a_k| r^k \geq |a_{n+1}| r^{n+1} \to 0 \; (n \to \infty)$$
that is, $f_n \to f$. □

Lemma 1.19. *Take $r > 0$ and take polynomials $f(z), g(z) = \sum_{n=0}^{k} b_n z^n \in \kappa[z]$ such that $\mu(r, g) = |b_k| r^k$. Let $Q(z)$ and $R(z)$ be the quotient and the remainder dividing $f(z)$ by $g(z)$ so that*
$$f(z) = g(z)Q(z) + R(z), \quad \deg(R) < k.$$
Then one has
$$\mu(r, f) = \max\{\mu(r, g)\mu(r, Q), \mu(r, R)\}.$$

Proof. By theorem 1.14, we obtain
$$\mu(r, f) \leq \max\{\mu(r, gQ), \mu(r, R)\} = \max\{\mu(r, g)\mu(r, Q), \mu(r, R)\}.$$

To prove the inverse inequality, we first consider the case $r = 1$. W. l. o. g., we may assume
$$|b_k| = 1, \quad \max\{\mu(r, Q), \mu(r, R)\} = 1.$$

Thus, we need to prove $\mu(1, f) \geq 1$. Assume, on the contrary, that every coefficient of $f(z)$ belongs to the valuation ideal $\kappa(0; 1)$. Using bars to denote reduction modulo $\kappa(0; 1)$, one has the equation
$$0 = \bar{f}(z) = \bar{g}(z)\bar{Q}(z) + \bar{R}(z).$$
Since $|b_k| = 1$, then
$$\deg(\bar{g}) = k > \deg(R) \geq \deg(\bar{R}),$$
which forces $\bar{Q}(z) \equiv 0$, and hence $\bar{R}(z) \equiv 0$, which contradicts $\max\{\mu(r, Q), \mu(r, R)\} = 1$. Thus, it must be $\mu(1, f) \geq 1$.

Next we assume that $r \neq 1$, but $r \in |\kappa|$. Thus, we can choose an element $a \in \kappa$ with $|a| = r$. Let h be one of the polynomials f, g, Q and R and set $h_a(z) = h(az)$. Then one has $\mu(1, h_a) = \mu(r, h)$ and
$$f_a(z) = g_a(z)Q_a(z) + R_a(z).$$
Hence, the lemma follows from the first case.

Finally, assume that $r \notin |\kappa|$. Note that the set $|\kappa|$ is dense in $\mathbb{R}_+$. Thus, there exists a sequence $\{r_i \mid i \geq 1\} \subset |\kappa|$ such that $r_i \to r$ as $i \to \infty$ so that
$$\lim_{i \to \infty} \mu(r_i, h) = \mu(r, h),$$

where h is one of the polynomials f, g, Q and R. Since one has

$$\mu(r_i, f) = \max\{\mu(r_i, g)\mu(r_i, Q), \mu(r_i, R)\}$$

in the second case, the lemma follows by taking the limit. $\qquad\qquad\square$

Lemma 1.20. *Take $f \in A_r(\kappa)$ for some $r > 0$ and take a polynomial $g(z) = \sum_{n=0}^{k} b_n z^n \in \kappa[z]$ such that $\mu(r, g) = |b_k| r^k$. Then there exist a power series $Q \in A_r(\kappa)$ and a polynomial $R \in \kappa[z]$ such that*

$$f(z) = g(z)Q(z) + R(z), \quad \deg(R) < k,$$

and such that

$$\mu(r, f) = \max\{\mu(r, g)\mu(r, Q), \mu(r, R)\}.$$

Proof. Let f_n be a sequence of polynomials converging to f. Let $Q_n(z)$ and $R_n(z)$ be the quotient and the remainder dividing $f_n(z)$ by $g(z)$ so that

$$f_n(z) = g(z)Q_n(z) + R_n(z), \quad \deg(R_n) < k.$$

By Lemma 1.19, one has

$$\mu(r, f_n) = \max\{\mu(r, g)\mu(r, Q_n), \mu(r, R_n)\}.$$

Applying Lemma 1.19 to the following equation

$$f_{n+1}(z) - f_n(z) = g(z)(Q_{n+1}(z) - Q_n(z)) + R_{n+1}(z) - R_n(z),$$

one obtains

$$\mu(r, f_{n+1} - f_n) = \max\{\mu(r, g)\mu(r, Q_{n+1} - Q_n), \mu(r, R_{n+1} - R_n)\},$$

which means that both $\{Q_n\}$ and $\{R_n\}$ are Cauchy sequences and hence convergent since $\{f_n\}$ is a Cauchy sequence and since $A_r(\kappa)$ is complete. Letting

$$Q(z) = \lim_{n\to\infty} Q_n(z), \quad R(z) = \lim_{n\to\infty} R_n(z),$$

which gives the equation we want, since each $R_n(z)$ is a polynomial of degree less than k, so is $R(z)$. Finally, the estimate on the norms are preserved obviously by passing to the limit.

$\qquad\qquad\square$

Now we can prove the Weierstrass preparation theorem.

Theorem 1.21. *Take $f \in A_r(\kappa) - \{0\}$ for $r > 0$. Then there exists a polynomial*

$$g(z) = b_0 + b_1 z + \cdots + b_\nu z^\nu \in \kappa[z]$$

of degree $\nu = \nu(r, f)$, and a power series $h = 1 + \sum_{n=1}^{\infty} c_n z^n$ with coefficients in κ, satisfying
   1) $f(z) = g(z)h(z)$;
   2) $\mu(r, g) = |b_\nu| r^\nu$;
   3) $h \in A_r(\kappa)$;
   4) $\mu(r, h - 1) < 1$; and
   5) $\mu(r, f - g) < \mu(r, f)$.
In particular, h has no zeros in $\kappa[0; r]$, and f just has ν zeros in $\kappa[0; r]$.

Proof. Write

$$f(z) = a_0 + a_1 z + a_2 z^2 + \cdots, \quad g_1(z) = \sum_{i=0}^{\nu} a_i z^i.$$

Then we can choose $\delta \in \mathbb{R}^+$ satisfying

$$0 \le \frac{\mu(r, f - g_1)}{\mu(r, f)} < \delta < 1.$$

Take $h_1(z) = 1$. Now by induction, one prove that there exist polynomials $g_i(z) = \sum_{j=0}^{\nu} b_{ij} z^j$ and h_i such that

i) $\mu(r, g_i) = |b_{i\nu}| r^{\nu}$;
ii) $\mu(r, f - g_i) \le \delta \mu(r, f)$, $\mu(r, h_i - 1) \le \delta$; and
iii) $\mu(r, f - g_i h_i) \le \delta^i \mu(r, f)$.

This has been proved for the case $i = 1$. If (i), (ii) and (iii) are true, then by Lemma 1.20, there exist a power series $Q_i \in \mathcal{A}_r(\kappa)$ and a polynomial $R_i \in \kappa[z]$ such that

$$f(z) - g_i(z) h_i(z) = g_i(z) Q_i(z) + R_i(z), \quad \deg(R_i) < \nu,$$

and such that

$$\mu(r, f - g_i h_i) = \max\{\mu(r, g_i)\mu(r, Q_i), \mu(r, R_i)\}.$$

Define

$$g_{i+1}(z) = g_i(z) + R_i(z) = \sum_{j=0}^{\nu} b_{i+1,j} z^j, \quad h_{i+1} = h_i + Q_i.$$

Note that (ii) implies $\mu(r, f) = \mu(r, g_i)$. Hence, one obtains

$$\mu(r, Q_i) \le \frac{\mu(r, f - g_i h_i)}{\mu(r, g_i)} = \frac{\mu(r, f - g_i h_i)}{\mu(r, f)} \le \delta^i$$

and

$$\mu(r, R_i) \le \mu(r, f - g_i h_i) \le \delta^i \mu(r, f).$$

Then one has

$$\mu(r, g_{i+1}) = \mu(r, g_i),$$

and hence (i) holds for $i + 1$ since $\deg(R_i) < \deg(g_i) = \nu$. The condition (ii) also holds for $i + 1$ since

$$\mu(r, f - g_{i+1}) \le \max\{\mu(r, f - g_i), \mu(r, R_i)\} \le \delta \mu(r, f)$$

and

$$\mu(r, h_{i+1} - 1) \le \max\{\mu(r, h_i - 1), \mu(r, Q_i)\} \le \delta.$$

Note that

$$f - g_{i+1} h_{i+1} = R_i(1 - h_i - Q_i).$$

Then

$$\mu(r, f - g_{i+1} h_{i+1}) \le \mu(r, R_i) \max\{\mu(r, h_i - 1), \mu(r, Q_i)\} \le \delta^{i+1} \mu(r, f),$$

and hence (iii) holds for $i + 1$.

Note that

$$\mu(r, g_{i+1} - g_i) = \mu(r, R_i) \leq \delta^i \mu(r, f)$$

and that

$$\mu(r, h_{i+1} - h_i) = \mu(r, Q_i) \leq \delta^i.$$

Since $\delta < 1$, both $\{g_i\}$ and $\{h_i\}$ are Cauchy sequences with respect to the norm $\mu(r, \cdot)$. Then

$$|b_{i+1,j} - b_{ij}|r^j \leq \mu(r, g_{i+1} - g_i) \leq \delta^i \mu(r, f), \quad 0 \leq j \leq \nu, \; i \geq 1,$$

i.e., $\{b_{ij}\}_{i \geq 1}$ is a Cauchy sequence for each j, and hence convergent. Letting

$$b_j = \lim_{i \to \infty} b_{ij}, \quad g(z) = \sum_{j=0}^{\nu} b_j z^j,$$

obviously one has $g_i \to g$ and $\mu(r, g) = |b_\nu| r^\nu$. Since $\mathcal{A}_r(\kappa)$ is complete, then $\{h_i\}$ converges so that there exists $h \in \mathcal{A}_r(\kappa)$ such that $h_i \to h$. It is easy to prove that g and h satisfy the conditions in Theorem 1.21.

Note that $\mu(\cdot, h - 1)$ increases. The condition (4) implies $\mu(t, h - 1) < 1$ for $0 \leq t \leq r$, and hence $|h(z)| = 1$ for $z \in \kappa[0; r]$. Thus h has no zeros in $\kappa[0; r]$. Let $z_1, ..., z_\nu$ be the zeros of g. Then

$$g(z) = b_\nu(z - z_1) \cdots (z - z_\nu).$$

Condition (2) implies $\mu(r, g/b_\nu) = r^\nu$, and hence

$$(\max\{r, |z_1|\}) \cdots (\max\{r, |z_\nu|\}) = r^\nu,$$

which yields

$$|z_j| \leq r, \quad j = 1, ..., \nu.$$

Therefore, just g has ν zeros in $\kappa[0; r]$, so does f. $\square$

Theorem 1.22. *If $f \in \mathcal{A}_\rho(\kappa) - \kappa$ $(\rho > 0)$ with $f(0) = 0$, then $\mu(r, f)$ increases strictly.*

Proof. Take $r_1, r_2 \in \mathbb{R}^+$ with $r_1 < r_2$. By Theorem 1.21, f has $\nu(r_2, f) - \nu(r_1, f)$ zeros in the annulus $\kappa[0; r_2] - \kappa[0; r_1]$. Thus, we can choose $r \in \mathbb{R}^+$ with $r_1 < r < r_2$ such that f has no zeros in $\kappa[0; r] - \kappa[0; r_1]$, and, therefore, Theorem 1.21 means $\nu(r, f) = \nu(r_1, f)$. Write

$$f(z) = \sum_{n=1}^{\infty} a_n z^n,$$

and note that $\nu(r_1, f) \geq 1$. Then we have

$$\begin{aligned}
0 < \mu(r_1, f) &= |a_{\nu(r_1, f)}| r_1^{\nu(r_1, f)} < |a_{\nu(r_1, f)}| r^{\nu(r_1, f)} \\
&= |a_{\nu(r, f)}| r^{\nu(r, f)} = \mu(r, f) \leq \mu(r_2, f).
\end{aligned}$$

$\square$

Definition 1.23. *An element in $\mathcal{A}(\kappa)$ is said to be an entire function on κ.*

We also denote by $\mathcal{A}(\mathbb{C})$ the set of entire functions on $\mathbb{C}$. To prove the decomposition theorem for entire functions, we need the following fact about convergence of infinite products:

Lemma 1.24 (cf. [26]). *Given a sequence $\{f_n(z)\}$ of power series in $\mathcal{A}_r(\kappa)$ $(r > 0)$, such that*

$$\lim_{n \to \infty} \mu(r, f_n - 1) = 0,$$

then the infinite product

$$f = \prod_{n=1}^{\infty} f_n \ .$$

is a power series in $\mathcal{A}_r(\kappa)$. If the hypotheses hold for all $r > 0$, then f is an entire function on κ.

Proof. Write

$$F_n = \prod_{k=1}^{n} f_k \in \mathcal{A}_r(\kappa).$$

Since $\mu(r, f_n - 1) \to 0$ as $n \to \infty$, there exists an n_0 such that for all $n \geq n_0$, $\mu(r, f_n - 1) < 1$ and hence $\mu(r, f_n) = 1$. Therefore when $n > n_0$, one has

$$\mu(r, F_n - F_{n-1}) = \mu(r, F_{n-1})\mu(r, f_n - 1) = \mu(r, F_{n_0})\mu(r, f_n - 1) \to 0,$$

that is $\{F_n\}$ is a Cauchy sequence, and hence convergent since $\mathcal{A}_r(\kappa)$ is complete, and the lemma follows. $\qquad\square$

Corollary 1.25. *For any sequence $\{z_n\}_{n \geq 1}$ in κ_* satisfying $|z_n| \to \infty$ as $n \to \infty$, then the infinite product*

$$f(z) = \prod_{n=1}^{\infty} \left(1 - \frac{z}{z_n}\right)$$

is an entire function on κ.

Theorem 1.26 (cf. [39]). *Let f be a power series defining an entire function on κ and assume that f is not a polynomial. Then f can be written as an infinite product*

$$f(z) = az^m \prod_{n=1}^{\infty} \left(1 - \frac{z}{z_n}\right),$$

where $a \in \kappa$, m is a non-negative integer, and z_n ranges through the nonzero roots of f, which form a sequence tending to infinity.

Proof. W. l. o. g., we may assume that $f(0) = 1$. Let $\{z_i\}_{i=1}^{\omega}$ be the zeros of f such that $|z_1| \leq |z_2| \leq \cdots$, where ω is the number of zeros (counting multiplicity) of f if it is finite, otherwise, $\omega = \infty$. If $\omega = \infty$, by Theorem 1.21, it must be $|z_i| \to \infty$. Take $\{a_i \mid i = 1, 2, ...\} \subset \kappa_*$ with

$$0 < |a_1| < |a_2| < \cdots, \quad r_n = |a_n| \to \infty \ (n \to \infty).$$

Let $f_n(z) = f(a_n z)$. Applying Theorem 1.21 in $\kappa[0; 1]$ gives

$$f_n(z) = g_n(z)(1 + c_1 z + c_2 z^2 + \cdots),$$

with g_n a polynomial and $|c_i| < 1$, and hence

$$f(z) = f_n(z/a_n) = g_n(z/a_n) \left(1 + c_1 \frac{z}{a_n} + c_2 \left(\frac{z}{a_n}\right)^2 + \cdots\right).$$

We can write $g_n(z/a_n)$ as follows:

$$g_n(z/a_n) = \prod_{|z_i| \leq r_n} \left(1 - \frac{z}{z_i}\right).$$

Fix a positive integer N, then when $n > N$,

$$\mu(r_N, h_n - 1) = \max_i |c_i| \left(\frac{r_N}{r_n}\right)^i < \frac{r_N}{r_n} \to 0 \ (n \to \infty),$$

that is, h_n converges uniformly to 1 on $\kappa[0; r_N]$. By Lemma 1.24, one has

$$g_n(z/a_n) \to \prod_{i=1}^{\omega} \left(1 - \frac{z}{z_i}\right), \quad z \in \kappa[0; r_N]$$

as $n \to \infty$, and hence

$$f(z) = \prod_{i=1}^{\omega} \left(1 - \frac{z}{z_i}\right)$$

on $\kappa[0; r_N]$. Since r_N can be sufficiently large, the equality holds for any $z \in \kappa$, which implies $\omega = \infty$ since f is not a polynomial, and the theorem follows. $\square$

Corollary 1.27. *If f is an entire function on κ, but is not a polynomial, then f has infinitely many zeros.*

Corollary 1.28. *If f is an entire function on κ that is never zero, then f is constant.*

Corollary 1.29. *Greatest common divisors exist for any finite set of $\mathcal{A}(\kappa)$.*

Proof. Let $f_1, ..., f_q$ be finite entire functions on κ. If any of the f_i is identically zero, then the corollary is trivial in this case. Assume some of the f_i are not identically zero. Then each $f_i \ (\not\equiv 0)$ can be written as a product

$$f_i(z) = a_i z^{m_i} \prod_{w \in S_i} \left(1 - \frac{z}{w}\right),$$

where $a_i \in \kappa_*$, m_i is a non-negative integer, and S_i is the set of nonzero roots of f_i (counting multiplicity). Set

$$m = \min_{f_i \not\equiv 0}\{m_i\}, \quad S = \bigcap_{f_i \not\equiv 0} S_i,$$

and define

$$G(z) = z^m \prod_{w \in S} \left(1 - \frac{z}{w}\right).$$

If S is finite, then G is a polynomial. Theorem 1.26 and Corollary 1.25 implies that G is an entire function if S is infinite. Obviously, G is a greatest common divisor of $f_1, ..., f_q$.   □

If $r \in |\kappa_*|$, then $\mathcal{A}_r(\kappa)$ is a unique factorization domain (see [9], Lemma 5.2.6), and so greatest common divisors exist.

1.5 Newton polygons

Let κ be an algebraically closed field of characteristic zero, complete for a non-trivial non-Archimedean absolute value $|\cdot|$. Take $f \in \mathcal{A}_{(\rho}(\kappa)$ $(0 < \rho \le \infty)$ and write

$$f(z) = \sum_{n=m}^{\infty} a_n z^n \quad (a_m \ne 0, \ m \ge 0).$$

The coefficient a_m also will be denoted by $f^*(0)$. Take $a \in \kappa$ and let $n\left(r, \frac{1}{f-a}\right)$ be the *counting function* of f for a, which denotes the number of zeros (counting multiplicity) of $f - a$ with absolute value $\le r$. In particular, we have

$$n\left(0, \frac{1}{f}\right) = m.$$

Fix a real ρ_0 with $0 < \rho_0 < \rho$. Define the *valence function* of f for a by

$$N\left(r, \frac{1}{f-a}\right) = \int_{\rho_0}^{r} \frac{n(t, \frac{1}{f-a})}{t} dt \quad (\rho_0 < r < \rho), \tag{1.6}$$

and write

$$N(r, f = a) = \int_{0}^{r} \frac{n(t, \frac{1}{f-a}) - n(0, \frac{1}{f-a})}{t} dt + n\left(0, \frac{1}{f-a}\right) \log r \quad (0 < r < \rho). \tag{1.7}$$

Then we have

$$N(r, f = a) - N(\rho_0, f = a) = N\left(r, \frac{1}{f-a}\right) \ge 0.$$

Theorem 1.21 shows that

$$n\left(r, \frac{1}{f}\right) = \nu(r, f).$$

Then Lemma 1.13 implies the *Jensen formula* (cf. [50]):

$$N(r, f = 0) = \log \mu(r, f) - \log |f^*(0)|, \tag{1.8}$$

which means

$$N\left(r, \frac{1}{f}\right) = \log \mu(r, f) - \log \mu(\rho_0, f). \tag{1.9}$$

We also denote the number of distinct zeros of $f - a$ on $\kappa[0; r]$ by $\bar{n}(r, \frac{1}{f-a})$ and define

$$\bar{N}\left(r, \frac{1}{f-a}\right) = \int_{\rho_0}^r \frac{\bar{n}(t, \frac{1}{f-a})}{t} dt \quad (\rho_0 < r < \rho).$$

Let v_p be the additive valuation associated with the absolute value $|\cdot|$ for a real constant $p > 1$. For each $n \geq m$, we define a straight line

$$\gamma_n(t) = v_p(a_n) - nt,$$

with slope $-n$, where we think of $\gamma_n = +\infty$ if $a_n = 0$. Note that $|a_n| r^n \to 0$ for $0 \leq r < \rho$ as $n \to \infty$. We can write this in terms of v_p as

$$v_p(a_n) - nt \to +\infty, \quad t < \tau = \log_p \rho,$$

and it follows that for every t there exists an n for which $v_p(a_n) - nt$ is minimal. Let $\gamma(t, f)$ denote the boundary of the intersection of all of the half-planes lying under the lines $\gamma_n(t)$. This line is what we call the *Newton polygon* of the function $f(z)$ (see [50]). Obviously, it satisfies

$$\mu(p^t, f) = p^{-\gamma(t,f)}, \quad t < \tau.$$

The points $t(< \tau)$ at which $\gamma(t, f)$ has vertices are called the *vertices* of $f(z)$. It is clear that if t is a vertex, then $v_p(a_n) - nt$ attains its minimum at least at two values of n. Note that there exists a $r_0 \in \mathbb{R}^+$ such that

$$|a_n| r^n < |a_m| r^m, \quad n > m, \ 0 < r < r_0,$$

that is

$$\gamma_n(t) > \gamma_m(t), \quad n > m, \ t < \log_p r_0.$$

Hence f has no vertices in $(-\infty, \log_p r_0)$. We also can prove that a finite segment $[\alpha, \beta] \subset (-\infty, \tau)$ contains only finitely many vertices since in $[\alpha, \beta]$ there are only finitely many $\gamma_n(t)$ which appear in $\gamma(t, f)$. A basic property of the Newton polygon is that, if $-v_p(z)$ is not a vertex for $z \in \kappa(0; \rho)$, then

$$|f(z)| = p^{-\gamma(-v_p(z),f)},$$

or

$$|f(z)| = \mu(|z|, f).$$

Note that $|\kappa|$ is dense in $\mathbb{R}_+$. Since $\mu(r, f)$ is continuous at r, one obtains the *maximum modulus principle*

$$\sup_{|z| \leq r} |f(z)| = \mu(r, f) \quad (0 \leq r < \rho). \tag{1.10}$$

Let $\{t_i\}_{i \geq 1}$ be the vertices of f with $t_1 < t_2 < \cdots$ and set $r_i = p^{t_i}$, $\nu_i = \nu(r_i, f)$ for all $i \geq 1$. Then one has

$$\gamma_n(t) > \gamma_m(t), \quad n > m, \ t < t_1,$$

and hence
$$\mu(r, f - a_m z^m) < |a_m| r^m = \mu(r, f), \quad 0 < r < r_1,$$
which implies that in $\kappa(0; r_1)$, f has no zeros if $m = 0$, and that $z = 0$ is the unique zero of f with multiplicity m if $m > 0$. Since t_1 is a vertex of f, then $\nu_1 > m$. By Theorem 1.21, there exists a polynomial $g(z) = \sum_{i=m}^{\nu_1} b_i z^i$ and a power series $h = 1 + \sum_{n=1}^{\infty} c_n z^n$ with coefficients in κ, satisfying

1) $f(z) = g(z)h(z)$;
2) $\mu(r_1, g) = |b_{\nu_1}| r_1^{\nu_1}$;
3) $h \in \mathcal{A}_{r_1}(\kappa)$;
4) $\mu(r_1, h - 1) < 1$; and
5) $\mu(r_1, f - g) < \mu(r_1, f)$.

In particular, h has no zeros in $\kappa[0; r_1]$, and f just has ν_1 zeros in $\kappa[0; r_1]$. Therefore f just has $\nu_1 - m$ zeros (counting multiplicity) on the circle $\kappa\langle 0; r_1 \rangle$. By induction, we can prove that f just has $\nu_i - \nu_{i-1}$ zeros (counting multiplicity) on the circle $\kappa\langle 0; r_i \rangle$ for any $i \geq 1$, where $\nu_0 = m$. Thus, we integrate by parts, and get

$$\begin{aligned}
N(r, f = 0) &= \int_0^r \log \frac{r}{t} d\left(n\left(t, \frac{1}{f}\right) - n\left(0, \frac{1}{f}\right)\right) + n\left(0, \frac{1}{f}\right) \log r \\
&= \sum_{r_i \leq r} (\nu_i - \nu_{i-1}) \log \frac{r}{r_i} + \nu_0 \log r \quad (0 < r < \rho).
\end{aligned} \tag{1.11}$$

Let $\{z_i\}_{i \geq 1}$ be the zeros (counting multiplicity) of f in $\kappa - \{0\}$. Then (1.11) can be expressed as follows (cf. [44]):

$$\begin{aligned}
N(r, f = 0) &= n\left(0, \frac{1}{f}\right) \log r + \sum_{|z_i| \leq r} (v_p(z_i) \log p + \log r) \\
&= n\left(r, \frac{1}{f}\right) \log r + \sum_{|z_i| \leq r} v_p(z_i) \log p,
\end{aligned} \tag{1.12}$$

or equivalently,

$$\frac{1}{\log p} N(p^t, f = 0) = n\left(0, \frac{1}{f}\right) t + \sum_{v_p(z_i) \geq -t} (v_p(z_i) + t). \tag{1.13}$$

Theorem 1.30 (cf. [32]). *Take $f \in \mathcal{A}_r(\kappa)$ and let $s = \mu(r, f - f(0))$. Then $f(\kappa[0; r]) = \kappa[f(0); s]$ and $\gamma(\log_p r, f - f(0)) = -\log_p s$.*

Proof. Obviously, $f(\kappa[0; r]) \subset \kappa[f(0); s]$. On the other hand, we take $b \in \kappa[f(0); s]$ and consider the function $g(z) = f(z) - b$. Theorem 1.30 is trivial when $s = 0$ so we may assume that $s > 0$. Thus we have $\nu(r, g) = \nu(r, f - f(0)) \geq 1$ since

$$|f(0) - b| \leq s = \mu(r, f - f(0)).$$

By Theorem 1.21, g admits at least one zero in $\kappa[0; r]$, and hence $b \in f(\kappa[0; r])$. Clearly, one has

$$\gamma(\log_p r, f - f(0)) = -\log_p \mu(r, f - f(0)) = -\log_p s.$$

$\square$

Theorem 1.31 (cf. [32]). *Let $f \in \mathcal{A}_r(\kappa)$ have k zeros in $\kappa[0; r]$ with $k \geq 1$ (taking multiplicities into account) and let $b \in f(\kappa[0; r])$. Then $f - b$ also admits k zeros in $\kappa[0; r]$ (counting multiplicity).*

Proof. Write $f(z) = \sum_{n=0}^{\infty} a_n z^n$. By Theorem 1.21, we have $k = \nu(r, f)$ and hence

$$|a_n|r^n \leq |a_k|r^k \ (n < k), \quad |a_n|r^n < |a_k|r^k \ (n > k).$$

By hypothesis and by Theorem 1.30, one has

$$|a_0 - b| \leq \sup_{n \geq 1} |a_n|r^n \leq |a_k|r^k,$$

and hence

$$\nu(r, f - b) = k = \nu(r, f).$$

Now the theorem follows from Theorem 1.21 again. $\qquad\qquad\qquad\square$

Corollary 1.32. *Assume that $f \in \mathcal{A}_{(\rho}(\kappa)$ $(0 < \rho \leq \infty)$ is unbounded. Then for any $b \in \kappa$, we have*

$$N\left(r, \frac{1}{f - b}\right) = N\left(r, \frac{1}{f}\right) + O(1) \ (r \to \rho).$$

Proof. Note that f and $f - b$ all have at least one zero since $f - b$ also is unbounded. Thus there is a r' $(< \rho)$ such that f has at least one zero in $\kappa[0; r']$ and such that $b \in f(\kappa[0; r'])$. By Theorem 1.31, one obtains

$$n\left(r, \frac{1}{f - b}\right) = n\left(r, \frac{1}{f}\right) \quad (r \geq r').$$

Therefore when $r \geq r'$, we have

$$
\begin{aligned}
N\left(r, \frac{1}{f - b}\right) &= N\left(r', \frac{1}{f - b}\right) + \int_{r'}^{r} \frac{n(t, \frac{1}{f - b})}{t} dt \\
&= N\left(r', \frac{1}{f - b}\right) + \int_{r'}^{r} \frac{n(t, \frac{1}{f})}{t} dt \\
&= N\left(r', \frac{1}{f - b}\right) - N\left(r', \frac{1}{f}\right) + N\left(r, \frac{1}{f}\right),
\end{aligned}
$$

and the corollary follows. $\qquad\qquad\qquad\square$

Corollary 1.33. *Assume that f is a non-constant entire function. Then for any $b \in \kappa$, we have*

$$N\left(r, \frac{1}{f - b}\right) = N\left(r, \frac{1}{f}\right) + O(1).$$

H. H. Khoái [44] was the fist to begin a systematic development of one variable Nevanlinna theory over non-Archimedean fields. He gave some preliminary definitions for the Nevanlinna functions and took the first steps toward recognizing that the information coming from the Newton polygon could be used to prove a non-Archimedean analog of the Jensen formula. These valence functions, which appear in the works of Khoái [44], Corrales-Rodrigáñez [28], Khoái-Quang [50], and Boutabaa [10], are exact analogs of those in classical Nevanlinna theory. In the several variable situation, Khoái [46] also obtain partial results. A several variable Jensen formula is proved by Cherry and Ye [26].

1.6 Non-Archimedean meromorphic functions

Let κ be an algebraically closed field of characteristic zero, complete for a non-trivial non-Archimedean absolute value $|\cdot|$. We will discuss analytic functions on κ in this section. To do so, we begin with the following fact (see [39]):

Lemma 1.34. *Take $b_{jn} \in \kappa$ and suppose that*
 1) $\lim_{n \to \infty} b_{jn} = 0$ *for every j, and*
 2) $\lim_{j \to \infty} b_{jn} = 0$ *uniformly in n.*
Then both series

$$\sum_{j=0}^{\infty} \sum_{n=0}^{\infty} b_{jn}, \quad \sum_{n=0}^{\infty} \sum_{j=0}^{\infty} b_{jn}$$

converge, and their sums are equal.

Proposition 1.35 (cf. [39]). *Given a power series*

$$f(z) = \sum_{n=0}^{\infty} a_n (z - z_0)^n, \quad (z_0, a_n \in \kappa),$$

and consider the power series

$$g(z) = \sum_{n=0}^{\infty} b_n (z - \alpha)^n, \quad b_n = \sum_{j \geq n} \binom{j}{n} a_j (\alpha - z_0)^{j-n}$$

for a point $\alpha \in \kappa - \{z_0\}$. If $f(\alpha)$ converges, then
 1) the series defining b_n converges for every n, so that b_n are well-defined;
 2) the power series $f(z)$ and $g(z)$ have the same region of convergence, that is, $f(z)$ converges if and only if $g(z)$ converges; and
 3) $f(z) = g(z)$ for any z in the region of convergence.

Proof. With n fixed, note that

$$\left| \binom{j}{n} a_j (\alpha - z_0)^{j-n} \right| \leq \left| a_j (\alpha - z_0)^{j-n} \right| = |\alpha - z_0|^{-n} |a_j(\alpha - z_0)^j| \to 0.$$

The convergence in (1) is derived by Corollary 1.12.
 To prove (2) and (3), we consider

$$b_{jn} = \begin{cases} \binom{j}{n} a_j (\alpha - z_0)^{j-n} (z - \alpha)^n & : \ j \geq n \\ 0 & : \ j < n. \end{cases}$$

Since both z and α are in the region D of the convergence of f, we can choose a real $r > 0$ such that $z, \alpha \in \kappa[z_0; r] \subset D$. For any n, we obtain an estimate

$$|b_{jn}| \leq |a_j (\alpha - z_0)^{j-n} (z - \alpha)^n| \leq |a_j| r^j \to 0 \ (j \to \infty),$$

which means that b_{jn} tends to zero uniformly in n. Obviously, for each j one has $b_{jn} \to 0$ as $n \to \infty$. By using Lemma 1.34, we have

$$
\begin{aligned}
f(z) &= \sum_{j=0}^{\infty} a_j (z - \alpha + \alpha - z_0)^j = \sum_{j=0}^{\infty} \sum_{n=0}^{j} \binom{j}{n} a_j (\alpha - z_0)^{j-n} (z - \alpha)^n \\
&= \sum_{j=0}^{\infty} \sum_{n=0}^{\infty} b_{jn} = \sum_{n=0}^{\infty} \sum_{j=0}^{\infty} b_{jn} \\
&= \sum_{n=0}^{\infty} b_n (z - \alpha)^n = g(z),
\end{aligned}
$$

that is, $g(z)$ converges and is equal to $f(z)$. Noting that we can switch the roles of g and f in the argument, it shows that in fact the regions of convergence are identical. $\qquad\square$

Proposition 1.35 shows that the power series $f(z)$, which admits a disc $\kappa[z_0; r]$ for the disc of convergence, may not be extended outside its convergence disc as it is in complex analysis, by means of a change of origin. However, Runge's theorem shows that an analytic function in a compact subset D of $\mathbb{C}$ is equal to the limit of a sequence of rational functions with respect to the uniform convergence on D. According to this fact, Marc Krasner [78] introduced the analytic elements on a subset D of κ as follows:

Definition 1.36. *Let D be an infinite set in κ and let $R(D)$ be the set of the rational functions $h(z) \in \kappa(z)$ with no poles in D. For every $h \in R(D)$, we put*

$$
\|h\|_D = \sup_{z \in D} |h(z)|.
$$

Denote by $\mathscr{H}(D)$ the completion of $R(D)$ for the topology of uniform convergence on D. The elements of $\mathscr{H}(D)$ are called the analytic elements on D.

Definition 1.36 is a notion of global non-Archimedean analyticity. According to this definition, the identity theorem holds, i.e., two analytic elements which agree, say, on a disc, must agree everywhere (see [79]). In this book, the analytic elements of $\mathscr{H}(D)$ will be called the *global analytic functions* on D. Obviously, $\mathscr{H}(D)$ is a κ-vector space, and every $f \in \mathscr{H}(D)$ defines a function on D which is the uniform limit of a sequence $\{h_n\}_{n \geq 1}$ of $R(D)$ on D. If D is closed and bounded, then the product of two global analytic functions on D is a global analytic function on D (cf. [32]). Given two infinite sets D and D' with $D \subset D'$, the restriction to D of elements of $\mathscr{H}(D')$ belongs to $\mathscr{H}(D)$.

Theorem 1.37 (cf. [32]). *For $r \in \mathbb{R}^+$, one has $\mathscr{H}(\kappa[0; r]) = \mathcal{A}_r(\kappa)$.*

Proof. By Lemma 1.18, we know that $\mathcal{A}_r(\kappa) \subset \mathscr{H}(\kappa[0; r])$. To prove $\mathscr{H}(\kappa[0; r]) \subset \mathcal{A}_r(\kappa)$, we first show that $R(\kappa[0; r])$ is included in $\mathcal{A}_r(\kappa)$. For this, we have to show that given any $a \in \kappa - \kappa[0; r]$, $k \in \mathbb{Z}^+$,

$$
\left(\frac{1}{z - a} \right)^k = \left(-\frac{1}{a} \sum_{n=0}^{\infty} \left(\frac{z}{a} \right)^n \right)^k = \left(-\frac{1}{a} \right)^k \sum_{n=0}^{\infty} b_n \left(\frac{z}{a} \right)^n \in \mathcal{A}_r(\kappa),
$$

where $b_n \in \mathbb{Z}^+$. Note that

$$\left|\frac{b_n}{a^n}\right| r^n \le \left(\frac{r}{|a|}\right)^n \to 0$$

as $n \to \infty$, since $|a| > r$. This shows that $\left(\frac{1}{z-a}\right)^k \in \mathcal{A}_r(\kappa)$, and therefore we have the inclusion $R(\kappa[0;r]) \subset \mathcal{A}_r(\kappa)$. By (1.10), we have

$$\|f\|_{\kappa[0;r]} = \mu(r, f), \quad f \in \mathcal{A}_r(\kappa).$$

By using Lemma 1.17, $\mathcal{A}_r(\kappa)$ is complete with respect to the norm $\mu(r, \cdot)$, and hence it also is complete with respect to the norm $\|\cdot\|_{\kappa[0;r]}$. Therefore, we finally have $\mathcal{A}_r(\kappa) = \mathcal{H}(\kappa[0;r])$. $\square$

Definition 1.38. *Suppose that a subset $D \subset \kappa$ has no isolated points. A function $f : D \longrightarrow \kappa$ is called locally analytic if for every $a \in D$ there exist $r \in \mathbb{R}^+$ and $a_n \in \kappa$ such that*

$$f(z) = \sum_{n=0}^{\infty} a_n(z-a)^n, \quad z \in D \cap \kappa[a;r].$$

Denote by Hol(D) *the set of local analytic functions on D.*

Some properties of local analytic functions are discussed in [76], for example, a local analytic function is differentiable in the usual sense. If D is an open set, by Theorem 1.37 global non-Archimedean analyticity implies local non-Archimedean analyticity, that is,

$$\mathcal{H}(D) \subset \text{Hol}(D).$$

Assume that f is locally analytic in D. Let $z_0 \in D$ and $r \in \mathbb{R}^+$ be such that $\kappa[z_0;r] \subset D$. By Theorem 1.37, f is equal to a power series of the form

$$f(z) = \sum_{n=0}^{\infty} a_n(z-z_0)^n, \quad z \in \kappa[z_0;r].$$

If $f(z_0) = 0$ and if f is not identically zero in $\kappa[z_0;r]$, then there exists a unique integer q such that $a_n = 0$ for every $n < q$ and $a_q \neq 0$. The number q will be named the *multiplicity of z_0*, and z_0 is called a *zero of multiplicity q*. Thus f can be expressed as

$$f(z) = (z-z_0)^q g(z) \ (g(z_0) \neq 0),$$

where g is a power series whose radius of convergence is superior or equal to r. Note that g is continuous in $\kappa[z_0;r]$. Since $g(z_0) \neq 0$, the continuity of g shows that z_0 is an isolated zero of f. Theorem 1.15 implies the following result:

Proposition 1.39. *Let f be a local analytic function in an open set D. Then for every n, the n-th derivative $f^{(n)}$ exists in D. If z_0 is a zero of multiplicity q of f, then we have $f^{(n)}(z_0) = 0$ for every $n < q$, and $f^{(q)}(z_0) \neq 0$.*

Theorem 1.40 ([32]). *Take* $r \in \mathbb{R}^+$ *and let*

$$f(z) = \sum_{n=0}^{\infty} a_n z^n \in \mathcal{A}_r(\kappa), \quad s = \sup_{n \geq 1} |a_n| r^{n-1} > 0.$$

Then the following statements are equivalent:

1) $|a_1| > |a_n| r^{n-1}$ *whenever* $n > 1$;
2) $|f(x) - f(y)| = |x - y||a_1|$ *whenever* $x, y \in \kappa[0; r]$;
3) f *is injective in* $\kappa[0; r]$ *and* $f'(z) \neq 0$ *whenever* $z \in \kappa[0; r]$.

Proof. $(1) \Rightarrow (2)$. Since $|a_n| r^n \to 0$ as $n \to \infty$, then the condition (1) implies

$$|a_1| > \max_{n \geq 2} |a_n| r^{n-1}.$$

Note that

$$f(x) - f(y) = (x - y) \left(a_1 + \sum_{n=2}^{\infty} a_n \sum_{j=0}^{n-1} x^j y^{n-1-j} \right).$$

Since $|x^j y^{n-1-j}| \leq r^{n-1}$, then

$$|a_1| > \max_{n \geq 2} |a_n| r^{n-1} \geq \left| \sum_{n=2}^{\infty} a_n \sum_{j=0}^{n-1} x^j y^{n-1-j} \right|,$$

and hence $|f(x) - f(y)| = |x - y||a_1|$.

$(2) \Rightarrow (3)$. Since $s > 0$, then f is not a constant, and hence (2) means that $|a_1| \neq 0$. Again (2) means that $f(x) \neq f(y)$ whenever $x \neq y$, that is, f is injective in $\kappa[0; r]$, and that $|f'(x)| = |a_1| \neq 0$ whenever $x \in \kappa[0; r]$ by letting $y \to x$.

$(3) \Rightarrow (1)$. Since f is injective in $\kappa[0; r]$, then $f(z) - a_0$ has a unique zero at $z = 0$. By the Weierstrass preparation theorem, it follows that $\nu(r, f) = 1$, and hence (1) is satisfied. $\square$

Thus, if one of the conditions in Theorem 1.40 is satisfied, then the inverse function f^{-1} of f exists on $f(\kappa[0; r])$. Further, f^{-1} is globally analytic on $f(\kappa[0; r])$ (see [32]). In particular, one has the following inverse function theorem:

Theorem 1.41. *Let D be open set in κ with $0 \in D$ and take $f \in \text{Hol}(D)$ with $f'(0) \neq 0$. Then there exists a $r \in \mathbb{R}^+$ such that f is one-to-one in $\kappa[0; r]$, and such that f^{-1} is globally analytic in $f(\kappa[0; r])$.*

The following fact can be proved easily (see [32]):

Lemma 1.42. *Take $r \in \mathbb{R}^+$ and let $f(z) = \sum a_n z^n$ be a power series with coefficients in κ. The following conditions are equivalent:*

1) $f \in \mathcal{A}_{(r}(\kappa)$;
2) $f \in \bigcap_{s < r} \mathcal{A}_s(\kappa)$;
3) *The series f is convergent in all of $\kappa(0; r)$.*

Let $\mathcal{A}_{(r}^*(\kappa)$ be the set of the bounded power series convergent in $\kappa(0; r)$. Thus we have

$$\mathcal{A}_{(r}^*(\kappa) \subset \mathcal{A}_{(r}(\kappa).$$

Note that $\mathcal{A}_{(r}^*(\kappa)$ is complete with respect to the norm $\|\cdot\|_{\kappa(0;r)}$. By the proof of Theorem 1.37, we also see

$$R(\kappa(0; r)) \subset \mathcal{A}_{(r}^*(\kappa),$$

and therefore (see A. Escassut [32])

$$\mathcal{H}(\kappa(0; r)) \subset \mathcal{A}_{(r}^*(\kappa).$$

A. Escassut [32] also shows that $\mathcal{H}(\kappa(0; r))$ is much smaller than $\mathcal{A}_{(r}^*(\kappa)$ so that its field of fractions also is much smaller than the field of fractions of $\mathcal{A}_{(r}^*(\kappa)$. The latter field is denoted by $\mathcal{M}_{(r}^*(\kappa)$, and is usually referred to the Nevanlinna class. Its elements have simple properties, however, functions in the field of fractions of $\mathcal{A}_{(r}(\kappa)$ are very complicated. For example, by the Jensen formula, each of $\mathcal{A}_{(r}(\kappa) - \mathcal{A}_{(r}^*(\kappa)$ has infinitely many zeros in $\kappa(0; r)$. Thus we have to define meromorphic functions as a variety.

Definition 1.43. *Suppose that a subset $D \subset \kappa$ has no isolated points. A function $f : D \longrightarrow \kappa \cup \{\infty\}$ will be said to be globally meromorphic if there exists a (finite or infinite) numerable subset S of D such that S has no limit points in D, and such that f is an analytic element of $\mathcal{H}(D - S)$. Denote by $\mathcal{M}(D)$ the set of globally meromorphic functions on D.*

Let $R_0(D)$ be the subset of $R(D)$ that consists of the $h \in R(D)$ such that

$$\lim_{z \in D, |z| \to \infty} h(z) = 0.$$

A. Escassut [32] shows that

$$\mathcal{H}(D) = \mathcal{H}(\overline{D}) \oplus R_0(\kappa - (\overline{D} - D))$$

holds, where $\overline{D}$ is the *closure* of D. Thus if D is open, then for each subset S of D, which has the property in Definition 1.43, one has

$$\mathcal{H}(D - S) = \mathcal{H}(\overline{D}) \oplus R_0(\kappa - \{(\overline{D} - D) \cup S\}),$$

and hence each $f \in \mathcal{M}(D)$ has a decomposition

$$f = f_1 + f_0, \quad f_1 \in \mathcal{H}(\overline{D}), \quad f_0 \in R(\kappa - \overline{D}).$$

Therefore f has only finite many poles in D since f_0 does. The multiplicities of poles of f_0 in D are also said to be those of f. Further, if D is bounded, then (see [32])

$$f = \frac{g}{Q}, \quad g \in \mathcal{H}(\overline{D}), \quad Q \in \kappa[z].$$

Definition 1.44. *Suppose that a subset* $D \subset \kappa$ *has no isolated points. A function* $f :$ $D \longrightarrow \kappa \cup \{\infty\}$ *is called locally meromorphic if for every* $a \in D$ *there exist* $r \in \mathbb{R}^+$, $q \in \mathbb{Z}_+$, *and* $a_n \in \kappa$ *such that*

$$f(z) = \sum_{n=-q}^{\infty} a_n (z-a)^n, \quad z \in D \cap \kappa[a; r].$$

Denote by Mer(D) *the set of locally meromorphic functions on* D.

By the definition 1.44, if $a_{-q} \neq 0$ with $q > 0$, then it is said that f has a *pole of multiplicity* q at a. Obviously, all poles of f are isolated.

Definition 1.45. *Suppose that a subset* $D \subset \kappa$ *is open. A function* $f : D \longrightarrow \kappa$ *is called analytic at a point* $a \in D$ *if there exist* $\rho \in \mathbb{R}^+ \cup \{\infty\}$ *and* $a_n \in \kappa$ *such that* $\kappa(a; \rho) \subset D$, *but* $\kappa[a; \rho'] - D \neq \emptyset$ *for any* $\rho' > \rho$, *and such that*

$$f(z) = \sum_{n=0}^{\infty} a_n (z-a)^n, \quad z \in \kappa(a; \rho).$$

If f *is analytic at every point of* D, *then* f *is said to be analytic in* D. *Denote by* $\mathcal{H}(D)$ *the set of analytic functions on* D.

The disc $\kappa(a; \rho)$ in Definition 1.45 will be called a *maximal analytic disc of* f at a. Analytic functions on D may be referred to have *maximal analyticity* on D. By the identity theorem, we easily obtain

$$\mathscr{H}(D) \subset \mathcal{H}(D) \subset \text{Hol}(D).$$

The field of fractions of $\mathcal{H}(D)$ will is denoted by $\mathcal{M}(D)$. An element f in the set $\mathcal{M}(D)$ will be called a *meromorphic function* on D. If f has no poles in D, then f is also called *holomorphic*. Take $\rho \in \mathbb{R}^+$. If $f \in \mathcal{H}(\kappa(0; \rho))$, then the maximal analytic disc of f at each $a \in \kappa(0; \rho)$ is just $\kappa(0; \rho)$, particularly, $f \in \mathcal{A}_{(\rho)}(\kappa)$. By Proposition 1.35, we have

$$\mathcal{H}(\kappa(0; \rho)) = \mathcal{A}_{(\rho)}(\kappa),$$

and hence

$$\mathcal{M}(\kappa(0; \rho)) = \left\{ \frac{g}{h} \mid g, h \in \mathcal{A}_{(\rho)}(\kappa), \; h \not\equiv 0 \right\}.$$

We will write

$$\mathcal{M}_{(\rho)}(\kappa) = \mathcal{M}(\kappa(0; \rho)).$$

Thus we see

$$\mathcal{M}(\kappa(0; \rho)) = \bigcap_{r < \rho} \mathcal{M}(\kappa[0; r]).$$

In particular, elements in the set

$$\mathcal{M}_{(\infty)}(\kappa) = \mathcal{M}(\kappa(0; \infty)) = \mathcal{M}(\kappa)$$

are called *meromorphic functions on* κ. We also denote by $\mathcal{M}(\mathbb{C})$ the set of meromorphic functions on $\mathbb{C}$. Obviously, $\mathcal{M}(\kappa)$ contains the set $\kappa(z)$ of rational functions. The elements in $\mathcal{M}(\kappa) - \kappa(z)$ will be called *transcendental*. It is easy to prove that the identity theorem holds for elements of $\mathcal{M}(\kappa)$. In value distribution theory, we mainly study the functions in $\mathcal{M}(\kappa)$.

Take $f \in \mathcal{M}_{(\rho}(\kappa)$ $(0 < \rho \leq \infty)$. Then there are $g, h \in \mathcal{A}_{(\rho}(\kappa)$ with $f = \frac{g}{h}$. Since (3) in Theorem 1.14 holds, we can uniquely extend μ to the meromorphic function $f = \frac{g}{h}$ by defining

$$\mu(r, f) = \frac{\mu(r, g)}{\mu(r, h)} \quad (0 \leq r < \rho).$$

In particular, we have

$$\mu\left(r, \frac{1}{f}\right) = \frac{1}{\mu(r, f)}.$$

Theorem 1.14 can be transferred as follows:

Theorem 1.46. *For $0 < r < \rho$, the function $\mu(r, \cdot) : \mathcal{M}_{(\rho}(\kappa) \longrightarrow \mathbb{R}_+$ satisfies the following properties:*
1) $\mu(r, f) = 0$ if and only if $f \equiv 0$;
2) $\mu(r, f_1 + f_1) \leq \max\{\mu(r, f_1), \mu(r, f_2)\}$; and
3) $\mu(r, f_1 f_2) = \mu(r, f_1)\mu(r, f_2)$.

Also set

$$\gamma(t, f) = \gamma(t, g) - \gamma(t, h),$$

where $\gamma(t, g)$ and $\gamma(t, h)$ are Newton polygons of g and h, respectively. Then we have

$$\gamma\left(t, \frac{1}{f}\right) = -\gamma(t, f),$$

and

$$\mu(p^t, f) = p^{-\gamma(t,f)}.$$

The vertices of g and h is also said to be *vertices* of f. It is clear that, if $-v_p(z)$ is not a vertex for $f(z)$, then

$$|f(z)| = p^{-\gamma(-v_p(z),f)} = \mu(|z|, f).$$

If $a : [\rho_0, \rho) \longrightarrow \mathbb{R}$ and $b : \kappa(0; \rho) \longrightarrow \mathbb{R}$ are real valued functions, then

$$\| \quad a(|z|) \leq b(z)$$

means that for any finite positive number $\rho_0 < R < \rho$, there is a finite set E in $|\kappa| \cap [\rho_0, R]$ such that

$$a(|z|) \leq b(z), \quad |z| \in |\kappa| \cap [\rho_0, R] - E.$$

By using this notation, we have

$$\| \quad \mu(|z|, f) = |f(z)|,$$

for a non-Archimedean meromorphic function f on $\kappa(0; \rho)$. In the sequel, we will take $\rho = \infty$ when we deal with meromorphic (or entire) functions on κ.

Chapter 2

Nevanlinna theory

In this chapter, we will introduce the basic value distribution theory of meromorphic functions defined on a non-Archimedean algebraically closed field of characteristic zero, say, two main theorems and the defect relation.

2.1 Characteristic functions

Let κ be an algebraically closed field of characteristic zero, complete for a non-trivial non-Archimedean absolute value $|\cdot|$. Fix $0 < \rho_0 < r < \rho \leq \infty$ and take $f \in M_{(\rho)}(\kappa)$. Then there are $f_0, f_1 \in A_r(\kappa)$ with $f = \frac{f_1}{f_0}$ such that f_0 and f_1 have not common factors in the ring $A_r(\kappa)$. Take $a \in \kappa \cup \{\infty\}$ and define the *counting function* $n\left(r, \frac{1}{f-a}\right)$ of f for a by

$$n\left(r, \frac{1}{f-a}\right) = \begin{cases} n(r,f) = n\left(r, \frac{1}{f_0}\right) & : \quad a = \infty \\ n\left(r, \frac{1}{f_1-af_0}\right) & : \quad a \neq \infty, \end{cases}$$

define the *valence function* $N\left(r, \frac{1}{f-a}\right)$ of f for a by

$$N\left(r, \frac{1}{f-a}\right) = \begin{cases} N(r,f) = N\left(r, \frac{1}{f_0}\right) & : \quad a = \infty \\ N\left(r, \frac{1}{f_1-af_0}\right) & : \quad a \neq \infty, \end{cases}$$

and write

$$N(r, f = a) = \begin{cases} N(r, f_0 = 0) & : \quad a = \infty \\ N(r, f_1 - af_0 = 0) & : \quad a \neq \infty. \end{cases}$$

Similarly, we can define $\bar{n}(r,f)$, $\bar{N}(r,f)$, $\bar{n}\left(r, \frac{1}{f-a}\right)$ and $\bar{N}\left(r, \frac{1}{f-a}\right)$. Then applying (1.8) to f_0 and f_1, we obtain the *Jensen formula*:

$$N(r, f = 0) - N(r, f = \infty) = \log \mu(r, f) - \log |f^*(0)|, \tag{2.1}$$

and

$$N\left(r, \frac{1}{f}\right) - N(r,f) = \log \mu(r,f) - \log \mu(\rho_0, f), \tag{2.2}$$

33

where $f^*(0)$ is defined as follows: there exists an integer m such that

$$f^*(0) = \lim_{z \to 0} \frac{f(z)}{z^m} \in \kappa_*.$$

Let $\{z_i\}_{i \geq 1}$ and $\{w_i\}_{i \geq 1}$ be the zeros and the poles (counting multiplicity) of f in κ_*, respectively. Then by (1.13), (2.1) can be expressed as follows:

$$\gamma(t, f) - v_p(f^*(0)) = \sum_{v_p(w_i) \geq -t} (v_p(w_i) + t) - \sum_{v_p(z_i) \geq -t} (v_p(z_i) + t)$$
$$+ \left(n(0, f) - n\left(0, \frac{1}{f}\right) \right) t. \tag{2.3}$$

The following fact is trivial:

Proposition 2.1. *Take $f_i \in M_{(\rho)}(\kappa)$ $(i = 1, 2, ..., k)$. Then for $r > 0$, we have*

$$N\left(r, \sum_{i=1}^{k} f_i\right) \leq \sum_{i=1}^{k} N(r, f_i), \quad N\left(r, \prod_{i=1}^{k} f_i\right) \leq \sum_{i=1}^{k} N(r, f_i).$$

Define the *compensation function* by

$$m(r, f) = \log^+ \mu(r, f) = \max\{0, \log \mu(r, f)\}.$$

Then we have

$$\frac{1}{\log p} m\left(p^t, \frac{1}{f}\right) = \gamma^+(t, f) = \max\{0, \gamma(t, f)\},$$

and the following fact:

Proposition 2.2. *Take $f_i \in M_{(\rho)}(\kappa)$ $(i = 1, 2, ..., k)$. Then for $r > 0$, we have*

$$m\left(r, \sum_{i=1}^{k} f_i\right) \leq \max_{1 \leq i \leq k} m(r, f_i), \quad m\left(r, \prod_{i=1}^{k} f_i\right) \leq \sum_{i=1}^{k} m(r, f_i).$$

As usual, we define the *characteristic function*:

$$T(r, f) = m(r, f) + N(r, f) \quad (\rho_0 < r < \infty).$$

Note that

$$\log \mu(r, f) = \log^+ \mu(r, f) - \log^+ \frac{1}{\mu(r, f)}$$
$$= m(r, f) - m\left(r, \frac{1}{f}\right).$$

Then the Jensen formula (2.2) can be rewritten as

$$T\left(r, \frac{1}{f}\right) = T(r, f) - \log \mu(\rho_0, f). \tag{2.4}$$

If f is a non-constant meromorphic function on κ, then f has zeros or poles so that $N\left(r, \frac{1}{f}\right) \to \infty$ or $N(r, f) \to \infty$. For any case, we always have $T(r, f) \to \infty$. Also we have the following fact:

Proposition 2.3. *Take $f, f_i \in M_{(\rho}(\kappa)$ $(i = 1, 2, ..., k)$. Then for $r > 0$, we have*

$$T\left(r, \sum_{i=1}^{k} f_i\right) \le \sum_{i=1}^{k} T(r, f_i), \quad T\left(r, \prod_{i=1}^{k} f_i\right) \le \sum_{i=1}^{k} T(r, f_i),$$

and $T(r, f)$ is an increasing function of r.

Proof. The first part of the proposition follows from Proposition 2.1 and 2.2. Suppose $r_2 > r_1$. If $m(r_1, f) = 0$, then

$$T(r_1, f) = N(r_1, f) \le N(r_2, f) \le T(r_2, f).$$

If $m(r_1, f) > 0$, we obtain $m\left(r_1, \frac{1}{f}\right) = 0$, and hence

$$T\left(r_1, \frac{1}{f}\right) = N\left(r_1, \frac{1}{f}\right) \le N\left(r_2, \frac{1}{f}\right) \le T\left(r_2, \frac{1}{f}\right).$$

Then $T(r_1, f) \le T(r_2, f)$ follows from (2.4). $\square$

Example 2.4. *Given a polynomial*

$$A(z) = \sum_{j=0}^{k} a_j z^j \quad (a_k \neq 0),$$

now we estimate $T(r, A)$. Set

$$r(A) = \max_{0 \le j < k} \left\{ \left(\frac{1}{|a_k|}\right)^{\frac{1}{k}}, \left(\frac{|a_j|}{|a_k|}\right)^{\frac{1}{k-j}} \right\}.$$

If $r > r(A)$, we have $|a_k| r^k > |a_j| r^j$ $(0 \le j < k)$, and hence $\mu(r, A) = |a_k| r^k > 1$. Therefore

$$T(r, A) = m(r, A) = \log \mu(r, A) = k \log r + \log |a_k| \quad (r > r(A)).$$

Take $f \in M_{(\rho}(\kappa)$ again and write $f = f_1/f_0$, where $f_0, f_1 \in A_{(\rho}(\kappa)$. Then

$$\tilde{f} = (f_0, f_1) : \kappa(0; \rho) \longrightarrow \kappa^2$$

is called a *representation* of f. If f_0 and f_1 have no common factors, then $\tilde{f}$ is called a *reduced representation* of f. Write

$$|\tilde{f}(z)| = \max_{k} |f_k(z)|, \quad \mu(r, \tilde{f}) = \max_{k} \mu(r, f_k).$$

Note that

$$\| \quad \mu(|z|, f_k) = |f_k(z)| \quad (k = 0, 1).$$

Then we have

$$\| \quad \mu(|z|, \tilde{f}) = |\tilde{f}(z)|.$$

Assume that $\tilde{f}$ is a reduced representation of f. Noting that

$$
\begin{aligned}
\log \mu(r, \tilde{f}) &= \max\{\log \mu(r, f_0), \log \mu(r, f_1)\} \\
&= \max\left\{0, \log \frac{\mu(r, f_1)}{\mu(r, f_0)}\right\} + \log \mu(r, f_0) \\
&= \max\{0, \log \mu(r, f)\} + \log \mu(r, f_0) \\
&= m(r, f) + \log \mu(r, f_0),
\end{aligned}
$$

and by the Jensen formula

$$
N(r, f) = N\left(r, \frac{1}{f_0}\right) = \log \mu(r, f_0) - \log \mu(\rho_0, f_0),
$$

we obtain

$$
T(r, f) = \log \mu(r, \tilde{f}) - \log \mu(\rho_0, f_0), \tag{2.5}
$$

or equivalently

$$
T(r, f) = \log \mu(r, \tilde{f}) - \log \mu(\rho_0, \tilde{f}) + m(\rho_0, f). \tag{2.6}
$$

We have

$$
\| \quad T(|z|, f) = \log |\tilde{f}(z)| - \log \mu(\rho_0, f_0). \tag{2.7}
$$

From (2.5), we can prove easily the following fact:

Proposition 2.5. *Take $f \in \mathcal{M}_{(\rho)}(\kappa)$. Then $f \in \mathcal{M}^*_{(\rho)}(\kappa)$ if and only if $T(r, f)$ is bounded.*

The inverse proposition also is noticed by Boutabaa-Escassut [18]. If $T(r, f)$ is bounded, the function f also is said to be *bounded*.

Example 2.6. *Given two coprime polynomials*

$$
A(z) = \sum_{j=0}^{k} a_j z^j \ (a_k \neq 0), \quad B(z) = \sum_{j=0}^{q} b_j z^j \ (b_q \neq 0),
$$

now we estimate $T(r, R)$ for the rational function $R = B/A$. According to Example 2.4, when $r > \max\{r(A), r(B)\}$ we have

$$
\log \mu(r, A) = k \log r + O(1), \quad \log \mu(r, B) = q \log r + O(1).
$$

Therefore by (2.5), we have

$$
T(r, R) = \max\{\log \mu(r, A), \log \mu(r, B)\} - \log \mu(\rho_0, A) = \deg(R) \log r + O(1),
$$

where, by definition,

$$
\deg(R) = \max\{k, q\}.
$$

Conversely, if a meromorphic function f on κ satisfies

$$
T(r, f) = O(\log r) \ (r \to \infty),
$$

then f is a rational function. In fact, noting that

$$T\left(r, \frac{1}{f}\right) = T(r, f) + O(1) = O(\log r) \quad (r \to \infty),$$

we have

$$\lim_{r \to \infty} n(r, f) = \lim_{r \to \infty} \frac{N(r, f)}{\log r} \leq \liminf_{r \to \infty} \frac{T(r, f)}{\log r} < \infty,$$

$$\lim_{r \to \infty} n\left(r, \frac{1}{f}\right) = \lim_{r \to \infty} \frac{N\left(r, \frac{1}{f}\right)}{\log r} \leq \liminf_{r \to \infty} \frac{T\left(r, \frac{1}{f}\right)}{\log r} < \infty,$$

which mean that f has only finitely many zeros and poles. Thus f has to be a rational function.

By Example 2.6, we obtain the following properties:

Corollary 2.7. *A meromorphic function f on κ is transcendental if and only if*

$$\lim_{r \to \infty} \frac{T(r, f)}{\log r} = \infty.$$

Corollary 2.8. *Take $a \in \kappa \cup \{\infty\}$. Then a non-constant rational function R on κ satisfies*

$$N\left(r, \frac{1}{R - a}\right) = T(r, R) + O(1),$$

except for at most one value of a.

Proof. We use the symbols in Example 2.6. If $k \neq q$, we have

$$N\left(r, \frac{1}{R - a}\right) = \begin{cases} q \log r + O(1) & : \quad a = 0 \\ k \log r + O(1) & : \quad a = \infty \\ \deg(R) \log r + O(1) & : \quad a \neq 0, \infty . \end{cases}$$

If $k = q$, then

$$N\left(r, \frac{1}{R - a}\right) = \begin{cases} \deg(R) \log r + O(1) & : \quad a = 0 \\ \deg(R) \log r + O(1) & : \quad a = \infty \\ \deg(R) \log r + O(1) & : \quad a \neq b_q/a_k, \infty , \end{cases}$$

but

$$N\left(r, \frac{1}{R - a}\right) \leq (\deg(R) - 1) \log r + O(1)$$

when $a = b_q/a_k$. Thus the corollary follows from Example 2.6. □

For more detail on the definitions and basic properties of the Nevanlinna functions, please refer to the works of Khoái [44], Corrales-Rodrigáñez [28], Khoái-Quang [50], Boutabaa [10], [11], and Cherry and Ye [26].

2.2   Growth estimates of meromorphic functions

Let κ be an algebraically closed field of characteristic zero, complete for a non-trivial non-Archimedean absolute value $|\cdot|$. Let K denote the field κ or $\mathbb{C}$. Take $f, a_j \in \mathcal{M}(K)$ ($j = 0, 1, ..., k$) with $a_k \not\equiv 0$ and define

$$A(z, w) = \sum_{j=0}^{k} a_j(z) w^j. \tag{2.8}$$

For $a \in K \cup \{\infty\}$, let $\mu_f^a(z_0)$ denote the *a-valued multiplicity* of f at z_0, that is, $\mu_f^a(z_0) = m$ if and only if

$$f(z) = \begin{cases} a + (z - z_0)^m h(z) & : \quad a \neq \infty \\ \dfrac{h(z)}{(z - z_0)^m} & : \quad a = \infty \end{cases}$$

with $h(z_0) \neq 0, \infty$. By the definition, we obtain a function

$$\mu_f^a : K \longrightarrow \mathbb{Z}_+$$

with $\mu_f^a(z_0) > 0$ for some $z_0 \in K$ if and only if $f(z_0) = a$.

Lemma 2.9 ([138]). *If $f \in \mathcal{M}(\kappa)$ is non-constant, then*

$$N(r, A \circ f) = kN(r, f) + O\left(\sum_{j=0}^{k} \left\{ N(r, a_j) + N\left(r, \frac{1}{a_j}\right) \right\}\right), \tag{2.9}$$

where $A \circ f$ is defined by $A \circ f(z) = A(z, f(z))$.

Proof. Obviously, we have

$$\mu_{A \circ f}^\infty \leq k \mu_f^\infty + \sum_{j=0}^{k} \mu_{a_j}^\infty,$$

and hence

$$N(r, A \circ f) \leq kN(r, f) + \sum_{j=0}^{k} N(r, a_j). \tag{2.10}$$

We claim that the following inequality

$$\mu_{A \circ f}^\infty \geq k \mu_f^\infty - k \sum_{j=0}^{k} (\mu_{a_j}^\infty + \mu_{a_j}^0) \tag{2.11}$$

holds. Write

$$b_j(z) = a_j(z) f(z)^j \quad j = 0, ..., k,$$

and take $z \in \kappa$. The assertion holds clearly if $\mu_f^\infty(z) = 0$. So we may assume that $\mu_f^\infty(z) > 0$. We distinguish three cases to prove it. If

$$\mu_{b_j}^\infty(z) < \mu_{b_k}^\infty(z) \quad (j < k),$$

then the claim follows from

$$\mu_{A\circ f}^{\infty}(z) = \mu_{b_k}^{\infty}(z) \geq k\mu_f^{\infty}(z) + \mu_{a_k}^{\infty}(z) - \mu_{a_k}^{0}(z) \geq k\mu_f^{\infty}(z) - \mu_{a_k}^{0}(z).$$

If there exists $l < k$ such that

$$\mu_{b_j}^{\infty}(z) < \mu_{b_l}^{\infty}(z) \quad (j \neq l),$$

then for $j = k$

$$k\mu_f^{\infty}(z) + \mu_{a_k}^{\infty}(z) - \mu_{a_k}^{0}(z) < l\mu_f^{\infty}(z) + \mu_{a_l}^{\infty}(z) - \mu_{a_l}^{0}(z)$$

which yields

$$\mu_f^{\infty}(z) \leq (k-l)\mu_f^{\infty}(z) \leq \mu_{a_l}^{\infty}(z) + \mu_{a_k}^{0}(z),$$

and the claim is proved. Finally, if $\mu_{b_j}^{\infty}(z) = \mu_{b_l}^{\infty}(z)$ for some $j > l$, the claim follows from

$$\mu_f^{\infty}(z) \leq (j-l)\mu_f^{\infty}(z) \leq \mu_{a_l}^{\infty}(z) + \mu_{a_j}^{0}(z).$$

Consequently, (2.11) implies

$$N(r, A \circ f) \geq kN(r, f) - k\sum_{j=0}^{k}\left(N(r, a_j) + N\left(r, \frac{1}{a_j}\right)\right). \tag{2.12}$$

Now (2.9) follows clearly from (2.10) and (2.12). $\qquad\square$

Lemma 2.10 ([138]). *If $f \in \mathcal{M}(\kappa)$ is non-constant, then*

$$m(r, A \circ f) = km(r, f) + O\left(\sum_{j=0}^{k} m(r, a_j) + m\left(r, \frac{1}{a_k}\right)\right). \tag{2.13}$$

Proof. Since

$$\mu(r, A \circ f) \leq \max_{0 \leq j \leq k}\{\mu(r, a_j)\mu(r, f)^j\}$$

holds for all $r > 0$, thus we have

$$m(r, A \circ f) \leq km(r, f) + \max_{0 \leq j \leq k} m(r, a_j). \tag{2.14}$$

We first prove the converse for $z \in \kappa$ with

$$f(z) \neq 0, \infty; \quad a_j(z) \neq 0, \infty \quad (0 \leq j \leq k).$$

Write

$$a(z) = \max_{0 \leq j < k}\left\{1, \left(\frac{|a_j(z)|}{|a_k(z)|}\right)^{\frac{1}{k-j}}\right\}.$$

We distinguish two cases. If $|f(z)| > a(z)$, then

$$|a_j(z)||f(z)|^j \leq |a_k(z)|a(z)^{k-j}|f(z)|^j < |a_k(z)||f(z)|^k,$$

and so

$$|A(z, f(z))| = |a_k(z)||f(z)|^k.$$

Hence we obtain

$$\mu(r, f)^k = \frac{\mu(r, A \circ f)}{\mu(r, a_k)},$$

where $r = |z|$. If $|f(z)| \le a(z)$, we have

$$\mu(r, f)^k \le \max_{0 \le j < k} \left\{ 1, \left(\frac{\mu(r, a_j)}{\mu(r, a_k)} \right)^{\frac{k}{k-j}} \right\}.$$

Therefore for any case,

$$\| \ \mu(r, f)^k \le \max_{0 \le j < k} \left\{ 1, \frac{\mu(r, A \circ f)}{\mu(r, a_k)}, \left(\frac{\mu(r, a_j)}{\mu(r, a_k)} \right)^{\frac{k}{k-j}} \right\}$$

holds for the choice of r ($= |z|$), which also is true for any r by continuity of the functions μ. The inequality gives

$$km(r, f) \le m(r, A \circ f) + km\left(r, \frac{1}{a_k}\right) + k \max_{0 \le j < k} m(r, a_j). \qquad (2.15)$$

Thus (2.13) follows from (2.14) and (2.15).   □

The following result follows from Lemma 2.9 and Lemma 2.10:

Theorem 2.11 ([138]). *If $f \in \mathcal{M}(\kappa)$ is non-constant, then*

$$T(r, A \circ f) = kT(r, f) + O\left(\sum_{j=0}^{k} T(r, a_j) \right). \qquad (2.16)$$

Further take $\{b_0, ..., b_q\} \subset \mathcal{M}(\kappa)$ with $b_q \not\equiv 0$ such that

$$B(z, w) = \sum_{j=0}^{q} b_j(z) w^j \qquad (2.17)$$

and $A(z, w)$ are coprime polynomials in w. Define

$$R(z, w) = \frac{A(z, w)}{B(z, w)}. \qquad (2.18)$$

Theorem 2.12 ([138]). *If $f \in \mathcal{M}(\kappa)$ is non-constant, then*

$$T(r, R \circ f) = \max\{k, q\}T(r, f) + O\left(\sum_{j=0}^{k} T(r, a_j) + \sum_{j=0}^{q} T(r, b_j) \right). \qquad (2.19)$$

Proof. W.l.o.g, we may assume $\deg(A) = k \geq q = \deg(B)$. Using the algorithm of division,

$$
\begin{aligned}
A(z,w) &= P_1(z,w)B(z,w) + Q_1(z,w), \\
&\quad \deg(P_1) = k - q, \quad \deg(Q_1) = t_1 < q,
\end{aligned}
$$

$$
\begin{aligned}
B(z,w) &= P_2(z,w)Q_1(z,w) + Q_2(z,w), \\
&\quad \deg(P_2) = q - t_1, \quad \deg(Q_2) = t_2 < t_1,
\end{aligned}
$$

$$
\cdots \cdots
$$

$$
\begin{aligned}
Q_{m-2}(z,w) &= P_m(z,w)Q_{m-1}(z,w) + Q_m(z), \\
&\quad \deg(P_m) = t_{m-2} - t_{m-1}, \quad \deg(Q_m) = t_m = 0.
\end{aligned}
$$

Then $Q_m(z) \not\equiv 0$ since $A(z,w)$ and $B(z,w)$ are coprime, and so

$$
A(z,w)Q(z,w) + B(z,w)P(z,w) = 1, \tag{2.20}
$$

where $P(z,w)$ and $Q(z,w)$ are polynomials in w such that

$$
\deg(P) \leq k - 1, \quad \deg(Q) \leq q - 1,
$$

and such that coefficients are rational functions of $\{a_j(z)\}$ and $\{b_j(z)\}$. Hence

$$
k + \deg(Q) = q + \deg(P), \quad k \geq q.
$$

We thus have $\deg(Q) \leq \deg(P)$. By Theorem 2.11 and the Jensen formula,

$$
\begin{aligned}
T(r,R) &= T\left(r, \frac{A \circ f}{B}\right) \leq T(r,P_1) + T\left(r, \frac{Q_1}{B}\right) \\
&= T(r,P_1) + T\left(r, \frac{B}{Q_1}\right) + O(1) \\
&\leq T(r,P_1) + \cdots + T(r,P_m) + T\left(r, \frac{Q_{m-1}}{Q_m}\right) + O(1) \\
&= (k-q)T(r,f) + (q-t_1)T(r,f) + \cdots (t_{m-1} - t_m)T(r,f) \\
&\quad + O\left(\sum_{j=0}^{k} T(r,a_j) + \sum_{j=0}^{q} T(r,b_j)\right) \\
&= kT(r,f) + O\left(\sum_{j=0}^{k} T(r,a_j) + \sum_{j=0}^{q} T(r,b_j)\right), \tag{2.21}
\end{aligned}
$$

where we simply write $R = R \circ f$, and so on.

Now we use induction to prove the converse inequality. For the case $q = 0$, Theorem 2.12 follows from Theorem 2.11. Assume that Theorem 2.12 hold for rational functions of f with the degree of denominators $\leq q - 1$. By (2.20),

$$T\left(r, \frac{Q}{P} + \frac{B}{A}\right) = T\left(r, \frac{1}{AP}\right) = T(r, AP) + O(1)$$

$$= (k + \deg(P))T(r, f) + O\left(\sum_{j=0}^{k} T(r, a_j) + \sum_{j=0}^{q} T(r, b_j)\right),$$

and by the assumption of the induction,

$$T\left(r, \frac{Q}{P} + \frac{B}{A}\right) \leq T\left(r, \frac{P}{Q}\right) + T(r, R) + O(1)$$

$$\leq (\deg(P))T(r, f) + T(r, R)$$

$$+ O\left(\sum_{j=0}^{k} T(r, a_j) + \sum_{j=0}^{q} T(r, b_j)\right).$$

Therefore, we obtain

$$kT(r, f) \leq T(r, R) + O\left(\sum_{j=0}^{k} T(r, a_j) + \sum_{j=0}^{q} T(r, b_j)\right),$$

and hence (2.19) follows from this and (2.21).   □

Theorem 2.12 was proved by Gackstatter and Laine [37], and Mokhon'ko [93] for meromorphic functions on $\mathbb{C}$ (Also see He-Xiao [54]). For several variables, see Hu-Yang [66],[67]. A weak version of Theorem 2.12 is proved by Boutabaa [10], where the coefficients $\{a_j\}$ and $\{b_j\}$ of the rational function $R(z, w)$ on w are rational functions.

2.3   Two main theorems

Here we introduce some basic facts that will be used in the following sections. Let κ be an algebraically closed field of characteristic zero, complete for a non-trivial non-Archimedean absolute value $|\cdot|$. Fix ρ_0 and ρ with $0 < \rho_0 < \rho \leq \infty$. First of all, we prove the non-Archimedean analogue of the *first main theorem* (cf. [11], [28], [44], [50]).

Theorem 2.13. *Let f be a non-constant meromorphic function in $\kappa(0; \rho)$. Then for every $a \in \kappa$ we have*

$$m\left(r, \frac{1}{f - a}\right) + N\left(r, \frac{1}{f - a}\right) = T(r, f) + O(1) \quad (r \to \rho).$$

Proof. By (2.4), we have

$$m\left(r, \frac{1}{f - a}\right) + N\left(r, \frac{1}{f - a}\right) = T\left(r, \frac{1}{f - a}\right) = T(r, f - a) - \log \mu(\rho_0, f - a).$$

By Proposition 2.3, we obtain

$$T(r, f - a) \leq T(r, f) + \log^+ |a|,$$

$$T(r, f) \leq T(r, f - a) + \log^+ |a|,$$

and hence the theorem follows. □

The following fact usually is referred as the *lemma of logarithmic derivative* (see [1], [11], [26], [50]):

Lemma 2.14. *Let f be a nonconstant meromorphic function in $\kappa(0; \rho)$. Then for any integer $k > 0$,*

$$\mu\left(r, \frac{f^{(k)}}{f}\right) \leq \frac{1}{r^k},$$

and in particular,

$$m\left(r, \frac{f^{(k)}}{f}\right) \leq k \log^+ \frac{1}{r}.$$

Proof. If $f \in \mathcal{A}_{(\rho)}(\kappa)$, then Theorem 1.15 implies

$$\mu\left(r, \frac{f'}{f}\right) = \frac{\mu(r, f')}{\mu(r, f)} \leq \frac{1}{r},$$

and hence

$$\mu\left(r, \frac{f^{(k)}}{f}\right) = \mu\left(r, \prod_{i=1}^{k} \frac{f^{(i)}}{f^{(i-1)}}\right) = \prod_{i=1}^{k} \mu\left(r, \frac{f^{(i)}}{f^{(i-1)}}\right) \leq \frac{1}{r^k},$$

where $f^{(0)} = f$. Now let $f = g/h \in \mathcal{M}_{(\rho)}(\kappa)$. Then

$$\mu\left(r, \frac{f'}{f}\right) = \mu\left(r, \frac{hg' - gh'}{h^2} \cdot \frac{h}{g}\right) = \mu\left(r, \frac{g'}{g} - \frac{h'}{h}\right)$$

$$\leq \max\left\{\mu\left(r, \frac{g'}{g}\right), \mu\left(r, \frac{h'}{h}\right)\right\} \leq \frac{1}{r},$$

and similarly, we can obtain

$$\mu\left(r, \frac{f^{(k)}}{f}\right) \leq \frac{1}{r^k}.$$

□

For a non-constant meromorphic function f in $\kappa(0; \rho)$, define the *ramification term* by

$$N_{\text{Ram}}(r, f) = 2N(r, f) - N(r, f') + N\left(r, \frac{1}{f'}\right).$$

Now we prove the non-Archimedean analogue of the *second main theorem* (cf. [11], [26], [28], [29], [50]).

Theorem 2.15. *Let f be a non-constant meromorphic function in $\kappa(0;\rho)$ and let $a_1,...,a_q$ be distinct numbers of κ. Define*

$$\delta = \min_{i \neq j}\{1, |a_i - a_j|\}, \quad A = \max_i\{1, |a_i|\}.$$

Then for $0 < r < \rho$,

$$(q-1)T(r,f) \ \leq \ N(r,f) + \sum_{j=1}^{q} N\left(r, \frac{1}{f - a_j}\right) - N_{\mathrm{Ram}}(r,f) - \log r + S_f$$

$$\leq \ \overline{N}(r,f) + \sum_{j=1}^{q} \overline{N}\left(r, \frac{1}{f - a_j}\right) - \log r + S_f,$$

where

$$S_f = \sum_{j=1}^{q} \log \mu(\rho_0, f - a_j) - \log \mu(\rho_0, f') + (q-1)\log\frac{A}{\delta}.$$

Proof. Take any $r' \in |\kappa|$ with $\rho_0 < r' < \rho$. Write $f = f_1/f_0$, where $f_0, f_1 \in \mathcal{A}_{r'}(\kappa)$ have no common factors, and set

$$F_0 = f_0, \quad F_i = f_1 - a_i f_0 \ (i = 1, 2, ..., q).$$

Then

$$|f_k(z)| \leq A \max_i\{|F_0(z)|, |F_i(z)|\} \ (k = 0, 1).$$

We will use $\mathbf{W} = \mathbf{W}(f_0, f_1)$ to denote the Wronskian of f_0 and f_1. Then

$$\mathbf{W}_i = \mathbf{W}(F_0, F_i) = \mathbf{W}.$$

Next we fix $z \in \kappa[0; r'] - \kappa[0; \rho_0]$ such that

$$\mathbf{W}(z), f_1(z), F_i(z) \neq 0, \quad i = 0, 1, ..., q.$$

Then there is a $j \in \{1, 2, ..., q\}$ such that

$$|F_j(z)| = \min_{1 \leq i \leq q} |F_i(z)|.$$

Note that

$$|f_0(z)| = \frac{|F_i(z) - F_j(z)|}{|a_j - a_i|} \leq \frac{1}{\delta}|F_i(z)| \ (i \neq j).$$

Thus we can take distinct indices $\beta_1, ..., \beta_{q-1}$ with $\beta_l \neq j$ $(l = 1, 2, ..., q-1)$ such that

$$0 < \max\{\delta|f_0(z)|, |F_j(z)|\} \leq |F_{\beta_1}(z)| \leq \cdots \leq |F_{\beta_{q-1}}(z)| < \infty.$$

Therefore, we have

$$|f_k(z)| \leq \frac{A}{\delta} \max\{\delta|f_0(z)|, |F_j(z)|\} \leq \frac{A}{\delta}|F_{\beta_1}(z)|$$

for $k = 0, 1; l = 1, 2, ..., q - 1$. Hence we obtain

$$|\tilde{f}(z)| = \max_k |f_k(z)| \leq \frac{A}{\delta} |F_{\beta_l}(z)|, \quad l = 0, ..., q - 1,$$

where

$$\tilde{f} = (f_0, f_1) : \kappa \longrightarrow \kappa^2$$

is a representation of f. By $\mathbf{W}_j = \mathbf{W}$, we obtain

$$\log \frac{|F_0(z) \cdots F_q(z)|}{|\mathbf{W}(z)|} = \log |F_{\beta_1}(z) \cdots F_{\beta_{q-1}}(z)| - \log D_j(z),$$

where

$$D_j = \frac{|\mathbf{W}_j|}{|F_0 F_j|} = \left| \frac{F_j'}{F_j} - \frac{F_0'}{F_0} \right|.$$

Then

$$\log |F_{\beta_1}(z) \cdots F_{\beta_{q-1}}(z)| \leq \log \frac{|F_0(z) \cdots F_q(z)|}{|\mathbf{W}(z)|} + \log D_j(z).$$

Therefore, we have

$$(q - 1) \log |\tilde{f}(z)| \leq \log \frac{|F_0(z) \cdots F_q(z)|}{|\mathbf{W}(z)|} + \log D_j(z) + (q - 1) \log \frac{A}{\delta}. \tag{2.22}$$

Set $r = |z|$. By using Theorem 1.15, we obtain

$$D_j(z) \leq \max \left\{ \frac{|F_0'(z)|}{|F_0(z)|}, \frac{|F_j'(z)|}{|F_j(z)|} \right\} \leq \frac{1}{r},$$

and hence

$$\log D_j(z) \leq -\log r.$$

By using the Jensen formula again,

$$\begin{aligned}
\log |F_0(z)| &= \log \mu(r, F_0) = \log \mu(r, f_0) \\
&= N\left(r, \frac{1}{f_0}\right) + \log \mu(\rho_0, f_0) \\
&= N(r, f) + \log \mu(\rho_0, f_0),
\end{aligned}$$

$$\begin{aligned}
\log |\mathbf{W}(z)| &= \log \mu(r, \mathbf{W}) = \log \mu(r, f_0 f_1' - f_1 f_0') \\
&= N\left(r, \frac{1}{\mathbf{W}}\right) + \log \mu(\rho_0, \mathbf{W}) \\
&= N\left(r, \frac{1}{\mathbf{W}}\right) + \log \mu(\rho_0, f') + 2 \log \mu(\rho_0, f_0),
\end{aligned}$$

$$\begin{aligned}
\log |F_i(z)| &= \log \mu(r, F_i) = \log \mu(r, f_1 - a_i f_0) \\
&= N\left(r, \frac{1}{f - a_i}\right) + \log \mu(\rho_0, f - a_i) + \log \mu(\rho_0, f_0),
\end{aligned}$$

for $i = 1, 2, ..., q$, and noting that

$$\log|\tilde{f}(z)| = T(r, f) + \log\mu(\rho_0, f_0),$$

we obtain

$$(q-1)T(r,f) \;\leq\; N(r,f) + \sum_{j=1}^{q} N\left(r, \frac{1}{f-a_j}\right)$$
$$-N\left(r, \frac{1}{\mathbf{W}}\right) - \log r + S_f. \tag{2.23}$$

Note that

$$\mathbf{W} = f_0 f_1' - f_1 f_0' = f_0^2 f'.$$

We can easily prove

$$n\left(r, \frac{1}{\mathbf{W}}\right) = 2n(r,f) - n(r,f') + n\left(r, \frac{1}{f'}\right),$$

which implies

$$N\left(r, \frac{1}{\mathbf{W}}\right) = N_{\mathrm{Ram}}(r,f)$$

and

$$n(r,f) + \sum_{j=1}^{q} n\left(r, \frac{1}{f-a_j}\right) - n\left(r, \frac{1}{\mathbf{W}}\right) \leq \overline{n}(r,f) + \sum_{j=1}^{q} \overline{n}\left(r, \frac{1}{f-a_j}\right).$$

Thus, the inequalities in theorem follows. Note that the set of r for which (2.23) holds is dense in $(\rho_0, r']$. Thus (2.23) also holds for all $\rho_0 < r \leq r'$ by continuity of the functions contained in the inequality. Since r' is arbitrary, hence the theorem is proved. $\square$

Remark. Write

$$n\left(r, \frac{1}{f'}; a_1, ..., a_q\right) := n\left(r, \frac{1}{f'}\right) + \sum_{j=1}^{q} \overline{n}\left(r, \frac{1}{f-a_j}\right) - \sum_{j=1}^{q} n\left(r, \frac{1}{f-a_j}\right), \tag{2.24}$$

and define

$$N\left(r, \frac{1}{f'}; a_1, ..., a_q\right) = \int_{\rho_0}^{r} \frac{n\left(t, \frac{1}{f'}; a_1, ..., a_q\right)}{t} dt. \tag{2.25}$$

Then we have

$$0 \leq n\left(r, \frac{1}{f'}; a_1, ..., a_q\right) \leq n\left(r, \frac{1}{f'}\right),$$

and the inequality in Theorem 2.15 can be expressed as follows

$$(q-1)T(r,f) \leq \overline{N}(r,f) + \sum_{j=1}^{q} \overline{N}\left(r, \frac{1}{f-a_j}\right) - N\left(r, \frac{1}{f'}; a_1, ..., a_q\right) - \log r + S_f.$$

2.4 Notes on the second main theorem

Let κ be an algebraically closed field of characteristic zero, complete for a non-trivial non-Archimedean absolute value $|\cdot|$. Let f be a non-constant meromorphic function in κ. Note that $T(r, f) \to \infty$ as $r \to \infty$. We can define the *defect* of f for a by

$$\delta_f(a) = \liminf_{r \to \infty} \frac{m\left(r, \frac{1}{f-a}\right)}{T(r, f)} = 1 - \limsup_{r \to \infty} \frac{N\left(r, \frac{1}{f-a}\right)}{T(r, f)},$$

$$\Theta_f(a) = 1 - \limsup_{r \to \infty} \frac{\overline{N}\left(r, \frac{1}{f-a}\right)}{T(r, f)},$$

with $0 \le \delta_f(a) \le \Theta_f(a) \le 1$. If $a = \infty$, here we note

$$\delta_f(\infty) = \liminf_{r \to \infty} \frac{m(r, f)}{T(r, f)} = 1 - \limsup_{r \to \infty} \frac{N(r, f)}{T(r, f)},$$

$$\Theta_f(\infty) = 1 - \limsup_{r \to \infty} \frac{\overline{N}(r, f)}{T(r, f)}.$$

Thus Theorem 2.15 directly yields the *defect relation* as follows (cf. [11], [26], [28], [29], [50]):

Theorem 2.16. *Let f be a non-constant meromorphic function in κ. Then*

$$\sum_{a \in \kappa \cup \{\infty\}} \delta_f(a) \le \sum_{a \in \kappa \cup \{\infty\}} \Theta_f(a) \le 2.$$

However, the defect relation is not sharp. Now we examine it more closely. Let f be a non-constant entire function in κ. Then

$$N\left(r, \frac{1}{f}\right) = \log \mu(r, f) - \log \mu(\rho_0, f) \to +\infty$$

as $r \to \infty$, and hence $\mu(r, f) > 1$ when r is sufficiently large. Therefore,

$$T(r, f) = m(r, f) = \log \mu(r, f)$$

if r is sufficiently large, and hence

$$N\left(r, \frac{1}{f}\right) = T(r, f) + O(1).$$

Corollary 1.33 implies

$$N\left(r, \frac{1}{f-a}\right) = T(r, f) + O(1) \tag{2.26}$$

for all $a \in \kappa$. Thus, we obtain the following result:

Theorem 2.17. *Let f be a non-constant entire function in κ. Then $\delta_f(a) = 0$ for all $a \in \kappa$, and $\delta_f(\infty) = 1$.*

Theorem 2.18 ([26]). *If f is a non-constant meromorphic function in κ, then there exists at most one element $a \in \kappa \cup \{\infty\}$ such that $\delta_f(a) > 0$.*

Proof. Assume, to the contrary, that there are two distinct elements $a, b \in \kappa \cup \{\infty\}$ such that $\delta_f(a) > 0$, $\delta_f(b) > 0$. W. l. o. g., let $a = 0$ and $b = \infty$. Otherwise, it is sufficient to consider the following functions:

$$F = f - a \ (a \neq \infty, \ b = \infty) \text{ or } F = \frac{f-a}{f-b} \ (a, b \neq \infty).$$

By assumption,

$$\delta_f(0) = \liminf_{r \to \infty} \frac{m\left(r, \frac{1}{f}\right)}{T(r,f)} > 0, \quad \delta_f(\infty) = \liminf_{r \to \infty} \frac{m(r,f)}{T(r,f)} > 0,$$

which mean that for all r sufficiently large,

$$m\left(r, \frac{1}{f}\right) = \log\frac{1}{\mu(r,f)} > 0, \quad m(r,f) = \log\mu(r,f) > 0.$$

Clearly this is absurd.   □

Corollary 2.19. *Let f be a non-constant meromorphic function in κ. Then*

$$\sum_{a \in \kappa \cup \{\infty\}} \delta_f(a) \leq 1.$$

In fact, we can obtain a strong result. By (2.5) and the Jensen formula, the following formula

$$T(r,f) = \max\left\{N(r,f), N\left(r, \frac{1}{f}\right)\right\} + O(1)$$

holds for a non-constant meromorphic function f in κ. Thus it is easy to prove that the following formula

$$T(r,f) = \max\left\{N\left(r, \frac{1}{f-a}\right), N\left(r, \frac{1}{f-b}\right)\right\} + O(1) \qquad (2.27)$$

holds for any two distinct elements $a, b \in \kappa \cup \{\infty\}$.

Definition 2.20. *Let f be a non-constant meromorphic function in κ. An element $a \in \kappa \cup \{\infty\}$ is said to be a Picard exceptional value of f if $a \notin f(\kappa)$.*

Entire functions in κ have unique Picard exceptional value ∞. Theorem 2.18 shows that any meromorphic function in κ has at most one Picard exceptional value. Next we make a few of notes on the second main theorem. We start by proving the non-Archimedean analogue of a result due to Nevanlinna.

Theorem 2.21. *Let f be a non-constant meromorphic function in κ and let a_1, a_2, a_3 be distinct meromorphic functions in κ. Then*

$$T(r, f) \le \sum_{j=1}^{3} \overline{N}\left(r, \frac{1}{f - a_j}\right) - \log r + S(r), \qquad (2.28)$$

where

$$S(r) = 4T(r, a_1) + 4T(r, a_2) + 5T(r, a_3) + O(1).$$

Proof. Setting

$$\Psi = \frac{f - a_1}{f - a_3} \cdot \frac{a_2 - a_3}{a_2 - a_1}$$

and applying Theorem 2.15 to Ψ with $a_1 = 0$, $a_2 = 1$, we have

$$T(r, \Psi) \le \overline{N}(r, \Psi) + \overline{N}\left(r, \frac{1}{\Psi}\right) + \overline{N}\left(r, \frac{1}{\Psi - 1}\right) - \log r + O(1). \qquad (2.29)$$

Note that

$$
\begin{aligned}
T(r, f) \ &\le\ T(r, f - a_3) + T(r, a_3) \\
&=\ T\left(r, \frac{1}{f - a_3}\right) + T(r, a_3) + O(1) \\
&\le\ T\left(r, \frac{a_3 - a_1}{f - a_3}\right) + T\left(r, \frac{1}{a_3 - a_1}\right) + T(r, a_3) + O(1) \\
&\le\ T\left(r, 1 + \frac{a_3 - a_1}{f - a_3}\right) + T(r, a_1) + 2T(r, a_3) + O(1) \\
&=\ T\left(r, \frac{f - a_1}{f - a_3}\right) + T(r, a_1) + 2T(r, a_3) + O(1) \\
&\le\ T(r, \Psi) + T\left(r, \frac{a_2 - a_1}{a_2 - a_3}\right) \\
&\quad + T(r, a_1) + 2T(r, a_3) + O(1) \\
&\le\ T(r, \Psi) + 2T(r, a_1) \\
&\quad + 2T(r, a_2) + 3T(r, a_3) + O(1).
\end{aligned}
$$

Note that the equations $\Psi(z) = 0, 1, \infty$ have roots only if either $f(z) - a_j(z) = 0$ $(j = 1, 2, 3)$ or if two of the functions a_j become equal. Thus

$$
\begin{aligned}
\overline{N}(r, \Psi) \ &+\ \overline{N}\left(r, \frac{1}{\Psi}\right) + \overline{N}\left(r, \frac{1}{\Psi - 1}\right) \\
&\le\ \sum_{j=1}^{3} \overline{N}\left(r, \frac{1}{f - a_j}\right) + \overline{N}\left(r, \frac{1}{a_1 - a_2}\right) \\
&\quad + \overline{N}\left(r, \frac{1}{a_2 - a_3}\right) + \overline{N}\left(r, \frac{1}{a_3 - a_1}\right) \\
&\le\ \sum_{j=1}^{3} \overline{N}\left(r, \frac{1}{f - a_j}\right) + 2T(r, a_1) \\
&\quad + 2T(r, a_2) + 2T(r, a_3) + O(1).
\end{aligned}
$$

Now (2.28) follows from (2.29).   □

For a non-constant meromorphic function f in κ, define the *ramification term of order 2* by

$$N_{2,\text{Ram}}(r, f) = 3N(r, f) - N(r, f'') + N\left(r, \frac{1}{f''}\right).$$

Now we give a form of the *second main theorem*.

Theorem 2.22. *Let f be a non-constant meromorphic function in κ and let $a_1, ..., a_q$ be distinct numbers of κ. Define*

$$\delta = \min_{i \neq j}\{1, |a_i - a_j|\}, \quad A = \max_i\{1, |a_i|\}.$$

Then

$$
\begin{aligned}
(q-1)T(r, f) \;\leq\; & 2N(r, f) + \sum_{j=1}^{q} N\left(r, \frac{1}{f - a_j}\right) \\
& - N_{2,\text{Ram}}(r, f) - 2\log r + S_f
\end{aligned}
$$

where

$$S_f = \sum_{j=1}^{q} \log \mu(\rho_0, f - a_j) - \log \mu(\rho_0, f'') + (q - 1) \log \frac{A}{\delta}.$$

Proof. Write $f = f_1/f_0$, where $f_0, f_1 \in \mathcal{A}(\kappa)$ have no common factors, and set

$$F_0 = f_0, \quad F_i = f_1 - a_i f_0 \ (i = 1, 2, ..., q),$$

$$\mathbf{W}_2(f_0, f_1) = \begin{vmatrix} 0 & f_0 & f_1 \\ f_0 & f_0' & f_1' \\ 2f_0' & f_0'' & f_1'' \end{vmatrix} = -f_0^3 f''. \tag{2.30}$$

Then

$$|f_k(z)| \leq A \max_i\{|F_0(z)|, |F_i(z)|\} \ (k = 0, 1).$$

We will abbreviate $\mathbf{W}_2 = \mathbf{W}_2(f_0, f_1)$, $\mathbf{W}_{2i} = \mathbf{W}_2(f_0, F_i)$. Then

$$\mathbf{W}_{2i} = \mathbf{W}_2, \quad i = 1, 2, ..., q.$$

Next we fix $z \in \kappa - \kappa[0; \rho_0]$ such that

$$\mathbf{W}_2(z), f_1(z), F_i(z) \neq 0, \quad i = 0, 1, ..., q.$$

According to the proof of Theorem 2.15, we can similarly obtain

$$(q - 1) \log |\tilde{f}(z)| \leq \log \frac{|F_0(z) \cdots F_q(z)|}{|\mathbf{W}_2(z)|} + \log |f_0(z)| + \log D_{2j}(z) + (q - 1) \log \frac{A}{\delta},$$

where

$$D_{2j} = \frac{|\mathbf{W}_{2j}|}{|f_0^2 F_j|} = \left| \frac{F_j''}{F_j} - \frac{f_0''}{f_0} - 2\frac{f_0'}{f_0}\frac{F_j'}{F_j} + 2\left(\frac{f_0'}{f_0}\right)^2 \right|.$$

Set $r = |z|$. By using Theorem 1.15, we obtain

$$D_{2j}(z) \le \max_j \left\{ \left| \frac{F_j''(z)}{F_j(z)} \right|, \left| \frac{f_0''(z)}{f_0(z)} \right|, |2| \left| \frac{f_0'(z)}{f_0(z)} \right| \cdot \left| \frac{F_j'(z)}{F_j(z)} \right|, |2| \left| \frac{f_0'(z)}{f_0(z)} \right|^2 \right\} \le \frac{1}{r^2},$$

and hence

$$\log D_{2j}(z) \le -2 \log r.$$

Now the proof can be completed according to the proof of Theorem 2.15. □

Similarly, for a non-constant meromorphic function f in κ, define the *ramification term of order k* by

$$N_{k,\mathrm{Ram}}(r, f) = (k+1)N(r, f) - N\left(r, f^{(k)}\right) + N\left(r, \frac{1}{f^{(k)}}\right).$$

By considering the following auxiliary function

$$\mathbf{W}_k(f_0, f_1) = (-1)^{[k/2]} f_0^{k+1} \left(\frac{f_1}{f_0}\right)^{(k)},$$

where $[k/2]$ denotes the integral part of $[k/2]$, we can show

$$N_{k,\mathrm{Ram}}(r, f) = N\left(r, \frac{1}{\mathbf{W}_k}\right),$$

where $f = f_1/f_0$, and we can obtain a form of the *second main theorem*:

Theorem 2.23. *Let f be a non-constant meromorphic function in κ and let $a_1, ..., a_q$ be distinct numbers of κ. Define*

$$\delta = \min_{i \ne j}\{1, |a_i - a_j|\}, \quad A = \max_i\{1, |a_i|\}.$$

Then for $k \ge 1$,

$$(q-1)T(r, f) \le kN(r, f) + \sum_{j=1}^q N\left(r, \frac{1}{f - a_j}\right)$$
$$- N_{k,\mathrm{Ram}}(r, f) - k \log r + S_f$$

where

$$S_f = \sum_{j=1}^q \log \mu(\rho_0, f - a_j) - \log \mu\left(\rho_0, f^{(k)}\right) + (q-1)\log\frac{A}{\delta}.$$

Take $a \in \kappa \cup \{\infty\}$. For a positive integer k, define a function

$$\mu_{f,k}^a(z) = \begin{cases} \mu_f^a(z) & : \ \mu_f^a(z) \le k \\ k & : \ \mu_f^a(z) > k. \end{cases}$$

Write

$$n_k\left(r, \frac{1}{f-a}\right) = \sum_{|z|\leq r} \mu^a_{f,k}(z)$$

and

$$N_k\left(r, \frac{1}{f-a}\right) = \int_{\rho_0}^r n_k\left(t, \frac{1}{f-a}\right)\frac{dt}{t}.$$

Write

$$n\left(r, \frac{1}{f^{(k)}}; a_1, ..., a_q\right) := n\left(r, \frac{1}{f^{(k)}}\right) + \sum_{j=1}^q n_k\left(r, \frac{1}{f-a_j}\right) - \sum_{j=1}^q n\left(r, \frac{1}{f-a_j}\right), \quad (2.31)$$

and define

$$N\left(r, \frac{1}{f^{(k)}}; a_1, ..., a_q\right) = \int_{\rho_0}^r \frac{n\left(t, \frac{1}{f^{(k)}}; a_1, ..., a_q\right)}{t} dt. \quad (2.32)$$

We can easily prove

$$kn(r, f) + \sum_{j=1}^q n\left(r, \frac{1}{f-a_j}\right) - n\left(r, \frac{1}{\mathbf{W}_k}\right) \leq k\overline{n}(r, f) + \sum_{j=1}^q n_k\left(r, \frac{1}{f-a_j}\right).$$

Then we have

$$0 \leq n\left(r, \frac{1}{f^{(k)}}; a_1, ..., a_q\right) \leq n\left(r, \frac{1}{f^{(k)}}\right),$$

and the inequality in Theorem 2.23 can be expressed as follows

$$(q-1)T(r, f) \leq k\overline{N}(r, f) + \sum_{j=1}^q N_k\left(r, \frac{1}{f-a_j}\right)$$
$$-N\left(r, \frac{1}{f^{(k)}}; a_1, ..., a_q\right) - k\log r + S_f. \quad (2.33)$$

For two entire functions a ($\neq 0$) and b ($\neq 0$), we suggest the following relation:

$$\mathbf{W}_k(a,b) = \begin{vmatrix} 0 & 0 & 0 & \cdots & 0 & a & b \\ 0 & 0 & 0 & \cdots & a & a' & b' \\ 0 & 0 & 0 & \cdots & \binom{2}{1}a' & a'' & b'' \\ \cdots\cdots\cdots\cdots\cdots\cdots\cdots\cdots \\ 0 & a & \binom{k-2}{1}a' & \cdots & \binom{k-2}{k-3}a^{(k-3)} & a^{(k-2)} & b^{(k-2)} \\ a & \binom{k-1}{1}a' & \binom{k-2}{2}a'' & \cdots & \binom{k-1}{k-2}a^{(k-2)} & a^{(k-1)} & b^{(k-1)} \\ \binom{k}{1}a' & \binom{k}{2}a'' & \binom{k}{3}a^{(3)} & \cdots & \binom{k}{k-1}a^{(k-1)} & a^{(k)} & b^{(k)} \end{vmatrix}. \quad (2.34)$$

We can prove that it is true for $k \leq 5$.

2.5 'abc' conjecture over function fields

Let κ be an algebraically closed field of characteristic zero, complete for a non-trivial non-Archimedean absolute value $|\cdot|$. The following result can be referred to the 'abc' conjecture over $\mathcal{A}(\kappa)$.

Theorem 2.24 ([72]). *Let $a(z)$, $b(z)$ and $c(z)$ be entire functions in κ without common zeros and not all constants such that $a + b = c$. Then*

$$\max\{T(r,a), T(r,b), T(r,c)\} \leq \overline{N}\left(r, \frac{1}{abc}\right) - \log r + O(1).$$

Proof. Write

$$f = \frac{a}{c}, \quad g = \frac{b}{c}.$$

Then f and g all are not constants by our assumptions, and satisfy $f + g = 1$. By the second main theorem (Theorem 2.15), and noting that

$$\overline{N}\left(r, \frac{1}{f-1}\right) = \overline{N}\left(r, \frac{1}{g}\right) = \overline{N}\left(r, \frac{1}{b}\right),$$

we obtain

$$
\begin{aligned}
T(r,f) &\leq \overline{N}(r,f) + \overline{N}\left(r, \frac{1}{f}\right) + \overline{N}\left(r, \frac{1}{f-1}\right) - \log r + O(1) \\
&= \overline{N}\left(r, \frac{1}{c}\right) + \overline{N}\left(r, \frac{1}{a}\right) + \overline{N}\left(r, \frac{1}{b}\right) - \log r + O(1) \\
&= \overline{N}\left(r, \frac{1}{abc}\right) - \log r + O(1).
\end{aligned}
$$

Similarly, we have

$$T(r,g) \leq \overline{N}\left(r, \frac{1}{abc}\right) - \log r + O(1).$$

By (2.27) and (2.26),

$$
\begin{aligned}
T(r,f) &= \max\left\{N(r,f), N\left(r, \frac{1}{f}\right)\right\} + O(1) \\
&= \max\left\{N\left(r, \frac{1}{c}\right), N\left(r, \frac{1}{a}\right)\right\} + O(1) \\
&= \max\left\{T(r,c), T(r,a)\right\} + O(1).
\end{aligned}
$$

Similarly,

$$T(r,g) = \max\left\{T(r,c), T(r,b)\right\} + O(1),$$

and, hence, the theorem follows from the above estimates. $\square$

Corollary 2.25 (Mason's theorem, cf. [89],[90],[91], or [84]). *Let* $a(z)$, $b(z)$, $c(z)$ *be relatively prime polynomials in* κ *and not all constants such that* $a + b = c$. *Then*

$$\max\{\deg(a), \deg(b), \deg(c)\} \leq \bar{n}\left(\infty, \frac{1}{abc}\right) - 1,$$

where

$$\bar{n}\left(\infty, \frac{1}{abc}\right) = \lim_{r \to \infty} \bar{n}\left(r, \frac{1}{abc}\right).$$

Next we extend Theorem 2.24. First of all, we prove a non-Archimedean analogue of a result due to Nevanlinna[98].

Lemma 2.26. *Let* $f_j(j = 1, \cdots, k)$ *be linearly independent meromorphic functions on* κ *such that*

$$f_1 + \cdots + f_k = 1 \ (k \geq 2). \tag{2.35}$$

Then

$$\begin{aligned}
T(r, f_j) \ &< \ \sum_{i=1}^{k} N\left(r, \frac{1}{f_i}\right) - \sum_{i \neq j} N(r, f_i) + N(r, \mathbf{W}) \\
&\quad -N\left(r, \frac{1}{\mathbf{W}}\right) - \frac{k(k-1)}{2} \log r + O(1), \quad 1 \leq j \leq k
\end{aligned} \tag{2.36}$$

where $\mathbf{W}$ *is the Wronskian of* $f_1, ..., f_k$.

Proof. Since $f_1, \cdots, f_k$ are linearly independent, then the Wronskian $\mathbf{W} \not\equiv 0$. By (2.35) and

$$f_1^{(i)} + \cdots + f_k^{(i)} = 0 \quad i = 1, \cdots, k - 1,$$

we have $f_j = \frac{g_j}{g}, j = 1, \cdots, k$, where

$$g = \frac{\mathbf{W}}{f_1 \cdots f_k} = \det\left(\frac{f_j^{(i)}}{f_j}\right),$$

and g_j is the minor of the j-th term on the first row of g. Write $f_j = f_{j,1}/f_{j,0}$, where $f_{j,0}, f_{j,1} \in \mathcal{A}(\kappa)$ have no common factors. Then

$$\begin{aligned}
T(r, f_j) \ &= \ \max\{\log \mu(r, f_{j,0}), \log \mu(r, f_{j,1})\} + O(1) \\
&= \ \max\left\{\log \mu(r, f_{j,0}), \log \mu\left(r, f_{j,0}\frac{g_j}{g}\right)\right\} + O(1) \\
&= \ \log \mu(r, f_{j,0}) + \max\left\{0, \log \frac{\mu(r, g_j)}{\mu(r, g)}\right\} + O(1) \\
&= \ N(r, f_j) - \log \mu(r, g) + \max\{\log \mu(r, g_j), \log \mu(r, g)\} + O(1), \quad (2.37)
\end{aligned}$$

where we use Jensen's formula which also gives

$$
\begin{aligned}
- \log \mu(r, g) &= -\log \mu(r, \mathbf{W}) + \sum_{i=1}^{k} \log \mu(r, f_i) \\
&= N(r, \mathbf{W}) - N\left(r, \frac{1}{\mathbf{W}}\right) \\
&\quad + \sum_{i=1}^{k} \left\{ N\left(r, \frac{1}{f_i}\right) - N(r, f_i) \right\} + O(1).
\end{aligned}
\tag{2.38}
$$

By using Lemma 2.14, we obtain

$$
\max\{\log \mu(r, g_j), \log \mu(r, g)\} \leq -\frac{k(k-1)}{2} \log r.
\tag{2.39}
$$

Hence (2.36) follows from (2.37), (2.38), and (2.39). □

By using Lemma 2.26, we can obtain a non-Archimedean version of a result due to Li and Yang [87] as follows.

Corollary 2.27. *Assume that the conditions of Lemma 2.26 hold. Then*

$$
T(r, f_j) < \sum_{i=1}^{k} N_{k-1}\left(r, \frac{1}{f_i}\right) + \vartheta_k \sum_{i=1}^{k} \overline{N}(r, f_i) - \frac{k(k-1)}{2} \log r + O(1)
$$

and

$$
T(r, f_j) < \sum_{i=1}^{k} N_{k-1}\left(r, \frac{1}{f_i}\right) + (k-1) \sum_{i \neq j} \overline{N}(r, f_i) - \frac{k(k-1)}{2} \log r + O(1),
$$

hold for $j = 1, ..., k$, where

$$
\vartheta_k = \begin{cases} \frac{1}{2} & : \quad k = 2, \\ \frac{2k-\frac{2}{3}}{3} & : \quad k = 3, 4, 5, \\ \frac{2k+1-2\sqrt{2k}}{2} & : \quad k \geq 6. \end{cases}
$$

Proof. Here we follow the main idea of Li and Yang [87]. W. l. o. g., we consider only the case $j = 1$. For $f \in \mathcal{M}(\kappa)$ and $a \in \kappa \cup \{\infty\}$, define a function $\overline{\mu}_f^a$ by

$$
\overline{\mu}_f^a(z) = \begin{cases} 1 & : \quad \mu_f^a(z) > 0 \\ 0 & : \quad \text{others}. \end{cases}
$$

Clearly, Corollary 2.27 follows from Lemma 2.26 and the following inequalities

$$
\begin{aligned}
u &= \sum_{i=1}^{k} \mu_{f_i}^0 - \sum_{i=2}^{k} \mu_{f_i}^\infty + \mu_{\mathbf{W}}^\infty - \mu_{\mathbf{W}}^0 \\
&\leq \sum_{i=1}^{k} \mu_{f_i, k-1}^0 + \vartheta_k \sum_{i=1}^{k} \overline{\mu}_{f_i}^\infty = v,
\end{aligned}
\tag{2.40}
$$

and

$$u \le \sum_{i=1}^{k} \mu^0_{f_i,k-1} + (k-1) \sum_{i=2}^{k} \overline{\mu}^\infty_{f_i} = w. \tag{2.41}$$

Take $z \in \kappa$. We distinguish several cases to show $u(z) \le \max\{v(z), w(z)\}$. If z is not a pole of f_i for any $i = 1, ..., k$, obviously we have

$$\mu^0_{f_i^{(j)}}(z) \ge \mu^0_{f_i}(z) - \mu^0_{f_i,j}(z) \ge \mu^0_{f_i}(z) - \mu^0_{f_i,k-1}(z), \quad 1 \le i \le k, \ 1 \le j \le k-1,$$

and, hence,

$$\mu^0_{\mathbf{W}}(z) \ge \sum_{i=1}^{k} \{\mu^0_{f_i}(z) - \mu^0_{f_i,k-1}(z)\},$$

that is

$$u(z) = \sum_{i=1}^{k} \mu^0_{f_i}(z) - \mu^0_{\mathbf{W}}(z) \le \sum_{i=1}^{k} \mu^0_{f_i,k-1}(z) = v(z) = w(z).$$

Next, suppose that z is a pole of f_i for each $i = 1, ..., k$. Let $\mathbf{W}_i$ be the Wronskian of $f'_1, ..., f'_{i-1}, f'_{i+1}, ..., f'_k$. Then $\mathbf{W} = (-1)^{i+1} \mathbf{W}_i$. Since

$$\mu^\infty_{f_i^{(j)}}(z) = \mu^\infty_{f_i}(z) + j = \mu^\infty_{f_i}(z) + j\overline{\mu}^\infty_{f_i}(z), \quad 1 \le i \le k, \ 1 \le j \le k-1,$$

we have

$$\begin{aligned}
\mu^\infty_{\mathbf{W}}(z) &= \mu^\infty_{\mathbf{W}_1}(z) \le \sum_{i=2}^{k} \mu^\infty_{f_i}(z) + \frac{k(k-1)}{2} \\
&= \sum_{i=2}^{k} \mu^\infty_{f_i}(z) + \frac{k}{2} \sum_{i=2}^{k} \overline{\mu}^\infty_{f_i}(z) \\
&= \sum_{i=2}^{k} \mu^\infty_{f_i}(z) + \frac{k-1}{2} \sum_{i=1}^{k} \overline{\mu}^\infty_{f_i}(z),
\end{aligned}$$

which implies

$$\begin{aligned}
u(z) &= \mu^\infty_{\mathbf{W}}(z) - \sum_{i=2}^{k} \mu^\infty_{f_i}(z) \le \max\left\{ \frac{k-1}{2} \sum_{i=1}^{k} \overline{\mu}^\infty_{f_i}(z), \frac{k}{2} \sum_{i=2}^{k} \overline{\mu}^\infty_{f_i}(z) \right\} \\
&\le \max\{v(z), w(z)\}.
\end{aligned}$$

If $k = 2$, these are all possible cases, and therefore Corollary 2.27 is proved.

Assume $k \ge 3$. Except for above two cases, the following two cases may occur: Case 3: $\mu^\infty_{f_i}(z) > 0$, but $\mu^\infty_{f_i}(z) = 0$ for some i; or Case 4: $\mu^\infty_{f_1}(z) = 0$, but $\mu^\infty_{f_i}(z) > 0$ for some i. For Case 3, w. l. o. g., assume that $\mu^\infty_{f_i}(z) > 0$ for $i = 1, ..., l(< k)$, and $\mu^\infty_{f_i}(z) = 0$ for $i > l$. When $\mu^\infty_{\mathbf{W}}(z) > 0$, we have

$$\begin{aligned}
\mu^\infty_{\mathbf{W}}(z) &= \mu^\infty_{\mathbf{W}_1}(z) \le \sum_{i=2}^{l} \mu^\infty_{f_i}(z) + \sum_{i=1}^{l-1}(k-i) - \sum_{i=l+1}^{k} \{\mu^0_{f_i}(z) - \mu^0_{f_i,k-1}(z)\} \\
&= \sum_{i=2}^{l} \mu^\infty_{f_i}(z) + \frac{(l-1)(2k-l)}{2} - \sum_{i=l+1}^{k} \{\mu^0_{f_i}(z) - \mu^0_{f_i,k-1}(z)\},
\end{aligned}$$

which means

$$u(z) \leq \sum_{i=l+1}^{k} \mu_{f_i,k-1}^0(z) + \frac{(l-1)(2k-l)}{2}$$

$$= \sum_{i=l+1}^{k} \mu_{f_i,k-1}^0(z) + \frac{2k-l}{2} \sum_{i=2}^{l} \overline{\mu}_{f_i}^\infty(z)$$

$$= \sum_{i=l+1}^{k} \mu_{f_i,k-1}^0(z) + \frac{(l-1)(2k-l)}{2l} \sum_{i=1}^{l} \overline{\mu}_{f_i}^\infty(z)$$

$$\leq \max\{v(z), w(z)\}.$$

Otherwise, we have

$$\mu_{\mathbf{W}}^0(z) = \mu_{\mathbf{W}_1}^0(z) \geq \sum_{i=l+1}^{k} \{\mu_{f_i}^0(z) - \mu_{f_i,k-1}^0(z)\} - \sum_{i=2}^{l} \mu_{f_i}^\infty(z) - \frac{(l-1)(2k-l)}{2},$$

and, similarly, we have the inequality $u(z) \leq \max\{v(z), w(z)\}$.

Finally, we consider Case 4. W. l. o. g., assume that $\mu_{f_i}^\infty(z) = 0$ for $i = 1, ..., j (< k)$, and $\mu_{f_i}^\infty(z) > 0$ for $i > j$. When $\mu_{\mathbf{W}}^\infty(z) > 0$, we have

$$\mu_{\mathbf{W}}^\infty(z) = \mu_{\mathbf{W}_k}^\infty(z) \leq \sum_{i=j+1}^{k-1} \mu_{f_i}^\infty(z) + \sum_{i=1}^{k-j-1}(k-i) - \sum_{i=1}^{j}\{\mu_{f_i}^0(z) - \mu_{f_i,k-1}^0(z)\}$$

$$= \sum_{i=j+1}^{k-1} \mu_{f_i}^\infty(z) + \frac{(k-j-1)(k+j)}{2} - \sum_{i=1}^{j}\{\mu_{f_i}^0(z) - \mu_{f_i,k-1}^0(z)\},$$

which yields

$$u(z) \leq \sum_{i=1}^{j} \mu_{f_i,k-1}^0(z) + \frac{(k-j-1)(k+j)}{2}$$

$$= \sum_{i=1}^{j} \mu_{f_i,k-1}^0(z) + \frac{k+j}{2} \sum_{i=j+1}^{k-1} \overline{\mu}_{f_i}^\infty(z)$$

$$= \sum_{i=1}^{j} \mu_{f_i,k-1}^0(z) + \frac{(k-j-1)(k+j)}{2(k-j)} \sum_{i=j+1}^{k} \overline{\mu}_{f_i}^\infty(z)$$

$$\leq \max\{v(z), w(z)\}.$$

Otherwise,

$$\mu_{\mathbf{W}}^0(z) = \mu_{\mathbf{W}_k}^0(z) \geq \sum_{i=1}^{j}\{\mu_{f_i}^0(z) - \mu_{f_i,k-1}^0(z)\} - \sum_{i=j+1}^{k-1} \mu_{f_i}^\infty(z) - \frac{(k-j-1)(k+j)}{2},$$

and, similarly, we have the inequality $u(z) \leq \max\{v(z), w(z)\}$. $\square$

By estimating the multiplicities of zeros and poles of $\mathbf{W}$ at zeros of $f_0, f_1, ..., f_k$ carefully, from Lemma 2.26 one can obtain the following generalization of Theorem 2.24:

Corollary 2.28. *Let $f_j(j = 0, \cdots, k)$ be entire functions on κ such that f_j, f_0 have no common zeros for $j = 1, ..., k$, f_j/f_0 $(j = 1, \cdots, k)$ be linearly independent on κ and*

$$f_1 + \cdots + f_k = f_0. \tag{2.42}$$

Then

$$\max_{0 \le j \le k} \{T(r, f_j)\} \le \sum_{i=1}^{k} N_{k-1}\left(r, \frac{1}{f_i}\right) + \iota_k \overline{N}\left(r, \frac{1}{f_0}\right) - \frac{k(k-1)}{2} \log r + O(1),$$

where

$$\iota_k = \begin{cases} 1 & : \quad k = 2 \\ \frac{k(k-1)}{2} - 1 & : \quad k \ge 3. \end{cases}$$

Corollary 2.29. *Let $f_j(j = 0, \cdots, k)$ be polynomials on κ such that f_j, f_0 are relatively prime for $j = 1, ..., k$, f_j/f_0 $(j = 1, \cdots, k)$ be linearly independent on κ and*

$$f_1 + \cdots + f_k = f_0. \tag{2.43}$$

Then

$$\max_{0 \le j \le k} \{\deg(f_j)\} \le \sum_{i=1}^{k} n_{k-1}\left(\infty, \frac{1}{f_i}\right) + \iota_k \overline{n}\left(\infty, \frac{1}{f_0}\right) - \frac{k(k-1)}{2},$$

where

$$n_{k-1}\left(\infty, \frac{1}{f_i}\right) = \lim_{r \to \infty} n_{k-1}\left(r, \frac{1}{f_i}\right).$$

2.6   Waring's problem over function fields

Let κ be an algebraically closed field of characteristic zero, complete for a non-trivial non-Archimedean absolute value $|\cdot|$. By using Corollary 2.27, we can easily solve the *Waring's problem* for meromorphic functions. The following result is the non-Archimedean version of a result due to Gross [41]:

Theorem 2.30. *For $n \ge 3$, there do not exist two nonconstant meromorphic functions f and g on κ satisfying*

$$f^n + g^n = 1. \tag{2.44}$$

Proof. Assume, to the contrary, that there exist two nonconstant meromorphic functions f and g on κ satisfying (2.44) for some $n \ge 3$. Then f^n and g^n are linearly independent and satisfy

$$nT(r, f) = T(r, f^n) = T(r, 1 - g^n) = nT(r, g) + O(1).$$

By Corollary 2.27, we have

$$\begin{aligned} nT(r, f) &= T(r, f^n) \le \overline{N}\left(r, \frac{1}{f^n}\right) + \overline{N}\left(r, \frac{1}{g^n}\right) + \overline{N}(r, g^n) - \log r + O(1) \\ &= \overline{N}\left(r, \frac{1}{f}\right) + \overline{N}\left(r, \frac{1}{g}\right) + \overline{N}(r, g) - \log r + O(1) \\ &\le 3T(r, f) - \log r + O(1) \end{aligned}$$

which is impossible when $n \geq 3$. $\qquad\qquad\qquad\qquad\qquad\qquad\qquad\qquad$ $\square$

For the case $n = 2$, we can prove easily that there do not exist two nonconstant entire functions f and g on κ satisfying

$$f^2 + g^2 = 1. \tag{2.45}$$

In fact, if there are two f and g satisfying (2.45). Let a_1 and a_2 be the zeros of $z^2 + 1 = 0$ over κ. Then $a_1 + a_2 = 0$ and $a_1 a_2 = 1$. By (2.45), we have

$$(f - a_1 g)(f - a_2 g) = 1.$$

Thus $f - a_1 g$ and $f - a_2 g$ have to be constant, and so f and g are constant. Similarly, we can prove the following results:

Theorem 2.31. *For $n \geq 4$, there do not exist two unbounded meromorphic functions f and g on $\kappa(0; \rho)$ satisfying (2.44).*

Theorem 2.32. *For $\min\{m, n\} \geq 3$, there do not exist two nonconstant meromorphic functions f and g on κ satisfying*

$$f^m + g^n = 1. \tag{2.46}$$

If $\min\{m, n\} \geq 4$, there do not exist two unbounded meromorphic functions f and g on $\kappa(0; \rho)$ satisfying (2.46).

These results are also obtained independently by Boutabaa [12] and Boutabaa-Escassut [15]. In fact, they prove the following better result than above theorems:

Theorem 2.33. *For $\min\{m, n\} \geq 2$, $\max\{m, n\} \geq 3$, there do not exist two nonconstant meromorphic functions f and g on κ satisfying (2.46). If $\min\{m, n\} \geq 2$, $\max\{m, n\} \geq 3$, there do not exist two unbounded holomorphic functions f and g on $\kappa(0; \rho)$ satisfying (2.46). If $\min\{m, n\} \geq 3$, $\max\{m, n\} \geq 4$, there do not exist two unbounded meromorphic functions f and g on $\kappa(0; \rho)$ satisfying (2.46).*

Proof. Assume, to the contrary, that there exist two meromorphic functions f and g on $\kappa(0; \rho)$ (or κ) satisfying (2.46). W. l. o. g., we assume $m \geq n$. Let $a_1, ..., a_m$ be the zeros of $z^m - 1$ in κ. Then for each $j = 1, ..., m$, each zero of $f - a_j$ has order $\geq n$, and hence

$$\overline{N}\left(r, \frac{1}{f - a_j}\right) \leq \frac{1}{n} N\left(r, \frac{1}{f - a_j}\right) \leq \frac{1}{n} T(r, f) + O(1).$$

By using the second main theorem, one has

$$(m - 1)T(r, f) \leq \overline{N}(r, f) + \frac{m}{n} T(r, f) - \log r + O(1), \tag{2.47}$$

which implies

$$\frac{mn - 2n - m}{n} T(r, f) \leq -\log r + O(1). \tag{2.48}$$

Therefore $mn \leq 2n + m$ if f is unbounded on $\kappa(0; \rho)$. This inequality contradicts the hypothesis $\min\{m, n\} \geq 3$, $\max\{m, n\} \geq 4$. If $f \in A_{(\rho)}(\kappa)$ is unbounded, then (2.47) implies $n(m - 1) \leq m$, a contradiction with $\min\{m, n\} \geq 2$, $\max\{m, n\} \geq 3$.

If $f, g \in \mathcal{M}(\kappa) - \kappa$, it follows from (2.48) that $mn < 2n + m$. This is impossible if either $n \geq 3$ or $n = 2, m \geq 4$. For the special case $n = 2$, $m = 3$, each pole of f^3 is a pole of g^2, and therefore has an order at least 2. Hence

$$\overline{N}(r, f) \leq \frac{1}{2} N(r, f) \leq \frac{1}{2} T(r, f).$$

Thus by (2.47), we obtain

$$2T(r, f) \leq \left(\frac{1}{2} + \frac{3}{2}\right) T(r, f) - \log r + O(1),$$

which is impossible. The proof of the theorem is completed. $\square$

Generally, they obtain the following two results (see [15]):

Theorem 2.34. *Let A and B be two relatively prime polynomials over κ, let t be the number of distinct zeros of B, and let $g \in \mathcal{M}_{(\rho}(\kappa)$ be such that all poles of g have order $\geq m \geq 1$. Suppose that there exists a function $f \in \mathcal{M}_{(\rho}(\kappa)$ satisfying*

$$g(z) B(f(z)) = A(f(z)), \quad z \in \kappa(0; \rho).$$

1) Assume that f is unbounded. Then $mt \leq 2m + \deg(B)$. Moreover, if $\deg(A) > \deg(B)$, then $mt \leq \min\{m + \deg(A), 2m + \deg(B)\}$.

2) Assume $f \in \mathcal{M}(\kappa) - \kappa$. Then $mt < 2m + \deg(B)$. Moreover, if $\deg(A) > \deg(B)$, then $mt < \min\{m + \deg(A), 2m + \deg(B)\}$.

Proof. Set $k = \deg(A)$ and $q = \deg(B)$. The inequality $mt < 2m + q$ is trivial if $t < 2$. So we may suppose $t \geq 2$. Write

$$B(z) = (z - b_1)^{q_1} \cdots (z - b_t)^{q_t}.$$

Since A and B have no common zeros, each zero z_0 of $B(f(z))$ is a pole of $g(z)$. Hence z_0 is a zero of $B(f(z))$ of order at least m, that is, $q_j \mu_f^{b_j}(z_0) \geq m$ when $f(z_0) - b_j = 0$. Therefore

$$\overline{N}\left(r, \frac{1}{f - b_j}\right) \leq \frac{q_j}{m} N\left(r, \frac{1}{f - b_j}\right) \leq \frac{q_j}{m} T(r, f) + O(1),$$

where the last inequality is obtained by first main theorem, and hence

$$\sum_{j=1}^{t} \overline{N}\left(r, \frac{1}{f - b_j}\right) \leq \frac{q}{m} T(r, f) + O(1).$$

Further, by using the second main theorem, one has

$$(t - 1) T(r, f) \leq \overline{N}(r, f) + \frac{q}{m} T(r, f) - \log r + O(1), \tag{2.49}$$

which implies

$$\frac{mt - 2m - q}{m} T(r, f) \leq -\log r + O(1). \tag{2.50}$$

Therefore $mt \leq 2m + q$ if f is unbounded. If $f, g \in \mathcal{M}(\kappa) - \kappa$, it follows from (2.50) that $mt < 2m + q$ since $\log r \to +\infty$ as $r \to +\infty$.

Next suppose that $k > q$. Then each pole of f also is a pole of g with $\mu_g^\infty = (k - q)\mu_f^\infty$. Hence

$$\overline{N}(r, f) \leq \frac{k - q}{m} N(r, f) \leq \frac{k - q}{m} T(r, f).$$

Thus the inequality (2.49) yields

$$(t - 1)T(r, f) \leq \frac{k}{m} T(r, f) - \log r + O(1), \tag{2.51}$$

which further implies $t - 1 \leq k/m$ if f is unbounded. Hence we obtain

$$mt \leq \min\{m + k, 2m + q\}.$$

If $f, g \in \mathcal{M}(\kappa) - \kappa$, it follows from (2.51) that $mt < m + k$ since $\log r \to +\infty$ as $r \to +\infty$, and so

$$mt < \min\{m + k, 2m + q\}.$$

The theorem is proved. $\qquad\square$

Theorem 2.35. *Given two relatively prime polynomials*

$$A(z) = a \prod_{i=1}^{s} (z - a_i)^{k_i}, \quad B(z) = b \prod_{j=1}^{t} (z - b_j)^{q_j}$$

over κ, where all a_i and b_j are distinct. Take a positive integer m and suppose that there exist two functions $f, g \in \mathcal{M}_{(\rho}(\kappa)$ satisfying

$$(g(z))^m B(f(z)) = A(f(z)), \quad z \in \kappa(0; \rho).$$

Then
   1) both f and g are bounded on $\kappa(0; \rho)$ if either

$$s + t > 1 + \frac{1}{m} \left((m, |\deg(A) - \deg(B)|) + \sum_{i=1}^{s} (m, k_i) + \sum_{j=1}^{t} (m, q_j) \right),$$

or $f \in \mathcal{A}_{(\rho}(\kappa)$ and

$$s + t > 1 + \frac{1}{m} \left(\sum_{i=1}^{s} (m, k_i) + \sum_{j=1}^{t} (m, q_j) \right),$$

where (m, n) is the largest common factor of m and n, or
   2) both f and g are constant if $f, g \in \mathcal{M}(\kappa)$ and if

$$s + t \geq 1 + \frac{1}{m} \left((m, |\deg(A) - \deg(B)|) + \sum_{i=1}^{s} (m, k_i) + \sum_{j=1}^{t} (m, q_j) \right).$$

Proof. It is clear that if f is constant (or bounded) so is g. Suppose, to the contrary, that f is not constant (or bounded). Set

$$k = \deg(A) = \sum_{i=1}^{s} k_i, \quad q = \deg(B) = \sum_{j=1}^{t} q_j.$$

Obviously, we have

$$k_i \mu_f^{a_i} = m \mu_g^0, \quad i = 1, ..., s,$$

and

$$q_j \mu_f^{b_j} = m \mu_g^\infty, \quad j = 1, ..., t.$$

It follow that

$$\mu_f^{a_i} \geq \frac{m}{(m, k_i)}, \quad i = 1, ..., s,$$

and

$$\mu_f^{b_j} \geq \frac{m}{(m, q_j)}, \quad j = 1, ..., t.$$

Hence we obtain

$$\overline{N}\left(r, \frac{1}{f - a_i}\right) \leq \frac{(m, k_i)}{m} N\left(r, \frac{1}{f - a_i}\right) \leq \frac{(m, k_i)}{m} T(r, f) + O(1), \quad i = 1, ..., s,$$

and

$$\overline{N}\left(r, \frac{1}{f - b_j}\right) \leq \frac{(m, q_j)}{m} N\left(r, \frac{1}{f - b_j}\right) \leq \frac{(m, q_j)}{m} T(r, f) + O(1), \quad j = 1, ..., t.$$

By using the second main theorem, one has

$$(s + t - 1)T(r, f) \leq \sum_{i=1}^{s} \overline{N}\left(r, \frac{1}{f - a_i}\right) + \sum_{j=1}^{t} \overline{N}\left(r, \frac{1}{f - b_j}\right)$$
$$+ \overline{N}(r, f) - \log r + O(1)$$
$$\leq \frac{1}{m}\left(\sum_{i=1}^{s}(m, k_i) + \sum_{j=1}^{t}(m, q_j)\right) T(r, f)$$
$$+ \overline{N}(r, f) - \log r + O(1). \tag{2.52}$$

Note that if $\mu_f^\infty(z_0) > 0$,

$$(k - q)\mu_f^\infty(z_0) = m\left(\mu_g^\infty(z_0) - \mu_g^0(z_0)\right).$$

We have

$$\overline{N}(r, f) \leq \frac{(m, |k - q|)}{m} N(r, f) \leq \frac{(m, |k - q|)}{m} T(r, f),$$

which and (2.52) yield

$$(s + t - 1)T(r, f) \leq \frac{1}{m}\left((m, |k - q|) + \sum_{i=1}^{s}(m, k_i) + \sum_{j=1}^{t}(m, q_j)\right) T(r, f)$$
$$- \log r + O(1). \tag{2.53}$$

If f lies in $\mathcal{A}_{(\rho}(\kappa)$ and is unbounded, then (2.52) implies

$$s + t - 1 \leq \frac{1}{m}\left(\sum_{i=1}^{s}(m, k_i) + \sum_{j=1}^{t}(m, q_j)\right).$$

This is a contradiction with the hypothesis. Hence f is bounded.

Assume that $f \in \mathcal{M}_{(\rho}(\kappa)$ is unbounded. Then (2.53) implies

$$s + t - 1 \leq \frac{1}{m}\left((m, |k - q|) + \sum_{i=1}^{s}(m, k_i) + \sum_{j=1}^{t}(m, q_j)\right). \tag{2.54}$$

This is a contradiction with the hypothesis. Hence f is bounded.

Finally, if $f \in \mathcal{M}(\kappa)$ is not constant, then from (2.53), (2.54) is strict since $\log r \to +\infty$, which also contradicts with the hypothesis. Hence f (and so g) is constant. $\square$

Theorem 2.35 easily yields Picard-Berkovich's theorem (see [6]) as far as curves of genus 1 and 2 are concerned.

Corollary 2.36. *Let M be an algebraic curve of genus 1 or 2 on κ and let $f, g \in M(\kappa)$ be such that $(f(z), g(z)) \in M$ when $z \in \kappa$. Then f and g are constant.*

Corollary 2.37. *Let M be an algebraic curve of genus 2 on κ and let $f, g \in M_{(\rho}(\kappa)$ be such that $(f(z), g(z)) \in M$ when $z \in \kappa(0; \rho)$. Then f and g are bounded.*

In fact, according to Picard [105], every algebraic curve of genus 1 (resp. 2) is birationally equivalent to a smooth elliptic (resp. hyperelliptic) curve. Hence one can apply Theorem 2.35 with $m = 2, \deg(B) = 0, \deg(A) = s = 3$ in Corollary 2.36, and $\deg(A) \geq 4, s \geq 4$ in Corollary 2.36 and Corollary 2.37 (see [15]). Picard-Berkovich's theorem (see [6]) claims that the conclusion in Corollary 2.36 holds if the genus of M is not less than 1. Cherry [23] shows that it is also true if M is an Abelian variety, and obtains a non-Archimedean analogue of Bloch's conjecture. Further, in [24] Cherry proves that each Abelian variety over κ carries with a Kobayashi distance.

Theorem 2.38. *Take positive integers l, m and n satisfying*

$$\frac{1}{l} + \frac{1}{m} + \frac{1}{n} \leq \frac{1}{3}. \tag{2.55}$$

Then there do not exist three nonconstant meromorphic functions f, g and h on κ satisfying

$$f^l + g^m + h^n = 1. \tag{2.56}$$

Proof. Assume, on the contrary, that there exist three nonconstant meromorphic functions f, g and h on κ satisfying (2.56). Then f^l, g^m and h^n are linearly independent. Otherwise, there is $(a, b, c) \in \kappa^3 - \{0\}$ such that

$$af^l + bg^m + ch^n = 0.$$

W. l. o. g., we may assume $a \neq 0$. Thus

$$\left(1 - \frac{b}{a}\right) g^m + \left(1 - \frac{c}{a}\right) h^n = 1.$$

Since g and h are nonconstant, then $a \neq b$ and $a \neq c$. Since we can take $\alpha, \beta \in \kappa_*$ such that

$$\alpha^m = 1 - \frac{b}{a}, \quad \beta^n = 1 - \frac{c}{a},$$

then

$$(\alpha g)^m + (\beta h)^n = 1,$$

which is impossible by Theorem 2.32.

By Corollary 2.27, we have

$$
\begin{aligned}
lT(r,f) \; = \; & T(r, f^l) \leq N_2\left(r, \frac{1}{f^l}\right) + N_2\left(r, \frac{1}{g^m}\right) + N_2\left(r, \frac{1}{h^n}\right) \\
& + \overline{N}(r, f^l) + \overline{N}(r, g^m) + \overline{N}(r, h^n) - \log r + O(1) \\
\leq \; & 3T(r,f) + 3T(r,g) + 3T(r,h) - \log r + O(1).
\end{aligned}
$$

Similarly,

$$mT(r,g) \leq 3T(r,f) + 3T(r,g) + 3T(r,h) - \log r + O(1),$$

and

$$nT(r,h) \leq 3T(r,f) + 3T(r,g) + 3T(r,h) - \log r + O(1).$$

Therefore,

$$\left(1 - \frac{3}{l} - \frac{3}{m} - \frac{3}{n}\right)(T(r,f) + T(r,g) + T(r,h)) + \left(\frac{1}{l} + \frac{1}{m} + \frac{1}{n}\right)\log r \leq O(1),$$

which is impossible under the condition (2.55). $\qquad \square$

According to the proof of Theorem 2.38, the following fact is obvious.

Theorem 2.39. *Take positive integers l, m and n satisfying*

$$\frac{1}{l} + \frac{1}{m} + \frac{1}{n} \leq \frac{1}{2}. \tag{2.57}$$

Then there do not exist three nonconstant entire functions f, g and h on κ satisfying (2.56).

Theorem 2.38 and Theorem 2.39 imply the following non-Archimedean version of a result due to Hayman [53]:

Theorem 2.40. *For $n \geq 9$ (resp. $n \geq 6$), there do not exist three nonconstant meromorphic (resp. entire) functions f, g and h on κ satisfying*

$$f^n + g^n + h^n = 1. \tag{2.58}$$

Generally, by induction and Corollary 2.27, we can prove the following non-Archimedean version of two results due to Toda [130] and Yu-Yang [145], respectively:

Theorem 2.41. *Take positive integers* $k(\geq 3), n_1, ..., n_k$ *satisfying*

$$\frac{1}{n_1} + \frac{1}{n_2} + \cdots + \frac{1}{n_k} \leq \frac{1}{k-1} \left(resp. \ \frac{1}{k-1+\vartheta_k} \right). \tag{2.59}$$

Then there do not exist nonconstant entire (resp. meromorphic) functions $f_1, ..., f_k$ *on* κ
satisfying

$$f_1^{n_1} + f_2^{n_2} + \cdots + f_k^{n_k} = 1. \tag{2.60}$$

Theorem 2.41 immediately yields the following fact:

Theorem 2.42. *For* $k \geq 3$ *and* $n \geq k(k-1+\vartheta_k)$ *(resp.* $n \geq k(k-1)$*), there do not exist
nonconstant meromorphic (resp. entire) functions* $f_1, ..., f_k$ *on* κ *satisfying*

$$f_1^n + f_2^n + \cdots + f_k^n = 1. \tag{2.61}$$

An interesting problem is to find the minimal integer $c(k)$ such that when $n \geq c(k)$,
there do not exist nonconstant meromorphic functions $f_1, ..., f_k$ ($k \geq 2$) on κ satisfying
(2.61). For the complex variable case, $c(2) = 4$ (see [4], [41]), and $7 \leq c(3) \leq 9$ (see [43],
[53]).

2.7 Exponent of convergence of zeros

Let κ be an algebraically closed field of characteristic zero, complete for a non-trivial non-
Archimedean absolute value $|\cdot|$. Let f be a transcendental entire function in κ and denote
its zeros in κ_* by $z_1, z_2, ...$ (counting multiplicity). Put $|z_n| = r_n$ and assume $r_1 \leq r_2 \leq \cdots$.
The number

$$\lambda = \inf \left\{ \tau \mid \sum_{n=1}^{\infty} \frac{1}{r_n^\tau} < \infty, \tau > 0 \right\}$$

is said to be the *exponent of convergence of zeros* for f. Then for any $\varepsilon > 0$, by the definition,
the series $\sum_{n=1}^{\infty} \frac{1}{r_n^{\lambda+\varepsilon}}$ converges, but $\sum_{n=1}^{\infty} \frac{1}{r_n^{\lambda-\varepsilon}}$ diverges.

Theorem 2.43. *Let* f *be a transcendental entire function in* κ *with the exponent* λ *of
convergence of zeros. Then*

$$\limsup_{r \to \infty} \frac{\log n\left(r, \frac{1}{f-a}\right)}{\log r} = \limsup_{r \to \infty} \frac{\log N\left(r, \frac{1}{f-a}\right)}{\log r} = \limsup_{r \to \infty} \frac{\log T(r, f)}{\log r} = \lambda$$

holds for all $a \in \kappa$.

Proof. The equality

$$\limsup_{r \to \infty} \frac{\log N\left(r, \frac{1}{f-a}\right)}{\log r} = \limsup_{r \to \infty} \frac{\log T(r, f)}{\log r}$$

follows from (2.26). Note that

$$n\left(r, \frac{1}{f-a}\right) = \frac{n\left(r, \frac{1}{f-a}\right)}{\log 2} \int_r^{2r} \frac{dt}{t} \leq \frac{1}{\log 2} N\left(2r, \frac{1}{f-a}\right)$$

and

$$N\left(r, \frac{1}{f-a}\right) = \int_{\rho_0}^r \frac{n\left(t, \frac{1}{f-a}\right)}{t} dt \le n\left(r, \frac{1}{f-a}\right) \log \frac{r}{\rho_0}.$$

We obtain the another equality

$$\limsup_{r\to\infty} \frac{\log n\left(r, \frac{1}{f-a}\right)}{\log r} = \limsup_{r\to\infty} \frac{\log N\left(r, \frac{1}{f-a}\right)}{\log r}.$$

Now we prove that these equalities are equal to λ. To do this, write

$$\limsup_{r\to\infty} \frac{\log n\left(r, \frac{1}{f}\right)}{\log r} = c.$$

First we assume $c < \infty$. Then for any $\varepsilon > 0$, there is a $r_0 > 0$ such that when $r > r_0$

$$\frac{\log n\left(r, \frac{1}{f}\right)}{\log r} < c + \varepsilon,$$

or

$$n\left(r, \frac{1}{f}\right) < r^{c+\varepsilon}.$$

In particular,

$$n \le n\left(r_n, \frac{1}{f}\right) < r_n^{c+\varepsilon}$$

holds when $r_n > r_0$, which means

$$\frac{1}{r_n^{c+2\varepsilon}} < \frac{1}{n^{1+\delta}} \quad \left(\delta = \frac{\varepsilon}{c+\varepsilon} > 0\right).$$

Hence the series $\sum_{n=1}^\infty \frac{1}{r_n^{c+2\varepsilon}}$ converges, that is, $\lambda \le c + 2\varepsilon$, and therefore $\lambda \le c$ by letting $\varepsilon \to 0$.

On another hand, the series $\sum_{n=1}^\infty \frac{1}{r_n^{c-\varepsilon}}$ diverges. In fact, if it converges, then

$$\lim_{n\to\infty} \frac{n}{r_n^{c-\varepsilon}} = 0,$$

and hence $n < r_n^{c-\varepsilon}$ when n is sufficiently large. Thus there is a $n_0 \ge n\left(0, \frac{1}{f}\right)$ such that

$$n\left(r_n, \frac{1}{f}\right) = n + n\left(0, \frac{1}{f}\right) < 2r_n^{c-\varepsilon}$$

holds when $n > n_0$. For any $r > r_{n_0}$, we can choose $n > n_0$ with $r_n \le r < r_{n+1}$. Then

$$n\left(r, \frac{1}{f}\right) = n\left(r_n, \frac{1}{f}\right) < 2r_n^{c-\varepsilon} \le 2r^{c-\varepsilon}$$

yields

$$c = \limsup_{r \to \infty} \frac{\log n \left(r, \frac{1}{f}\right)}{\log r} \leq c - \varepsilon.$$

This is a contradiction. Thus our claim is true, and hence $c - \varepsilon \leq \lambda$, that is, $c \leq \lambda$ by letting $\varepsilon \to 0$. Therefore we obtain $c = \lambda$.

If $c = \infty$, the series $\sum_{n=1}^{\infty} \frac{1}{r_n^b}$ diverges for any $b > 0$. Otherwise, there is some b such that $\sum_{n=1}^{\infty} \frac{1}{r_n^b}$ converges, which implies $c \leq b < \infty$ according to the argument above. Hence it must be $\lambda = \infty$. Thus, by Corollary 1.33 and the equalities proved above,

$$\limsup_{r \to \infty} \frac{\log n \left(r, \frac{1}{f-a}\right)}{\log r} = \limsup_{r \to \infty} \frac{\log n \left(r, \frac{1}{f}\right)}{\log r} = \lambda$$

holds for all $a \in \kappa$. □

Finally we show the following result:

Theorem 2.44. *If w is a non-constant entire function and if $f \in M(\kappa) - \kappa(z)$, then*

$$\lim_{r \to \infty} \frac{T(r, f \circ w)}{T(r, w)} = +\infty.$$

Proof. Since $f \in M(\kappa) - \kappa(z)$, there exists some $c \in \kappa$ such that $f - c = 0$ has infinitely many zeros $a_1, a_2, \ldots$ with $|a_j - a_l| > 1$ $(j \neq l)$. Set

$$f(z) - c = (z - a_j) g_j(z), \quad j = 1, 2, \ldots .$$

Then for any positive integer ν, there exist positive constants K and $\delta (< \frac{1}{2})$ such that

$$|g_j(z)| \leq K, \quad |z - a_j| \leq \delta, \quad j = 1, \ldots, \nu.$$

Hence we have

$$\log^+ \frac{1}{|f(z) - c|} \geq \sum_{j=1}^{\nu} \log^+ \frac{\delta}{|z - a_j|} - \log^+(\delta K), \quad z \in \kappa,$$

which yields

$$m \left(r, \frac{1}{f \circ w - c}\right) \geq \sum_{j=1}^{\nu} m \left(r, \frac{1}{w - a_j}\right) - \nu \log^+ \frac{1}{\delta} - \log^+(\delta K).$$

Note that

$$\sum_{j=1}^{\nu} N \left(r, \frac{1}{w - a_j}\right) \leq N \left(r, \frac{1}{f \circ w - c}\right).$$

Adding the two inequalities above and by using the first main theorem, we have

$$\nu T(r, w) \leq T(r, f \circ w) + O(1).$$

Since $T(r, w) \to \infty$ as $r \to \infty$, the theorem follows. □

Corollary 2.45. *A meromorphic function f on κ is a rational function of degree d if and only if, for any non-constant entire function w on κ, we have*

$$\lim_{r\to\infty} \frac{T(r, f\circ w)}{T(r, w)} = d.$$

Corollary 2.46. *A meromorphic function f on κ is a rational function of degree d if and only if*

$$\lim_{r\to\infty} \frac{T(r, f)}{\log r} = d.$$

2.8   Value distribution of differential polynomials

Let κ be an algebraically closed field of characteristic zero, complete for a non-trivial non-Archimedean absolute value $|\cdot|$. Given a non-constant non-Archimedean meromorphic function f on κ, let $\mathcal{M}_f(\kappa)$ be the set of meromorphic functions g on κ satisfying

$$\lim_{r\to\infty} \frac{T(r, g)}{T(r, f)} = 0,$$

that is, elements in $\mathcal{M}_f(\kappa)$ grow slower than f, and usually are called *small functions* for f. Take a polynomial $\Omega \in \mathcal{M}_f(\kappa)[z_0, z_1, ..., z_n]$ with coefficients in $\mathcal{M}_f(\kappa)$ so that we can write

$$\Omega(z, z_0, z_1, ..., z_n) = \sum_{i\in I} c_i(z) z_0^{i_0} z_1^{i_1} \cdots z_n^{i_n}$$

where I is a finite set, $i = (i_0, ..., i_n) \in \mathbb{Z}_+^{n+1}$, and $c_i (\not\equiv 0) \in \mathcal{M}_f(\kappa)$. We will call

$$\Omega(z) := \Omega(z, f(z), f'(z), ..., f^{(n)}(z)) = \sum_{i\in I} c_i(z) f(z)^{i_0} f'(z)^{i_1} \cdots f^{(n)}(z)^{i_n}$$

a *differential polynomial* of f. For $i = (i_0, ..., i_n) \in \mathbb{Z}_+^{n+1}$, set

$$|i| = i_0 + \cdots + i_n.$$

Define

$$\deg(\Omega) = \max_{i\in I}\{|i|\}, \quad \deg_*(\Omega) = \min_{i\in I}\{|i|\}, \quad \gamma(\Omega) = \max_{i\in I}\left\{\sum_{k=1}^{n} k i_k\right\}.$$

Theorem 2.47 ([62]). *Take distinct points $\{a_1, ..., a_{q+1}\} \subset (\kappa_*) \cup \{\infty\}$ with $q \geq 1$. If the differential polynomial Ω of f is non-constant with $\deg_*(\Omega) \geq 1$, then*

$$q\deg_*(\Omega)T(r, f) \leq q\deg_*(\Omega)N\left(r, \frac{1}{f}\right) + \sum_{j=1}^{q+1} N\left(r, \frac{1}{\Omega - a_j}\right)$$

$$-(q-1)N\left(r, \frac{1}{\Omega}\right) - N_{\mathrm{Ram}}(r, \Omega) - \log r + qS(r), \qquad (2.62)$$

where

$$S(r) = \max_{i\in I} m(r, c_i) + \gamma(\Omega)\log^+ \frac{1}{r} + O(1).$$

Proof. Set $d = \deg(\Omega), \nu = \deg_*(\Omega)$ and note that

$$\mu\left(r, \frac{\Omega}{f^d}\right) \leq \max_{i \in I} \mu\left(r, \frac{c_i}{f^{d-|i|}}\left(\frac{f'}{f}\right)^{i_1} \cdots \left(\frac{f^{(n)}}{f}\right)^{i_n}\right)$$

$$= \max_{i \in I} \mu(r, c_i) \mu\left(r, \frac{1}{f}\right)^{d-|i|} \mu\left(r, \frac{f'}{f}\right)^{i_1} \cdots \mu\left(r, \frac{f^{(n)}}{f}\right)^{i_n}.$$

Thus Lemma 2.14 yields

$$m\left(r, \frac{\Omega}{f^d}\right) \leq (d - \nu) m\left(r, \frac{1}{f}\right) + S(r),$$

and hence

$$d \cdot T(r, f) = T(r, f^d) = m\left(r, \frac{1}{f^d}\right) + N\left(r, \frac{1}{f^d}\right) + O(1)$$

$$\leq m\left(r, \frac{1}{\Omega}\right) + m\left(r, \frac{\Omega}{f^d}\right) + d \cdot N\left(r, \frac{1}{f}\right) + O(1)$$

$$\leq m\left(r, \frac{1}{\Omega}\right) + (d - \nu) m\left(r, \frac{1}{f}\right) + d \cdot N\left(r, \frac{1}{f}\right) + S(r),$$

that is,

$$\nu T(r, f) \leq m\left(r, \frac{1}{\Omega}\right) + \nu N\left(r, \frac{1}{f}\right) + S(r). \tag{2.63}$$

By using the second main theorem (Theorem 2.15), we see

$$m\left(r, \frac{1}{\Omega}\right) + \sum_{j=1}^{q+1} m\left(r, \frac{1}{\Omega - a_j}\right) \leq 2T(r, \Omega) - N_{\text{Ram}}(r, \Omega) - \log r + O(1).$$

Hence, the first main theorem implies

$$m\left(r, \frac{1}{\Omega}\right) + \sum_{j=3}^{q+1} m\left(r, \frac{1}{\Omega - a_j}\right) \leq N\left(r, \frac{1}{\Omega - a_1}\right) + N\left(r, \frac{1}{\Omega - a_2}\right)$$

$$- N_{\text{Ram}}(r, \Omega) - \log r + O(1). \tag{2.64}$$

Therefore (2.63) and (2.64) give

$$\nu T(r, f) + (q - 1) T(r, \Omega) \leq \nu N\left(r, \frac{1}{f}\right) + \sum_{j=1}^{q+1} N\left(r, \frac{1}{\Omega - a_j}\right)$$

$$- N_{\text{Ram}}(r, \Omega) - \log r + S(r). \tag{2.65}$$

Note that (2.63) yields

$$\nu T(r, f) + N\left(r, \frac{1}{\Omega}\right) \leq T(r, \Omega) + \nu N\left(r, \frac{1}{f}\right) + S(r).$$

Then

$$
\begin{aligned}
\nu T(r,f) + (q-1)T(r,\Omega) \; &\geq \; \nu T(r,f) + (q-1)\{\nu T(r,f) \\
&\qquad + N\left(r,\frac{1}{\Omega}\right) - \nu N\left(r,\frac{1}{f}\right) - S(r)\} \\
&= \; \nu q T(r,f) + (q-1)N\left(r,\frac{1}{\Omega}\right) \\
&\qquad - \nu(q-1)N\left(r,\frac{1}{f}\right) - (q-1)S(r). \qquad (2.66)
\end{aligned}
$$

The theorem follows from (2.65) and (2.66).   □

In particular, take $q = 1, a_1 = 1, a_2 = \infty$ and let

$$
\Psi = c_0 f + c_1 f' + \cdots + c_n f^{(n)}, \qquad (2.67)
$$

and note that

$$
\overline{N}(r,\Psi) \leq \overline{N}(r,f) + \sum_{i=0}^{n} \overline{N}(r,c_i).
$$

We obtain an analogue of the Milloux's inequality [92]:

Theorem 2.48. *Let f be a non-constant meromorphic function on κ and define Ψ by (2.67) for $c_i \in M_f(\kappa)$ $(i = 0,1,...,n)$. If Ψ is non-constant, then*

$$
\begin{aligned}
T(r,f) \; &\leq \; N\left(r,\frac{1}{f}\right) + \overline{N}(r,f) + \overline{N}\left(r,\frac{1}{\Psi-1}\right) \\
&\qquad - N\left(r,\frac{1}{\Psi'};1\right) - \log r + S(r), \qquad (2.68)
\end{aligned}
$$

where

$$
S(r) = \sum_{i=0}^{n} \overline{N}(r,c_i) + \max_{0\leq i\leq n} m(r,c_i) + \log^+ \frac{1}{r} + O(1) = o(T(r,f)).
$$

Next we give an analogue of the Hayman's inequality [51]. Here we will consider a more general case (see [58]). To do so, we will assume that the differential polynomial Ψ is of the following form

$$
\Psi = c_0 f + c_1 f' + \cdots + c_l f^{(l)} + c_n f^{(n)} \; (0 \leq l \leq n-2; \; c_n \not\equiv 0). \qquad (2.69)
$$

We first introduce a few of symbols. Let K denote the field κ or $\mathbb{C}$. Take $a \in K \cup \{\infty\}$ and $f \in \mathcal{M}(K)$. For a positive integer k, define functions $\overline{\mu}_{f,k}^a$ and $\overline{\mu}_{f(k}^a$ respectively by

$$
\overline{\mu}_{f,k}^a(z) = \begin{cases} 1 & : \quad 0 < \mu_f^a(z) \leq k \\ 0 & : \quad \text{others}, \end{cases}
$$

and

$$
\overline{\mu}_{f(k}^a(z) = \begin{cases} 1 & : \quad \mu_f^a(z) \geq k \\ 0 & : \quad \text{others}. \end{cases}
$$

Thus we can define the following *counting functions*

$$\overline{n}_k\left(r, \frac{1}{f-a}\right) = \sum_{|z| \leq r} \overline{\mu}^a_{f,k}(z),$$

and

$$\overline{n}_{(k}\left(r, \frac{1}{f-a}\right) = \sum_{|z| \leq r} \overline{\mu}^a_{f(k}(z),$$

We also write

$$\overline{N}_k\left(r, \frac{1}{f-a}\right) = \int_{\rho_0}^r \overline{n}_k\left(t, \frac{1}{f-a}\right) \frac{dt}{t},$$

and

$$\overline{N}_{(k}\left(r, \frac{1}{f-a}\right) = \int_{\rho_0}^r \overline{n}_{(k}\left(t, \frac{1}{f-a}\right) \frac{dt}{t}.$$

Theorem 2.49. *Let f be a non-constant meromorphic function on κ and define Ψ by (2.69) for $c_i \in \mathcal{M}_f(\kappa)$ ($i = 0, ..., l$) and for a non-zero constant c_n. If Ψ is non-constant, then*

$$T(r,f) \leq \left(2 + \frac{1}{n-k}\right) N\left(r, \frac{1}{f}\right) + \left(2 + \frac{2}{n-k}\right) \overline{N}\left(r, \frac{1}{\Psi-1}\right)$$
$$-2N\left(r, \frac{1}{\Psi'}; 1\right) - \left(2 + \frac{2}{n-k}\right) \log r + S(r), \tag{2.70}$$

where

$$k = \begin{cases} l+1 & : \quad c_l \not\equiv 0 \\ l & : \quad c_l \equiv 0, \end{cases}$$

$$S(r) = \left(2 + \frac{1}{n-k}\right) \left(\max_{0 \leq i \leq l} m(r, c_i) + \log^- \frac{1}{r}\right)$$
$$+ \left(3 + \frac{2}{n-k}\right) \sum_{i=0}^l \overline{N}(r, c_i) + O(1) = o(T(r, f)).$$

Proof. We will follow Hayman's proof [52]. Set

$$g = \frac{\{\Psi'\}^{n+1}}{\{1 - \Psi\}^{n+2}}. \tag{2.71}$$

Suppose that z_0 is a simple pole of f, which is not a pole of any of $c_0, ..., c_l$. Then we may write

$$f(z) = \frac{b}{z - z_0} + O(1)$$

near $z = z_0$, for some $b \neq 0$. Thus the Laurent expansion of $1 - \Psi(z)$ about z_0 is

$$1 - \Psi(z) = \frac{(-1)^{n+1} n! b c_n}{(z - z_0)^{n+1}} \left\{1 + O((z - z_0)^{n+1-k})\right\}.$$

Differentiating Ψ, we deduce that

$$\Psi'(z) = \frac{(-1)^{n+1}(n+1)!bc_n}{(z-z_0)^{n+2}}\left\{1 + O((z-z_0)^{n+1-k})\right\}.$$

Thus,

$$g(z) = \frac{(-1)^{n+1}(n+1)^{n+1}}{bc_n n!}\left\{1 + O((z-z_0)^{n+1-k})\right\},$$

so that $g(z_0) \neq 0, \infty$, but $g'(z)$ has a zero of order at least $n-k$ at z_0, and hence, noting that $\overline{N}_1(r,f)$ counts the simple poles of f, we have

$$(n-k)\overline{N}_1(r,f) \leq N\left(r,\frac{1}{g'};0\right) + (n-k)\sum_{i=1}^{l}\overline{N}(r,c_i).$$

Now Jensen's formula applied to g'/g shows that

$$\begin{aligned}
N\left(r,\frac{g}{g'}\right) - N\left(r,\frac{g'}{g}\right) &= \log\mu\left(r,\frac{g'}{g}\right) - \log\mu\left(\rho_0,\frac{g'}{g}\right) \\
&= \log\mu(r,g') - \log\mu(\rho_0,g') \\
&\quad - \log\mu(r,g) + \log\mu(\rho_0,g) \\
&= N\left(r,\frac{1}{g'}\right) - N(r,g') \\
&\quad -N\left(r,\frac{1}{g}\right) + N(r,g) \\
&= N\left(r,\frac{1}{g'};0\right) - \overline{N}\left(r,\frac{1}{g}\right) - \overline{N}(r,g),
\end{aligned}$$

and hence

$$\begin{aligned}
(n-k)\overline{N}_1(r,f) &\leq N\left(r,\frac{1}{g'};0\right) + (n-k)\sum_{i=1}^{l}\overline{N}(r,c_i) \\
&\leq \overline{N}\left(r,\frac{1}{g}\right) + \overline{N}(r,g) - \log r \\
&\quad + (n-k)\sum_{i=1}^{l}\overline{N}(r,c_i) + O(1),
\end{aligned}$$

where Lemma 2.14 is used. Clearly zeros and poles of g can occur only at multiple poles of f, or zeros of $\Psi - 1$, or zeros of Ψ' other than zeros of $\Psi - 1$. Thus

$$\overline{N}\left(r,\frac{1}{g}\right) + \overline{N}(r,g) \leq \overline{N}\left(r,\frac{1}{\Psi-1}\right) + \overline{N}_{(2}(r,f) + N\left(r,\frac{1}{\Psi'};1\right),$$

where $\overline{N}_{(2}(r,f)$ denotes the valence function with respect to the multiple poles of f, each pole being counted only once. Therefore

$$\begin{aligned}
(n-k)\overline{N}_1(r,f) &\leq \overline{N}_{(2}(r,f) + \overline{N}\left(r,\frac{1}{\Psi-1}\right) + N\left(r,\frac{1}{\Psi'};1\right) \\
&\quad - \log r + (n-k)\sum_{i=1}^{l}\overline{N}(r,c_i) + O(1). \qquad (2.72)
\end{aligned}$$

By Theorem 2.48,

$$
\begin{aligned}
\overline{N}_1(r,f) + 2\overline{N}_{(2}(r,f) &\le N(r,f) \le T(r,f) \\
&\le N\left(r,\frac{1}{f}\right) + \overline{N}(r,f) + \overline{N}\left(r,\frac{1}{\Psi-1}\right) \\
&\quad - N\left(r,\frac{1}{\Psi'};1\right) - \log r + S_0(r),
\end{aligned}
$$

where

$$
S_0(r) = \sum_{i=0}^{l} \overline{N}(r,c_i) + \max_{0\le i\le l} m(r,c_i) + \log^+\frac{1}{r} + O(1).
$$

Since $\overline{N}(r,f) = \overline{N}_1(r,f) + \overline{N}_{(2}(r,f)$, this gives

$$
\overline{N}_{(2}(r,f) \le N\left(r,\frac{1}{f}\right) + \overline{N}\left(r,\frac{1}{\Psi-1}\right) - N\left(r,\frac{1}{\Psi'};1\right) - \log r + S_0(r). \tag{2.73}
$$

On combining this with (2.72) we deduce

$$
\begin{aligned}
(n-k)\overline{N}_1(r,f) &\le N\left(r,\frac{1}{f}\right) + 2\overline{N}\left(r,\frac{1}{\Psi-1}\right) \\
&\quad - 2\log r + (n-k)\sum_{i=1}^{l}\overline{N}(r,c_i) + S_0(r).
\end{aligned}
$$

Thus,

$$
\begin{aligned}
\overline{N}(r,f) &= \overline{N}_1(r,f) + \overline{N}_{(2}(r,f) \\
&\le \left(1+\frac{1}{n-k}\right)N\left(r,\frac{1}{f}\right) + \left(1+\frac{2}{n-k}\right)\overline{N}\left(r,\frac{1}{\Psi-1}\right) \\
&\quad - N\left(r,\frac{1}{\Psi'};1\right) - \left(1+\frac{2}{n-k}\right)\log r \\
&\quad + \sum_{i=1}^{l}\overline{N}(r,c_i) + \left(1+\frac{1}{n-k}\right)S_0(r).
\end{aligned}
$$

Upon substituting this inequality in Theorem 2.48, we obtain the inequality of Theorem 2.49.

□

Note that the differential polynomial Ψ in Theorem 2.49 has a gap. If this restriction is removed, we obtain an analogue of Langley's inequality [85]:

Theorem 2.50. *Let f be a non-constant meromorphic function on κ and define Ψ by (2.67) for $c_i \in \mathcal{M}_f(\kappa)$ $(i=0,1,...,n)$ with c_n not identically zero. Assume that Ψ is non-constant. Then either*

$$
\begin{aligned}
T(r,f) &\le 3N\left(r,\frac{1}{f}\right) + 4\overline{N}\left(r,\frac{1}{\Psi-1}\right) \\
&\quad - 2N\left(r,\frac{1}{\Psi'};1\right) - 3\log r + S(r)
\end{aligned} \tag{2.74}
$$

or Ψ satisfies the identity

$$(n+1)\frac{\Psi''}{\Psi'} - (n+2)\frac{\Psi'}{\Psi-1} = \frac{2c_{n-1}}{nc_n} - \frac{2c_n'}{c_n}, \tag{2.75}$$

where

$$S(r) = 5\sum_{i=0}^{n} N(r,c_i) + 2N\left(r,\frac{1}{c_n}\right) + m\left(r,\frac{c_{n-1}}{nc_n}\right)$$

$$+ \max_{0 \le i \le n} m(r,c_i) + 4\log^+\frac{1}{r} + O(1) = o(T(r,f)).$$

If $c_{n-1} = nc_n'$, then either (2.74) holds or $\Psi-1$ and Ψ' all have no zeros at all, and poles of Ψ have same multiplicity $n+1$ satisfying $\Theta_\Psi(\infty) \ge \frac{n}{n+1}$.

Proof. We will follow Langley's proof [85] and use the symbols in the proof of Theorem 2.49. Write

$$h(z) = (n+1)\frac{\Psi''}{\Psi'} - (n+2)\frac{\Psi'}{\Psi-1} - \frac{2c_{n-1}}{nc_n} + \frac{2c_n'}{c_n}.$$

Now the coefficient of $(z-z_0)^{-n-1}$ in the Laurent expansion of $\Psi(z) - 1$ about z_0 is $(-1)^n n! bc_n(z_0)$, where we assume $c_n(z_0) \ne 0$, while that of $(z-z_0)^{-n}$ is

$$(-1)^{n-1}(n-1)! bc_{n-1}(z_0) + (-1)^n n! bc_n'(z_0).$$

Thus, we obtain, near $z = z_0$,

$$\frac{\Psi'(z)}{\Psi(z)-1} = -\frac{n+1}{z-z_0} + \frac{c_n'(z_0)}{c_n(z_0)} - \frac{c_{n-1}(z_0)}{nc_n(z_0)} + O(z-z_0). \tag{2.76}$$

Similarly, we obtain, near $z = z_0$,

$$\frac{\Psi''(z)}{\Psi'(z)} = -\frac{n+2}{z-z_0} + \frac{c_n'(z_0)}{c_n(z_0)} - \frac{c_n'(z_0) + c_{n-1}(z_0)}{(n+1)c_n(z_0)} + O(z-z_0). \tag{2.77}$$

It follows that, near $z = z_0$,

$$(n+1)\frac{\Psi''(z)}{\Psi'(z)} - (n+2)\frac{\Psi'(z)}{\Psi(z)-1} = \frac{2c_{n-1}(z_0)}{nc_n(z_0)} - \frac{2c_n'(z_0)}{c_n(z_0)} + O(z-z_0).$$

If (2.75) is not true, i.e., $h(z) \not\equiv 0$, we have $h(z_0) = 0$, and thus

$$\overline{N}_1(r,f) \le N\left(r,\frac{1}{h}\right) + \overline{N}\left(r,\frac{1}{c_n}\right) + \sum_{i=0}^{n} \overline{N}(r,c_i).$$

Note that

$$\mu(r,h) = \mu\left(r,(n+1)\frac{\Psi''}{\Psi'} - (n+2)\frac{\Psi'}{\Psi-1} - \frac{2c_{n-1}}{nc_n} + \frac{2c_n'}{c_n}\right).$$

$$\le \max\left\{\mu\left(r,\frac{\Psi''}{\Psi'}\right), \mu\left(r,\frac{\Psi'}{\Psi-1}\right), \mu\left(r,\frac{c_n'}{c_n}\right), \mu\left(r,\frac{c_{n-1}}{nc_n}\right)\right\}$$

$$\le \max\left\{\frac{1}{r}, \mu\left(r,\frac{c_{n-1}}{nc_n}\right)\right\}.$$

Applying Jensen's formula to h,

$$
N\left(r, \frac{1}{h}\right) = \log \mu(r, h) + N(r, h) + O(1)
$$

$$
\leq N(r, h) + \log^+ \frac{1}{r} + m\left(r, \frac{c_{n-1}}{n c_n}\right) + O(1).
$$

Now h can have only simple poles at zeros and poles of Ψ' and $\Psi - 1$, which are not simple poles of f, and possibly other poles at zeros and poles of c_n and c_{n-1}, that is,

$$
N(r, h) \leq \overline{N}_{(2}(r, f) + \overline{N}\left(r, \frac{1}{\Psi - 1}\right) + N\left(r, \frac{1}{\Psi'}; 1\right)
$$

$$
+ N\left(r, \frac{1}{c_n}\right) + \sum_{i=0}^{n} N(r, c_i).
$$

Thus

$$
\overline{N}_1(r, f) \leq \overline{N}_{(2}(r, f) + \overline{N}\left(r, \frac{1}{\Psi - 1}\right) + N\left(r, \frac{1}{\Psi'}; 1\right) + S_1(r), \tag{2.78}
$$

where

$$
S_1(r) = m\left(r, \frac{c_{n-1}}{n c_n}\right) + 2N\left(r, \frac{1}{c_n}\right) + 2\sum_{i=0}^{n} N(r, c_i) + \log^+ \frac{1}{r} + O(1).
$$

Now (2.74) follows from Theorem 2.48, (2.73) and (2.78).

Finally, assume $c_{n-1} = n c_n'$ and assume that (2.75) holds. Then the meromorphic function g defined by (2.71) satisfies

$$
g' = g\left\{(n+1)\frac{\Psi''}{\Psi'} - (n+2)\frac{\Psi'}{\Psi - 1}\right\} = 0,
$$

that is, g is a non-zero constant. If $\Psi(z) - 1 = 0$ has at least a root z_1 with multiplicity $\nu \geq 1$, then it must satisfy $(\nu - 1)(n+1) = \nu(n+2)$. This is impossible. Thus $\Psi(z) - 1 = 0$ has no roots, and hence Ψ' has no roots since g is a non-zero constant. Obviously, all poles of Ψ must be of multiplicity $n + 1$, and hence $n(r, \Psi) = (n+1)\overline{n}(r, \Psi)$. Therefore

$$
\limsup_{r \to \infty} \frac{\overline{N}(r, \Psi)}{T(r, \Psi)} \leq \frac{1}{n+1} \limsup_{r \to \infty} \frac{N(r, \Psi)}{T(r, \Psi)} \leq \frac{1}{n+1},
$$

which means $\Theta_\Psi(\infty) \geq \frac{n}{n+1}$. $\qquad\qquad\qquad\qquad\qquad\qquad\square$

Chapter 3

Uniqueness of meromorphic functions

We will prove some uniqueness theorems of non-Archimedean meromorphic functions. These results can be regarded as the non-Archimedean analogue of corresponding theorems in value distribution theory of meromorphic functions of one complex variable, and the methods for proving those theorems also are similar and usually more simple for the non-Archimedean case. We will discuss only the functions in $\mathcal{M}(\kappa)$. Some results can be extended to the functions in $\mathcal{M}_{(\rho)}(\kappa)$ (see [17]).

3.1 Adams-Straus' uniqueness theorems

Let κ be an algebraically closed field of characteristic zero, complete for a non-trivial non-Archimedean absolute value $|\cdot|$. Let K denote the field κ or $\mathbb{C}$. Let f be a non-constant meromorphic function on K and take an element $a \in K \cup \{\infty\}$. We define the *a-valued divisor* of f by

$$E_f(a) = \{(\mu_f^a(z), z) \mid z \in K\},$$

and denote the *preimage* of a under f by

$$\overline{E}_f(a) = f^{-1}(a) = \{z \in K \mid \mu_f^a(z) > 0\}.$$

If a pair of non-constant meromorphic functions f and g on K satisfy $E_f(a) = E_g(a)$ (resp., $\overline{E}_f(a) = \overline{E}_g(a)$), it is said also that f and g *share the value* a CM (resp., IM), where CM (resp., IM) means counting multiplicities (resp., ignoring multiplicities).

Theorem 3.1 (cf. [1],[22],[62]). *Let f and g be two non-constant meromorphic functions on κ. Let a_1, a_2, a_3, a_4 be four different points in $\kappa \cup \{\infty\}$. Assume*

$$\overline{E}_f(a_j) = \overline{E}_g(a_j), \quad j = 1, ..., 4.$$

Then $f \equiv g$.

Proof. Assume that $f \not\equiv g$, on the contrary. By the second main theorem, we obtain

$$2T(r,f) + \log r \ \leq \ \sum_{j=1}^{4} \overline{N}\left(r, \frac{1}{f-a_j}\right) + O(1)$$

$$\leq \ N\left(r, \frac{1}{f-g}\right) + O(1) \leq T\left(r, \frac{1}{f-g}\right) + O(1)$$

$$= \ T(r, f-g) + O(1) \leq T(r,f) + T(r,g) + O(1),$$

and, similarly,

$$2T(r,g) + \log r \leq T(r,f) + T(r,g) + O(1).$$

Therefore

$$2\log r \leq O(1).$$

This is impossible. $\qquad\qquad\qquad\qquad\qquad\qquad\qquad\qquad\qquad\qquad\qquad\qquad\quad\square$

Consequently, if two non-constant entire functions f and g on κ share three distinct finite values IM, then $f = g$. However, Adams-Straus proved a strong result as follows (cf. [1], [22]):

Theorem 3.2. *Suppose that f and g are two non-constant entire functions on κ for which there exist two distinct finite values $a_1, a_2 \in \kappa$ such that*

$$\overline{E}_f(a_j) = \overline{E}_g(a_j), \quad j = 1, 2.$$

Then $f \equiv g$.

Proof. We will follow Adams-Straus's proof (also see [22]). Without loss of generality, we may assume that there is a sequence $\{z_n\}_{n\geq 1} \subset \kappa$ with $r_n = |z_n| \to \infty$ as $n \to \infty$ such that $|f(z_n)| \geq |g(z_n)|$. Since f is non-constant, we also assume $|f(z_n)| > \max\{|a_1|, |a_2|\}$ for all $n \geq 1$. Set

$$\Psi = \frac{f'(f-g)}{(f-a_1)(f-a_2)}.$$

Note that for every zero of the denominator, the numerator has a zero of multiplicity at least the same. Thus Ψ is an entire function on κ. By the non-Archimedean triangle inequality and by Lemma 2.14,

$$|\Psi(z_n)| = \frac{|f'(z_n)||f(z_n) - g(z_n)|}{|f(z_n) - a_1||f(z_n) - a_2|} \leq \frac{|f'(z_n)||f(z_n)|}{|f(z_n)||f(z_n)|} \leq \frac{1}{r_n} \to 0.$$

Therefore $\Psi = 0$, and hence $f = g$. $\qquad\qquad\qquad\qquad\qquad\qquad\qquad\qquad\qquad\quad\square$

It is well known that if f and g are two non-constant polynomials over an algebraically closed field κ of characteristic zero sharing two distinct elements a_1 and a_2 of κ IM, then $f = g$ (see [1]). Comparing this result with Theorem 3.2, W. Cherry proposed the following principle (see [23], [25]):

Cherry's principle *Any theorem which is true for polynomials (resp., rational functions) will also be true for non-Archimedean entire (resp., meromorphic) functions, unless it is obviously false.*

C. C. Yang [137] conjectured that for any two distinct values $a, b \in \mathbb{C}$, if two non-constant polynomials f and g of same degree on $\mathbb{C}$ are such that $(f-a)(f-b)$ and $(g-a)(g-b)$ share 0 IM, then $f = g$ or $f + g = a + b$. This conjecture and its generalization were proved by F. Pakovitch [104] and (I.V.) Ostrovskii-(F.) Pakovitch- (M.G.) Zaidenberg [103] , respectively. More recently, G. Schmieder [120] also gave an elementary proof of the conjecture. According to Cherry's principle, we propose the following problem:

Conjecture 3.3. *For any two distinct values* $a, b \in \kappa$, *if two non-constant transcendental entire functions* f *and* g *(or polynomials of the same degree) on* κ *are such that* $(f-a)(f-b)$ *and* $(g-a)(g-b)$ *share 0 IM, then* $f = g$ *or* $f + g = a + b$.

If $(f - a)(f - b)$ and $(g - a)(g - b)$ share 0 CM, it is true (see Theorem 3.31).

Theorem 3.4 ([62]). *Suppose that* f *and* g *are two non-constant meromorphic functions on* κ *for which there exist three distinct values* $a_1, a_2, a_3 \in \kappa \cup \{\infty\}$ *such that*

$$E_f(a_j) = E_g(a_j) \ (j = 1, 2); \quad \overline{E}_f(a_3) \cap \overline{E}_g(a_3) \neq \emptyset.$$

Then $f \equiv g$.

Proof. If $a_1, a_2 \in \kappa$, we consider the following functions

$$F = \frac{f - a_1}{f - a_2}, \quad G = \frac{g - a_1}{g - a_2}.$$

Then $\frac{F}{G}$ is a p-adic analytic function on κ that is never zero. Thus there exists $c \in \kappa_*$ such that $\frac{F}{G} = c$, that is,

$$\frac{f(z) - a_1}{f(z) - a_2} = c \frac{g(z) - a_1}{g(z) - a_2}. \tag{3.1}$$

Take $z_0 \in \overline{E}_f(a_3) \cap \overline{E}_g(a_3)$. Then $f(z_0) = g(z_0) = a_3$. Letting $z = z_0$ in (3.1), we obtain $c = 1$, and hence $f = g$.

If one of a_1 and a_2 is ∞, say, $a_2 = \infty$, then $\frac{f-a_1}{g-a_1}$ is an entire function on κ that is never zero. Similarly, we can prove $f = g$. $\qquad \square$

Corollary 3.5. *Suppose that* f *and* g *are two non-constant entire functions on* κ *for which there exist two finite distinct values* $a_1, a_2 \in \kappa$ *such that*

$$E_f(a_1) = E_g(a_1); \quad \overline{E}_f(a_2) \cap \overline{E}_g(a_2) \neq \emptyset.$$

Then $f \equiv g$.

In 1977, Rubel-Yang [116] proved that if f is a non-constant entire function on $\mathbb{C}$ such that $E_f(a_i) = E_{f'}(a_i)$ $(i = 1, 2)$ hold for two distinct finite complex numbers a_1 and a_2, then $f = f'$. After two years, Mues-Steinmetz [97] proved that if f is a non-constant entire function on $\mathbb{C}$ such that $\overline{E}_f(a_i) = \overline{E}_{f'}(a_i)$ $(i = 1, 2)$ hold for two distinct finite complex numbers a_1 and a_2, then $f = f'$. These results are not true for non-Archimedean case since the only meromorphic function solution of the equation $f = f'$ is zero. In fact, we can prove easily the following fact:

Proposition 3.6. *If* f *is a non-constant entire function on* κ, *then there is not any* $a \in \kappa$ *such that* $E_f(a) = E_{f'}(a)$, *and there is at most one* $a \in \kappa$ *such that* $\overline{E}_f(a) = \overline{E}_{f'}(a)$. *If* f *is a non-constant meromorphic function on* κ, *there are at most two distinct* $a \in \kappa$ *such that* $\overline{E}_f(a) = \overline{E}_{f'}(a)$.

3.2   Multiple values of meromorphic functions

In this section, we give the non-Archimedean analogues of uniqueness theorems of L. Yang and H. X. Yi dealing with multiple values of meromorphic functions. Let κ be an algebraically closed field of characteristic zero, complete for a non-trivial non-Archimedean absolute value $|\cdot|$. Let K denote the field κ or $\mathbb{C}$. Let f be a meromorphic function in K and take $a \in K \cup \{\infty\}$. For a positive integer k, define a set

$$\overline{E}_f(a, k) = \left\{ z \in K \mid \overline{\mu}^a_{f,k}(z) = 1 \right\},$$

and define a function

$$\mu^a_{f(k)}(z) = \begin{cases} \mu^a_f(z) & : \quad \mu^a_f(z) \geq k \\ 0 & : \quad \text{others.} \end{cases}$$

Write

$$n_{(k}\left(r, \frac{1}{f-a}\right) = \sum_{|z| \leq r} \mu^a_{f(k}(z)$$

and

$$N_{(k}\left(r, \frac{1}{f-a}\right) = \int_{\rho_0}^r n_{(k}\left(t, \frac{1}{f-a}\right) \frac{dt}{t}.$$

Lemma 3.7. *Let f be a non-constant meromorphic function in κ and let $a_1, ..., a_q$ be q distinct elements in $\kappa \cup \{\infty\}$. Then for $k_j \in \mathbb{Z}^+ \cup \{\infty\}$ $(j = 1, ..., q)$,*

$$\left(\sum_{j=1}^q \frac{k_j}{k_j+1} - 2 \right) T(r, f) < \sum_{j=1}^q \frac{k_j}{k_j+1} \overline{N}_{k_j}\left(r, \frac{1}{f-a_j}\right) - \log r + O(1). \tag{3.2}$$

Proof. By Theorem 2.15, we have

$$(q-2)T(r, f) \leq \sum_{j=1}^q \overline{N}\left(r, \frac{1}{f-a_j}\right) - \log r + O(1).$$

For $a \in \{a_1, ..., a_q\}$, $k \in \{k_1, ..., k_q\}$, note that

$$
\begin{aligned}
\overline{N}\left(r, \frac{1}{f-a}\right) &= \overline{N}_k\left(r, \frac{1}{f-a}\right) + \overline{N}_{(k+1}\left(r, \frac{1}{f-a}\right) \\
&\leq \frac{k}{k+1}\overline{N}_k\left(r, \frac{1}{f-a}\right) + \frac{1}{k+1}\overline{N}_k\left(r, \frac{1}{f-a}\right) \\
&\quad + \frac{1}{k+1}N_{(k+1}\left(r, \frac{1}{f-a}\right) \\
&\leq \frac{k}{k+1}\overline{N}_k\left(r, \frac{1}{f-a}\right) + \frac{1}{k+1}N\left(r, \frac{1}{f-a}\right) \tag{3.3} \\
&\leq \frac{k}{k+1}\overline{N}_k\left(r, \frac{1}{f-a}\right) + \frac{1}{k+1}T(r, f) + O(1), \tag{3.4}
\end{aligned}
$$

and hence (3.2) follows.   □

According to Yang [139], define

$$\Theta_f(a, k) = 1 - \limsup_{r \to \infty} \frac{\overline{N}_k\left(r, \frac{1}{f-a}\right)}{T(r, f)}$$

with $0 \leq \Theta_f(a, k) \leq 1$. We may think $\Theta_f(a) = \Theta_f(a, \infty)$. Now Lemma 3.7 means

$$\sum_{j=1}^{q} \frac{k_j}{k_j + 1} \Theta_f(a_j, k_j) \leq 2. \tag{3.5}$$

An element $a \in \kappa \cup \{\infty\}$ is said to be a *Borel exceptional value of order k* of f if

$$\limsup_{r \to \infty} \frac{\log^+ \overline{n}_k\left(r, \frac{1}{f-a}\right)}{\log r} < \limsup_{r \to \infty} \frac{\log^+ T(r, f)}{\log r}.$$

Note that

$$\limsup_{r \to \infty} \frac{\log^+ \overline{n}_k\left(r, \frac{1}{f-a}\right)}{\log r} = \limsup_{r \to \infty} \frac{\log^+ \overline{N}_k\left(r, \frac{1}{f-a}\right)}{\log r}.$$

Then Lemma 3.7 gives the following fact:

Corollary 3.8. *Let f be a non-constant meromorphic function in κ and let $a_1, ..., a_q$ be q distinct elements in $\kappa \cup \{\infty\}$. If there exist $k_j \in \mathbb{Z}^+ \cup \{\infty\}$ $(j = 1, ..., q)$ such that a_j is a Borel exceptional value of order k_j of f for each j, then*

$$\sum_{j=1}^{q} \frac{k_j}{k_j + 1} \leq 2.$$

Theorem 3.9. *Let f and g be non-constant meromorphic functions in κ. Let $a_1, ..., a_q$ be q distinct elements in $\kappa \cup \{\infty\}$ and take $k_j \in \mathbb{Z}^+ \cup \{\infty\}$ $(j = 1, ..., q)$ with*

$$k_1 \geq k_2 \geq \cdots \geq k_q, \quad \sum_{j=3}^{q} \frac{k_j}{k_j + 1} \geq 2. \tag{3.6}$$

Then $f = g$ if f and g satisfy

$$\overline{E}_f(a_j, k_j) = \overline{E}_g(a_j, k_j), \quad j = 1, 2, ..., q.$$

Proof. Assume, on the contrary, that $f \not\equiv g$. The condition (3.6) implies

$$1 \geq \frac{k_1}{k_1 + 1} \geq \frac{k_2}{k_2 + 1} \geq \cdots \geq \frac{k_q}{k_q + 1} \geq \frac{1}{2}.$$

By Lemma 3.7, we have

$$\left(\sum_{j=1}^{q} \frac{k_j}{k_j + 1} - 2\right) T(r, f) < \sum_{j=1}^{q} \frac{k_j}{k_j + 1} \overline{N}_{k_j}\left(r, \frac{1}{f - a_j}\right) - \log r + O(1)$$

$$\leq \frac{k_2}{k_2+1} \sum_{j=1}^{q} \overline{N}_{k_j}\left(r, \frac{1}{f-a_j}\right) - \log r + O(1)$$

$$+ \left(\frac{k_1}{k_1+1} - \frac{k_2}{k_2+1}\right) \overline{N}_{k_1}\left(r, \frac{1}{f-a_1}\right)$$

$$\leq \frac{k_2}{k_2+1} \sum_{j=1}^{q} \overline{N}_{k_j}\left(r, \frac{1}{f-a_j}\right) - \log r + O(1)$$

$$+ \left(\frac{k_1}{k_1+1} - \frac{k_2}{k_2+1}\right) T(r,f),$$

which gives

$$\left(\sum_{j=3}^{q} \frac{k_j}{k_j+1} + \frac{2k_2}{k_2+1} - 2\right) T(r,f) \; < \; \frac{k_2}{k_2+1} \sum_{j=1}^{q} \overline{N}_{k_j}\left(r, \frac{1}{f-a_j}\right) - \log r + O(1)$$

$$\leq \frac{k_2}{k_2+1} N\left(r, \frac{1}{f-g}\right) - \log r + O(1)$$

$$\leq \frac{k_2}{k_2+1}\left(T(r,f) + T(r,g)\right) - \log r + O(1).$$

Similarly,

$$\left(\sum_{j=3}^{q} \frac{k_j}{k_j+1} + \frac{2k_2}{k_2+1} - 2\right) T(r,g) \leq \frac{k_2}{k_2+1}\left(T(r,f) + T(r,g)\right) - \log r + O(1).$$

Therefore

$$\left(\sum_{j=3}^{q} \frac{k_j}{k_j+1} - 2\right)\left(T(r,f) + T(r,g)\right) + 2\log r \leq O(1).$$

This is a contradiction. Hence $f = g$.   □

Corollary 3.10. *Let f and g be non-constant entire functions in κ. Let $a_1, ..., a_q$ be q distinct elements in κ and take $k_j \in \mathbb{Z}^+ \cup \{\infty\}$ $(j = 1, ..., q)$ with*

$$k_1 \geq k_2 \geq \cdots \geq k_q, \quad \sum_{j=3}^{q} \frac{k_j}{k_j+1} \geq 1. \tag{3.7}$$

Then $f = g$ if f and g satisfy

$$\overline{E}_f(a_j, k_j) = \overline{E}_g(a_j, k_j), \quad j = 1, 2, ..., q.$$

Corollary 3.11. *Let f and g be non-constant meromorphic functions in κ. Let $a_1, ..., a_q$ be q distinct elements in $\kappa \cup \{\infty\}$ and take $k \in \mathbb{Z}^+ \cup \{\infty\}$ with $q \geq 4 + \frac{2}{k}$. Then $f = g$ if they satisfy*

$$\overline{E}_f(a_j, k) = \overline{E}_g(a_j, k), \quad j = 1, 2, ..., q.$$

Corollary 3.12. *Let f and g be non-constant entire functions in κ. Let $a_1, ..., a_q$ be q distinct elements in κ and take $k \in \mathbb{Z}^+ \cup \{\infty\}$ with $q \geq 3 + \frac{1}{k}$. Then $f = g$ if they satisfy*

$$\overline{E}_f(a_j, k) = \overline{E}_g(a_j, k), \quad j = 1, 2, ..., q.$$

Comparing Corollary 3.12 with Theorem 3.2, we suggest the following problem:

Conjecture 3.13. *Let f and g be non-constant entire functions in κ. Let $a_1, ..., a_q$ be q distinct elements in κ and take $k \in \mathbb{Z}^+$ with $q \geq 2 + \frac{1}{k}$. Then $f = g$ if they satisfy*

$$\overline{E}_f(a_j, k) = \overline{E}_g(a_j, k), \quad j = 1, 2, ..., q.$$

According to Cherry's principle, we also suggest a problem for polynomials:

Conjecture 3.14. *Let f and g be non-constant polynomials in $\mathbb{C}$. Let $a_1, ..., a_q$ be q distinct elements in $\mathbb{C}$ and take $k \in \mathbb{Z}^+$ with $q \geq 2 + \frac{1}{k}$. Then $f = g$ if they satisfy*

$$\overline{E}_f(a_j, k) = \overline{E}_g(a_j, k), \quad j = 1, 2, ..., q.$$

We also suggest to weakening the conditions in Theorem 3.1 and Theorem 3.2 as follows:

Conjecture 3.15. *Suppose that f and g are two non-constant meromorphic functions on κ and let a_1, a_2, a_3, a_4 be four different points in $\kappa \cup \{\infty\}$. Then there exists a $k \in \mathbb{Z}^+$ such that the condition*

$$\overline{E}_f(a_j) = \overline{E}_g(a_j) \ (j = 1, 2, 3), \quad \overline{E}_f(a_4, k) = \overline{E}_g(a_4, k)$$

imply $f \equiv g$.

Conjecture 3.16. *Suppose that f and g are two non-constant entire functions on κ and let a_1, a_2 be two distinct values in κ. Then there exists a $k \in \mathbb{Z}^+$ such that the condition*

$$\overline{E}_f(a_1) = \overline{E}_g(a_1), \quad \overline{E}_f(a_2, k) = \overline{E}_g(a_2, k)$$

yield $f \equiv g$.

Theorem 3.9 and Corollary 3.11 are respectively non-Archimedean versions of uniqueness theorems of H. X. Yi (cf. [143]) and L. Yang (cf. [139]) dealing with multiple values of meromorphic functions. Here we use their methods. In the complex case, Yi's theorem is a generalization of a Gopalakrishna-Bhoosnurmath's result [38].

3.3 Uniqueness polynomials of meromorphic functions

Let κ be an algebraically closed field of characteristic zero, complete for a non-trivial non-Archimedean absolute value $|\cdot|$. Let K denote the field κ or $\mathbb{C}$. For any $n \in \mathbb{Z}^+$, we denote the set of zeros of $z^n - 1$ in K by $\Omega_n(K)$. Obviously, $\Omega_n(K)$ contains n distinct elements. If n is a prime, then

$$\Omega_n(K) = \{1, w, w^2, ..., w^{n-1}\}$$

for any $w \in \Omega_n(K) - \{1\}$. Let (n, m) denote the largest common factor of the two integers n and m.

Lemma 3.17. *Take $n, m \in \mathbb{Z}^+$ with $n > m$ and set $d = (n, m)$. Then*

$$\Omega_n(K) \cap \Omega_m(K) = \Omega_d(K).$$

Proof. First of all, we assume that n and m are coprime. Then $d = 1$ and $\Omega_d(K) = \{1\}$. We will prove

$$\Omega_n(K) \cap \Omega_m(K) = \{1\}.$$

Assume, on the contrary, that there is a $z \in \Omega_n(K) \cap \Omega_m(K) - \{1\}$. Then there exists a positive integer $l \geq 2$ such that $z \in \Omega_l(K)$ but $z \notin \Omega_j(K)$ for $1 \leq j < l$. Write

$$n = pl + r, \quad m = ql + s \ (r, s \in \mathbb{Z}_+, \ 0 \leq r, s < l).$$

Since $z^n = 1 = z^l$, it follows $z^{n-l} = 1$. By induction, we have $z^r = z^{n-pl} = 1$ so that $r = 0$ according to our assumptions. Similarly, we also have $s = 0$, and therefore $l|n$, $l|m$. This is a contradiction.

Generally, n/d and m/d are coprime, and hence

$$\Omega_{n/d}(K) \cap \Omega_{m/d}(K) = \{1\}.$$

Write

$$\Omega_{n/d}(K) = \{1, u_1, ..., u_{n/d}\}, \quad \Omega_{m/d}(K) = \{1, w_1, ..., w_{m/d}\}.$$

Then $u_i \neq w_j$ for all i and j. Note that

$$z^n - 1 = (z^d - 1)(z^d - u_1) \cdots (z^d - u_{n/d}),$$

and

$$z^m - 1 = (z^d - 1)(z^d - w_1) \cdots (z^d - w_{m/d}).$$

We have

$$\Omega_d(K) \subset \Omega_n(K) \cap \Omega_m(K).$$

If $z_0 \in \Omega_n(K) \cap \Omega_m(K) - \Omega_d(K)$, then it must be $z_0^d - u_i = 0$ and $z_0^d - w_j = 0$ for some i and j, that is, $u_i = w_j$. This is a contradiction. Thus, Lemma 3.17 follows.   □

Consequently, we always have

$$\Omega_n(K) \cap \Omega_{n-1}(K) = \{1\} \ (n \geq 2).$$

Note that $1 \leq d \leq n - m$. We obtain an estimate

$$\#\{\Omega_n(K) \cap \Omega_m(K)\} \leq n - m.$$

According to Li-Yang [87], we introduce the following notation:

Definition 3.18. *Let P be a non-constant polynomial on K and let $\mathcal{F}$ be a family of $\mathcal{M}(K)$. If the condition $P(f) = P(g)$ implies $f = g$ for any $f, g \in \mathcal{F} - K$, then P is called a unique polynomial for $\mathcal{F}$.*

Next we introduce two classes of unique polynomials for meromorphic (or entire) functions. In the research on unique range sets of meromorphic functions on $\mathbb{C}$, the following polynomial

$$Y_{n,m}(z) = Y_{n,m}(a, b; z) = z^n - az^m + b \ (n, m \in \mathbb{Z}^+, \ n > m) \tag{3.8}$$

is often used (see, e. g., [143]), where we will let $a, b \in K_*$ with

$$\frac{a^n}{b^{n-m}} \neq \frac{n^n}{m^m(n-m)^{n-m}}. \tag{3.9}$$

The condition (3.9) makes $Y_{n,m}$ only has simple zeros. We will denote the set of zeros of $Y_{n,m}$ by $\dot{Y}_{n,m}$.

Theorem 3.19. *Let n and m be two coprime positive integers with $n > m$. If $n \geq 3$, then $Y_{n,m}$ is a unique polynomial for $\mathcal{A}(\kappa)$.*

Proof. Take $f, g \in \mathcal{A}(\kappa) - \kappa$ such that $Y_{n,m}(f) = Y_{n,m}(g)$. Set $h = f/g$. We see

$$g^{n-m} = a\frac{h^m - 1}{h^n - 1}.$$

Note that $\Omega_n(\kappa) \cap \Omega_m(\kappa) = \{1\}$. If h is not a constant, then the elements in $\Omega_n(\kappa) - \{1\}$ are all Picard exceptional values of h, and hence

$$\sum_{b \in \Omega_n(\kappa) - \{1\}} \delta_h(b) = n - 1 \geq 2.$$

This is a contradiction. Thus h is constant. If $h \neq 1$, we see that g is constant, which is contrary to our assumption. Hence $h = 1$, that is, $f = g$. □

Corollary 3.20. *Let n and m be two coprime positive integers with $n > m$. If $n \geq 3$, then $Y_{n,m}$ is a unique polynomial for $\kappa[z]$.*

Theorem 3.21. *For $m, n \in \mathbb{Z}^+$ with $n \geq m + 2$, $Y_{n,m}$ is a unique polynomial for $\mathcal{M}(K)$ if one of the following conditions holds:*
1) $(n, m) = 1$, $n > \frac{1}{2}\{5 + \sqrt{17 + 4m(m-3)}\}$;
2) $(n, m) = 1$, $m = n - 2 \geq 3$.

Proof. Take $f, g \in \mathcal{M}(K) - K$ such that $Y_{n,m}(f) = Y_{n,m}(g)$. Set $h = \frac{f}{g}$. We see

$$g^{n-m}(h^n - 1) = a(h^m - 1).$$

Write

$$z^n - 1 = (z - 1)(z - t_1) \cdots (z - t_{n-1}),$$
$$z^m - 1 = (z - 1)(z - u_1) \cdots (z - u_{m-1}).$$

Then

$$g^{n-m} = a\frac{(h - u_1) \cdots (h - u_{m-1})}{(h - t_1) \cdots (h - t_{n-1})}.$$

Note that

$$\#\{\Omega_n(K) \cap \Omega_m(K)\} = (n, m) = 1.$$

If h is not constant, then we have

$$\Theta_h(t_j) \geq 1 - \frac{1}{n-m} = \frac{n-m-1}{n-m} \quad (1 \leq j \leq n-1),$$

and

$$\Theta_h(u_j) \geq 1 - \frac{1}{n-m} = \frac{n-m-1}{n-m} \quad (1 \leq j \leq m-1),$$

and consequently,

$$2 \geq \sum_{j=1}^{n-1} \Theta_h(t_j) + \sum_{j=1}^{m-1} \Theta_h(u_j)$$

$$\geq \begin{cases} \frac{(n-m-1)(n+m-2)}{n-m} & : \quad (n,m) = 1 \\ n-2 & : \quad m = n-2, \ (n, n-2) = 1 \end{cases}$$

$$> 2.$$

This is a contradiction. Thus h is constant. If $h \neq 1$, we see that g is constant. Hence $h = 1$, i.e., $f \equiv g$. $\qquad\square$

Here we use the idea from Yi [143] in the proof of Theorem 3.19 and Theorem 3.21.

Problem 3.22. *Is $Y_{n,n-1}$ a unique polynomial for $\mathcal{M}(K)$ when $n \geq 5$?*

Finally, we consider the following polynomial

$$F_{n,b}(z) = \frac{(n-1)(n-2)}{2} z^n - n(n-2)z^{n-1} + \frac{n(n-1)}{2} z^{n-2} + b \ (b \in K_* - \{-1\}), \quad (3.10)$$

and denote the set of zeros of $F_{n,b}$ by $\dot{F}_{n,b}$. Note that $\dot{F}_{n,b}$ contains n distinct elements. G. Frank and M. Reinders [33] proved that $F_{n,b}$ is a unique polynomial for $\mathcal{M}(\mathbb{C})$ if $n \geq 11$.

Theorem 3.23. *If $n \geq 6$, then $F_{n,b}$ is a unique polynomial for $\mathcal{M}(K)$.*

Proof. Take $f, g \in \mathcal{M}(K) - K$ such that $F_{n,b}(f) = F_{n,b}(g)$. Set $h = \frac{f}{g}$. We see

$$\frac{(n-1)(n-2)}{2}(h^n - 1)g^2 - n(n-2)(h^{n-1} - 1)g + \frac{n(n-1)}{2}(h^{n-2} - 1) = 0. \quad (3.11)$$

If h is constant, (3.11) implies $h^n - 1 = 0$ and $h^{n-1} - 1 = 0$. It follows that $h = 1$ and hence $f = g$.

It remains to consider the case that h is not constant. We write (3.11) in the form

$$\left((h^n - 1)g - \frac{n}{n-1}(h^{n-1} - 1) \right)^2 = -\frac{n}{(n-1)^2(n-2)} \varphi(h), \quad (3.12)$$

where φ is defined by

$$\varphi(z) = (n-1)^2(z^n - 1)(z^{n-2} - 1) - n(n-2)(z^{n-1} - 1)^2.$$

If $\varphi(h) = 0$, that is,

$$(h^n - 1)(h^{n-2} - 1) = \frac{n(n-2)}{(n-1)^2}(h^{n-1} - 1)^2,$$

and noting that

$$\#\{\Omega_n(K) \cap \Omega_{n-2}(K)\} \leq 2,$$

i.e., $(z^n - 1)(z^{n-2} - 1)$ has at least $2n - 6$ simple roots $z_1, ..., z_{2n-6}$, say, then the second main theorem implies

$$\sum_{j=1}^{2n-6} \Theta_h(z_j) \geq \sum_{j=1}^{2n-6} \frac{1}{2} = \frac{2n-6}{2} = n - 3 \geq 3.$$

This is impossible. Thus we have $\varphi(h) \not\equiv 0$.

An elementary calculation gives

$$\varphi^{(k)}(1) = 0 \ (0 \leq k \leq 3), \quad \varphi^{(4)}(1) = 2n(n-1)^2(n-2) \neq 0.$$

Hence, we can write

$$\varphi(z) = (z - 1)^4(z - t_1)(z - t_2) \cdots (z - t_{2n-6}),$$

where $t_1, ..., t_{2n-6} \in K - \{1\}$. Note that

$$\varphi(z) = z^{2n-2} - (n-1)^2 z^n + 2n(n-2)z^{n-1} - (n-1)^2 z^{n-2} + 1, \tag{3.13}$$

$$\varphi'(z) = (n-1)z^{n-3}\{2z^n - n(n-1)z^2 + 2n(n-2)z - (n-1)(n-2)\}. \tag{3.14}$$

Now assume that

$$\varphi(z) = \varphi'(z) = 0,$$

for some $z \in K$. Obviously, we have $z \neq 0$. Thus by (3.14), z satisfies

$$2z^n - n(n-1)z^2 + 2n(n-2)z - (n-1)(n-2) = 0. \tag{3.15}$$

Solving (3.13) and (3.14), we obtain

$$z^{n-2}\{(n-1)(n-2)z^2 - 2n(n-2)z + n(n-1)\} = 2. \tag{3.16}$$

Solving (3.15) and (3.16), we obtain

$$4z^2 = \{n(n-1)z^2 - 2n(n-2)z + (n-1)(n-2)\}$$
$$\cdot\{(n-1)(n-2)z^2 - 2n(n-2)z + n(n-1)\}.$$

A simple calculation shows that z satisfies the following equation

$$z^4 - 4z^3 + 6z^2 - 4z + 1 = (z-1)^4 = 0,$$

that is, $z = 1$. Therefore $t_1, ..., t_{2n-6}$ all are simple zeros of φ. From (3.12), we see that

$$\Theta_h(t_j) \geq \frac{1}{2} \ (1 \leq j \leq 2n - 6).$$

Thus the second main theorem yields

$$2 \geq \sum_{j=1}^{2n-6} \Theta_h(t_j) \geq \frac{2n-6}{2} = n-3,$$

and hence $n \leq 5$ in contradiction to our assumption $n \geq 6$. This completes the proof of the theorem.   □

Problem 3.24. *Are $F_{3,b}$, $F_{4,b}$ and $F_{5,b}$ unique polynomials for $\mathcal{A}(\kappa)$?*

3.4   Unique range sets of meromorphic functions

Let κ be an algebraically closed field of characteristic zero, complete for a non-trivial non-Archimedean absolute value $|\cdot|$. Let K denote the field κ or $\mathbb{C}$ and set $\bar{K} = K \cup \{\infty\}$. Let f be a non-constant meromorphic function on K and take a non-empty set $S \subset \bar{K}$. We define the *S-valued divisor* of f by

$$E_f(S) = \bigcup_{a \in S} \{(\mu_f^a(z), z) \mid z \in K\} = \bigcup_{a \in S} E_f(a),$$

and denote the *preimage* of S under f

$$\overline{E}_f(S) = f^{-1}(S) = \bigcup_{a \in S} \{z \in K \mid \mu_f^a(z) > 0\}.$$

Definition 3.25. *Given a family $\mathcal{F}$ of $\mathcal{M}(K)$, a non-empty set S in $K \cup \{\infty\}$ is called a urscm (resp., ursim) for $\mathcal{F}$ if for any non-constant functions $f, g \in \mathcal{F}$ satisfying $E_f(S) = E_g(S)$ (resp., $\overline{E}_f(S) = \overline{E}_g(S)$), one has $f = g$.*

In the Definition 3.25, "urscm" (resp., "ursim") means a *unique range set for counting multiplicity* (resp., *unique range set for ignoring multiplicity*). If two functions $f, g \in \mathcal{F}$ satisfy $E_f(S) = E_g(S)$ (resp., $\overline{E}_f(S) = \overline{E}_g(S)$), we also say that f and g *share the set S CM* (resp., IM). These notations were introduced by Gross-Yang [42].

Generally, an n-tuple $S = (S_1, ..., S_n)$ of non-empty sets $S_1, S_2, ..., S_n$ in $K \cup \{\infty\}$ with $S_i \cap S_j = \emptyset$ ($i \neq j$) is called an *n-urscm* (resp., *n-ursim*) for $\mathcal{F}$ if for any non-constant functions $f, g \in \mathcal{F}$ satisfying $E_f(S) = E_g(S)$ (resp., $\overline{E}_f(S) = \overline{E}_g(S)$), one has $f = g$, where, by definition,

$$E_f(S) = (E_f(S_1), ..., E_f(S_n)), \quad \overline{E}_f(S) = (\overline{E}_f(S_1), ..., \overline{E}_f(S_n)).$$

If two functions $f, g \in \mathcal{F}$ satisfy $E_f(S) = E_g(S)$ (resp., $\overline{E}_f(S) = \overline{E}_g(S)$), we also say that f and g *share the n-tuple S CM* (resp., IM). For the n-tuple $S = (S_1, ..., S_n)$, define its *cardinal number* by

$$\#S = \#S_1 + \cdots + \#S_n,$$

where $\#S_i$ is the cardinal number of the set S_i. Thus we obtain two numbers

$$c_n(\mathcal{F}) = \min\{\#S \mid S \text{ is an } n\text{-urscm for } \mathcal{F}\}$$

and

$$i_n(\mathcal{F}) = \min\{\#S \mid S \text{ is an } n\text{-ursim for } \mathcal{F} \}.$$

Obviously, we have

$$n \leq c_n(\mathcal{F}) \leq i_n(\mathcal{F})$$

since an n-ursim must be an n-urscm. Given any positive integer $n > 1$, if S is an $(n-1)$-urscm or $(n-1)$-ursim, then for any set T in $K \cup \{\infty\}$, (S,T) is an n-urscm or n-ursim so that

$$c_n(\mathcal{F}) \leq c_{n-1}(\mathcal{F}) + 1 \leq c_1(\mathcal{F}) + n - 1$$

or

$$i_n(\mathcal{F}) \leq i_{n-1}(\mathcal{F}) + 1 \leq i_1(\mathcal{F}) + n - 1.$$

Theorem 3.1 and Theorem 3.2 mean respectively

$$c_n(\mathcal{M}(\kappa)) = i_n(\mathcal{M}(\kappa)) = n \ (n \geq 4), \quad c_n(\mathcal{A}(\kappa)) = i_n(\mathcal{A}(\kappa)) = n \ (n \geq 2).$$

Theorem 3.4 gives

$$c_3(\mathcal{M}(\kappa)) = 3.$$

Basic problems for studying n-urscm and n-ursim are the following

   (i) Find n-urscm and n-ursim.
   (ii) Find the sharp bound of $c_n(\mathcal{F})$ and $i_n(\mathcal{F})$.
   (iii) Characterize an n-urscm or n-ursim.

Let $\mathrm{Aut}(K)$ be the group of non-constant linear polynomials in $K[z]$, that is, $\sigma \in \mathrm{Aut}(K)$ if and only if $\sigma(z) = az + b$ with $a \neq 0$. Let id be the *identical mapping* in K. Note that a set S in K is an ursim for $\mathrm{Aut}(K)$ if and only if S is an urscm for $\mathrm{Aut}(K)$. Take $\sigma, \eta \in \mathrm{Aut}(K)$. We see that $\sigma^{-1}(S) = \eta^{-1}(S)$ if and only if $\eta \circ \sigma^{-1}(S) = S$. Thus S is an ursim for $\mathrm{Aut}(K)$ if and only if there exists no $\xi \in \mathrm{Aut}(K) - \{id\}$ such that $\xi(S) = S$.

Example 3.26. *Take* $\sigma, \eta \in \mathrm{Aut}(K)$ *defined by*

$$\sigma(z) = -z, \quad \eta(z) = wz \ (w \in \Omega_3(K) - \{1\}).$$

Obviously, one has

$$\sigma(\{-1,0,1\}) = \{-1,0,1\}, \quad \eta(\Omega_3(K)) = \Omega_3(K).$$

Hence $\{-1,0,1\}$ and $\Omega_3(K)$ are not ursim for $\mathrm{Aut}(K)$*. Note that for any distinct elements $a,b \in K$, the mapping $\xi \in \mathrm{Aut}(K)$ defined by*

$$\xi(z) = -z + a + b$$

satisfies $\xi(\{a,b\}) = \{a,b\}$. Hence an ursim for $\mathrm{Aut}(K)$ *contains at least 3 distinct elements. Therefore $c_1(\mathcal{A}(K)) \geq 3$.*

Example 3.27. *Take $S = \{a_1, a_2, a_3\} \subset K$ and set*

$$c = \frac{1}{3}(a_1 + a_2 + a_3), \quad a = c - a_1, \quad w \in \Omega_3(K) - \{1\}.$$

Then S is not an ursim for $\mathrm{Aut}(K)$ if and only if it is one of the following two forms (see [19]):

  (i) $S = \{c - a, c, c + a\}$;
  (ii) $S = \{c + a, c + wa, c + w^2 a\}$.
Two mappings in $\mathrm{Aut}(K)$ under which S are invariant are respectively

$$\sigma(z) = -z + 2c, \quad \eta(z) = wz + (1 - w)c.$$

Boutabaa, Escassut and Haddad [19] asked whether a finite set S in κ is an urscm for $\mathcal{A}(\kappa)$ if and only if S is an ursim for $\mathrm{Aut}(\kappa)$. They confirmed this case if either S has only three different points or $\mathcal{A}(\kappa)$ is replaced by $\kappa[x]$. The general case is proved by Cherry and Yang [25]:

Theorem 3.28. *A finite set S in κ is an urscm for $\mathcal{A}(\kappa)$ if and only if S is an ursim for $\mathrm{Aut}(\kappa)$.*

To prove Theorem 3.28, we will need the following result (cf. [25]):

Theorem 3.29. *Take $f, g \in \mathcal{A}(\kappa) - \kappa$ and let $F(x, y)$ be a polynomial in two variables with coefficients in κ. If $F(f, g) \equiv 0$, then there exist $h \in \mathcal{A}(\kappa)$, p, $q \in \kappa[z]$ such that $f = p(h)$ and $g = q(h)$.*

Proof. Here we will follow the proof of Cherry-Yang [25]. Let $F_0(x, y)$ be an irreducible factor of F with $F_0(f, g) \equiv 0$. By Berkovich's non-Archimedean Picard theorem ([6], Theorem 4.5.1, or see [23]), $F_0(x, y) = 0$ is a rational curve, and can therefore be rationally parametrized, that is, there exist rational functions $p(t)$, $q(t)$ and $R(x, y)$ such that $t = R(x, y)$, and $F_0(p(t), q(t)) \equiv 0$. Set $h = R(f, g)$. Then one has $f = p(h)$ and $g = q(h)$. For the rest, one shows that h can be chosen to be entire, and p, q polynomials. Since $f = p(h)$ and $g = q(h)$ are entire, then $h \in \mathcal{M}(\kappa)$ must omit $p^{-1}(\infty)$ and $q^{-1}(\infty)$. However, h omits at most one element in $\bar{\kappa}$, and so $p^{-1}(\infty) = q^{-1}(\infty)$ consists of exactly one element. Thus after making a linear change in coordinates, we may assume $p^{-1}(\infty) = q^{-1}(\infty)$, and so h omits ∞, that is, $h \in \mathcal{A}(\kappa)$ and p, $q \in \kappa[z]$.   □

The following lemma refers to [19] or [25]:

Lemma 3.30. *Take $h \in \mathcal{A}(\kappa) - \kappa$ and p, q, $P \in \kappa[z]$ satisfying*

$$n = \deg(P) \geq 1, \quad d = \deg(p) \geq \max\{1, \deg(q)\}.$$

Suppose that c is a non-zero constant such that $P(p(h)) = cP(q(h))$. Then $p(h) = Aq(h) + B$, where $A, B \in \kappa$ with $A^n = c$.

Proof. Write

$$p(z) = \sum_{j=0}^{d} a_j z^j, \quad q(z) = \sum_{j=0}^{d} b_j z^j, \quad P(z) = \sum_{j=0}^{n} c_j z^j,$$

and let κ_0 be the field generated by $\{a_j\}$, $\{b_j\}$ and $\{c_j\}$. By extending κ_0 if necessary, we may assume that κ contains at least an element ζ which is transcendental over κ_0. Since $h(\kappa) = \kappa$, we can choose $z_0 \in \kappa$ such that $h(z_0) = \zeta$. Then

$$P(p(\zeta)) = c_n(a_d\zeta^d)^{n-1}\sum_{j=1}^{d}a_j\zeta^j + P_1(\zeta),$$

and

$$cP(q(\zeta)) = cc_n(b_d\zeta^d)^{n-1}\sum_{j=1}^{d}b_j\zeta^j + P_2(\zeta),$$

where P_1, $P_2 \in \kappa[z]$ with

$$\deg(P_1) \leq d(n-1), \quad \deg(P_2) \leq d(n-1).$$

By comparing the coefficients of the terms with degrees in ζ larger than $d(n-1)$, we see that

$$a_j = Ab_j, \quad j = 1, ..., d,$$

for some constant A with $A^n = c$. Thus, one completes the proof of the theorem by setting $B = a_0 - Ab_0$. $\qquad\square$

Proof of Theorem 3.28. Write $S = \{s_1, ..., s_n\}$, and set

$$P(z) = (z - c_1)\cdots(z - s_n).$$

Let f, $g \in \mathcal{A}(\kappa) - \kappa$ share S CM. Then there exists a non-zero constant c such that $P(f) = cP(g)$. Applying Theorem 3.29 to $F(x,y) = P(x) - cP(y)$, there exist $h \in \mathcal{A}(\kappa)$, p, $q \in \kappa[z]$ such that $f = p(h)$ and $g = q(h)$. It follows $f = Ag + B$ from Lemma 3.30, where A and B are constant, and $A^n = c$. Since $S \subset g(\kappa) = \kappa$ and noting that $f^{-1}(S) = g^{-1}(S)$, then $g(g^{-1}(S)) = S$, and $f(g^{-1}(S)) = f(f^{-1}(S)) = S$. Write $\sigma(z) = Az + B$. Then

$$S = f(g^{-1}(S)) = (\sigma \circ g)(g^{-1}(S)) = \sigma(S).$$

If S is an urscm for $\mathcal{A}(\kappa)$, obviously S is an ursim for $\mathrm{Aut}(\kappa)$. Conversely, assume S is an ursim for $\mathrm{Aut}(\kappa)$. Then f, $g \in \mathcal{A}(\kappa) - \kappa$ sharing S CM means that $f = \sigma \circ g$ and $\sigma(S) = S$ according to the proof above, and hence $\sigma = id$. Thus we have $f = g$, and hence S is an urscm for $\mathcal{A}(\kappa)$. $\qquad\square$

Theorem 3.31. *For any two distinct values a, $b \in \kappa$, if two non-constant entire functions f and g on κ are such that $(f - a)(f - b)$ and $(g - a)(g - b)$ share 0 CM, then $f = g$ or $f + g = a + b$.*

Proof. W. l. o. g., we may assume $a = 0$ and $b = 1$. Set $P(z) = z(z - 1)$. Then there exists a non-zero constant c such that $P(f) = cP(g)$. Applying Theorem 3.29 to $F(x,y) = P(x) - cP(y)$, there exist $h \in \mathcal{A}(\kappa)$, p, $q \in \kappa[z]$ such that $f = p(h)$ and $g = q(h)$. It follows from Lemma 3.30 that $f = Ag + B$, where A and B are constant, and $A^2 = c$. Thus

$$cg^2 - cg = f(f - 1) = A^2g^2 + \{A(B - 1) + AB\}g + B(B - 1),$$

and hence $-c = A(B-1) + AB$ and $B(B-1) = 0$ since g is not a constant. We obtain either $B = 0$, $A = c = 1$ or $B = 1$, $A = -c = -1$, and the theorem follows. □

According to Cherry's principle, we suggest the following problem:

Conjecture 3.32 ([25]). *A finite set S in κ is an ursim for $\mathcal{A}(\kappa)$ if and only if S is an ursim for $\kappa[z]$.*

Let $\mathrm{Aut}(\bar{K})$ be the group of non-constant fractional linear functions in $K(z)$, that is, $\sigma \in \mathrm{Aut}(\bar{K})$ if and only if

$$\sigma(z) = \frac{az+b}{cz+d}, \quad a,b,c,d \in K, \quad ad-cb \neq 0.$$

An element of $\mathrm{Aut}(\bar{K})$ also is called a *Möbius transformation* on $\bar{K}$. Similarly, S is an ursim for $\mathrm{Aut}(\bar{K})$ if and only if there exists no $\xi \in \mathrm{Aut}(\bar{K}) - \{id\}$ such that $\xi(S) = S$. Given 3 points $a_1, a_2, a_3 \in \bar{K}$ and choosing $a_4 \in \bar{K}$ satisfying

$$a_3 \neq \frac{1}{2}(a_1 + a_2), \quad (a_1, a_2, a_3, a_4) = \frac{(a_1-a_3)(a_2-a_4)}{(a_1-a_4)(a_2-a_3)} = -1,$$

then the following transformation

$$\sigma(z) = \frac{(a_3+a_4)z - 2a_3a_4}{2z - (a_3+a_4)}$$

satisfies

$$\sigma(a_1) = a_2, \ \sigma(a_2) = a_1, \ \sigma(a_j) = a_j \ (j = 3,4).$$

In particular, $\sigma(\{a_1, a_2, a_3\}) = \{a_1, a_2, a_3\}$ so that $\{a_1, a_2, a_3\}$ is not an ursim for $\mathrm{Aut}(\bar{K})$. Thus an ursim for $\mathrm{Aut}(\bar{K})$ must have at least 4 points, and hence $c_1(\mathcal{M}(K)) \geq 4$. Here we suggest the following problem:

Problem 3.33. *A finite set S in κ is an urscm (resp., ursim) for $\mathcal{M}(\kappa)$ if and only if S is an urscm (resp., ursim) for $\kappa(z)$.*

3.5   The Frank-Reinders' technique

Let κ be an algebraically closed field of characteristic zero, complete for a non-trivial non-Archimedean absolute value $|\cdot|$. By using the Frank-Reinders' technique [33], we prove the following result:

Lemma 3.34. *Let F and G be non-constant meromorphic functions on κ sharing ∞ CM and let $a_1, ..., a_q$ be distinct elements of κ with $q \geq 2$. Then one of the following cases must occur*

1)

$$\left(q - \frac{3}{2}\right)\{T(r,F) + T(r,G)\} \leq \sum_{j=1}^{q}\left\{N_2\left(r, \frac{1}{F-a_j}\right) + N_2\left(r, \frac{1}{G-a_j}\right)\right\}$$
$$- \log r + O(1);$$

2) $G = AF + B$, where $A, B \in \kappa$ with $A \neq 0$, and

$$\#(\{a_1, ..., a_q\} \cap \{Aa_1 + B, ..., Aa_q + B\}) \geq 2.$$

Proof. Define

$$H = \frac{F''}{F'} - \frac{G''}{G'}.$$

First we suppose $H = 0$. Then $G = AF + B$ with $A, B \in \kappa$, $A \neq 0$. If (2) is not satisfied, then

$$\#\{a_1, ..., a_q, Aa_1 + B, ..., Aa_q + B\} \geq 2q - 1.$$

By the second main theorem, we have

$$
\begin{aligned}
(2q - 3)T(r, G) \leq{} & \sum_{j=1}^{q}\left\{\overline{N}\left(r, \frac{1}{G - a_j}\right) + \overline{N}\left(r, \frac{1}{G - (Aa_j + B)}\right)\right\} \\
& - N\left(r, \frac{1}{G'}; a_1, ..., a_q, Aa_1 + B, ..., Aa_q + B\right) \\
& - \log r + O(1) \\
={} & \sum_{j=1}^{q}\left\{\overline{N}\left(r, \frac{1}{G - a_j}\right) + \overline{N}\left(r, \frac{1}{F - a_j}\right)\right\} \\
& - N\left(r, \frac{1}{G'}; a_1, ..., a_q\right) - N\left(r, \frac{1}{F'}; a_1, ..., a_q\right) \\
& - \log r + O(1) \\
\leq{} & \sum_{j=1}^{q}\left\{N_2\left(r, \frac{1}{F - a_j}\right) + N_2\left(r, \frac{1}{G - a_j}\right)\right\} \\
& - \log r + O(1).
\end{aligned}
$$

Note that

$$T(r, G) = T(r, F) + O(1).$$

Then

$$
\left(q - \frac{3}{2}\right)(T(r, F) + T(r, G)) \leq \sum_{j=1}^{q}\left\{N_2\left(r, \frac{1}{F - a_j}\right) + N_2\left(r, \frac{1}{G - a_j}\right)\right\}
$$
$$
- \log r + O(1).
$$

Now suppose $H \not\equiv 0$. Note that poles of H can only occur at zeros of F' or G'. Then

$$
\begin{aligned}
N(r, H) \leq{} & \sum_{j=1}^{q}\left\{N_2\left(r, \frac{1}{F - a_j}\right) - \overline{N}\left(r, \frac{1}{F - a_j}\right)\right\} \\
& + \sum_{j=1}^{q}\left\{N_2\left(r, \frac{1}{G - a_j}\right) - \overline{N}\left(r, \frac{1}{G - a_j}\right)\right\} \\
& + N\left(r, \frac{1}{F'}; a_1, ..., a_q\right) + N\left(r, \frac{1}{G'}; a_1, ..., a_q\right).
\end{aligned}
$$

Note that H has a zero at every point where F and G have a simple pole. It follows that

$$\overline{N}(r, F) + \overline{N}(r, G) \le N\left(r, \frac{1}{H}\right) + \frac{1}{2}\{N(r, F) + N(r, G)\}.$$

By the first main theorem and the lemma of logarithmic derivative, we have

$$\overline{N}(r, F) + \overline{N}(r, G) \le N(r, H) + \frac{1}{2}\{T(r, F) + T(r, G)\} + O(1).$$

The second main theorem applied to F and G gives

$$(q - 1)\{T(r, F) + T(r, G)\} \le \sum_{j=1}^{q}\left\{\overline{N}\left(r, \frac{1}{F - a_j}\right) + \overline{N}\left(r, \frac{1}{G - a_j}\right)\right\}$$
$$+ \overline{N}(r, F) + \overline{N}(r, G)$$
$$- N\left(r, \frac{1}{F'}; a_1, ..., a_q\right) - N\left(r, \frac{1}{G'}; a_1, ..., a_q\right)$$
$$- 2\log r + O(1).$$

Hence

$$\left(q - \frac{3}{2}\right)\{T(r, F) + T(r, G)\} \le \sum_{j=1}^{q}\left\{N_2\left(r, \frac{1}{F - a_j}\right) + N_2\left(r, \frac{1}{G - a_j}\right)\right\}$$
$$- 2\log r + O(1).$$

Thus, we complete the proof of the lemma. $\qquad\square$

Frank-Reinders [33] remarked that if $q = 2$, then case (2) means that $F = G$ or $F + G = a_1 + a_2$. Thus Lemma 3.34 is the non-Archimedean analogue of Theorem 1 in Yi [142].

Lemma 3.35. *Let f and g be non-constant meromorphic (resp., entire) functions on κ satisfying*

$$E_f(\dot{F}_{n,b}) = E_g(\dot{F}_{n,b}),$$

and define

$$F = \frac{1}{F_{n,b}(f)}, \quad G = \frac{1}{F_{n,b}(g)}.$$

If $n \ge 10$ (resp., $n \ge 6$), then $G = AF + B$, where $A, B \in \kappa$ with $A \ne 0$, and

$$\#(\{a_1, a_2, a_3\} \cap \{Aa_1 + B, Aa_2 + B, Aa_3 + B\}) \ge 2,$$

where $a_1 = 0$, $a_2 = \frac{1}{6}$, $a_3 = \frac{1}{b+1}$.

Proof. Since

$$F'_{n,b}(z) = \frac{n(n - 1)(n - 2)}{2}z^{n-3}(z - 1)^2,$$

we have $F_{n,b}(1) = 1 + b$ with multiplicity 3 and $F_{n,b}(0) = b$ with multiplicity $n - 2$. Therefore, we can write

$$F_{n,b}(z) - b - 1 = (z - 1)^3 Q_1(z), \quad Q_1(1) \ne 0,$$
$$F_{n,b}(z) - b = z^{n-2}Q(z), \quad Q(0) \ne 0,$$

where $Q_1(z)$ is a polynomial of degree $n - 3$, having only simple zeros. For every $a \in \kappa - \{b, b+1\}$, $F_{n,b}(z) - a$ has only simple zeros. In particular, $F_{n,b}(z)$ has only simple zeros and thus $\dot{F}_{n,b}$ has exactly n elements. Noting that,

$$T(r, F) = nT(r, f) + O(1),$$

from the first main theorem we conclude that

$$
N_2\left(r, \frac{1}{F - a_1}\right) = N_2(r, F_{n,b}(f)) = 2\overline{N}(r, f),
$$

$$
N_2\left(r, \frac{1}{F - a_2}\right) = N_2\left(r, \frac{1}{F_{n,b}(f) - b}\right) \leq 2\overline{N}\left(r, \frac{1}{f}\right) + N_2\left(r, \frac{1}{Q(f)}\right)
$$
$$
\leq 4T(r, f) + O(1),
$$

$$
N_2\left(r, \frac{1}{F - a_3}\right) = N_2\left(r, \frac{1}{F_{n,b}(f) - b - 1}\right)
$$
$$
\leq 2\overline{N}\left(r, \frac{1}{f - 1}\right) + N_2\left(r, \frac{1}{Q_1(f)}\right)
$$
$$
\leq (n - 1)T(r, f) + O(1).
$$

It follows that

$$
\sum_{j=1}^{3} N_2\left(r, \frac{1}{F - a_j}\right) \leq (n + 3)T(r, f) + 2N(r, f) + O(1)
$$
$$
\leq \left(1 + \frac{5}{n}\right)T(r, F),
$$

and the same inequality holds with f and F replaced by g and G. Note that F and G share ∞ CM. Thus if (1) of Lemma 3.34 is true, we would have $3/2 < 1 + 5/n$, and hence $n < 10$, which is a contradiction to our assumptions. Therefore (2) of Lemma 3.34 must be satisfied.

$\square$

Theorem 3.36 ([63]). *For any integer $n \geq 10$, the set $\dot{F}_{n,b}$ is an urscm for $\mathcal{M}(\kappa)$.*

Proof. Write $\dot{F}_{n,b} = \{r_1, r_2, ..., r_n\}$ and define

$$
Q(z) = \frac{(n - 1)(n - 2)}{2}z^2 - n(n - 2)z + \frac{n(n - 1)}{2}.
$$

Let f and g be non-constant meromorphic functions on κ satisfying $E_f(\dot{F}_{n,b}) = E_g(\dot{F}_{n,b})$. By two main theorems, we have the estimate

$$
(n - 2)T(r, g) \leq \sum_{k=1}^{n} \overline{N}\left(r, \frac{1}{g - r_k}\right) - \log r + O(1)
$$
$$
= \sum_{k=1}^{n} \overline{N}\left(r, \frac{1}{f - r_k}\right) - \log r + O(1)
$$
$$
\leq nT(r, f) - \log r + O(1).
$$

Similarly, we can obtain the estimate

$$(n-2)T(r,f) \leq nT(r,g) - \log r + O(1).$$

Define

$$h_1 = -\frac{1}{b}f^{n-2}Q(f), \quad h_2 = \frac{h_3}{b}g^{n-2}Q(g), \quad h_3 = \frac{F_{n,b}(f)}{F_{n,b}(g)}.$$

Then we have

$$h_1 + h_2 + h_3 = 1.$$

Write $f = \frac{f_1}{f_2}$ and $g = \frac{g_1}{g_2}$, where pairs f_1, f_2 and g_1, g_2 are entire functions on κ without common factors, respectively. Then

$$h_3 = c\left(\frac{g_2}{f_2}\right)^n, \quad c = \frac{F_{n,b}(f)f_2^n}{F_{n,b}(g)g_2^n}.$$

Note that c is an entire function on κ which is never zero and hence is constant. Thus we have

$$\overline{N}(r,h_3) \leq \overline{N}(r,f), \quad \overline{N}\left(r,\frac{1}{h_3}\right) \leq \overline{N}(r,g).$$

In the following, we will prove $h_3 \equiv 1$.

First we prove that h_1 can not be expessed linearly by $\{1, h_3\}$. Assume that we have a linear expression

$$h_1 = a_1 h_3 + a_2, \quad a_1, a_2 \in \kappa.$$

Since h_1 is not constant, then $a_1 \neq 0$, and h_3 is not constant. If $a_2 \neq 0$, then the second main theorem implies

$$
\begin{aligned}
nT(r,f) &= T(r,h_1) + O(1) \\
&\leq \overline{N}\left(r,\frac{1}{h_1}\right) + \overline{N}(r,h_1) + \overline{N}\left(r,\frac{1}{h_1-a_2}\right) - \log r + O(1) \\
&\leq \overline{N}\left(r,\frac{1}{f}\right) + \overline{N}\left(r,\frac{1}{Q(f)}\right) + \overline{N}(r,f) + \overline{N}\left(r,\frac{1}{h_3}\right) - \log r + O(1) \\
&\leq 4T(r,f) + \overline{N}(r,g) - \log r + O(1) \\
&\leq 4T(r,f) + T(r,g) - \log r + O(1) \\
&\leq \left(4 + \frac{n}{n-2}\right)T(r,f) - \log r + O(1),
\end{aligned}
$$

which yields $n < 5 + \frac{2}{n-2}$, a contradiction! If $a_2 = 0$, setting

$$Q(z) = \frac{(n-1)(n-2)}{2}(z-s_1)(z-s_2),$$

then by $h_1 = a_1 c(\frac{g_2}{f_2})^n$, we see

$$N\left(r,\frac{1}{f}\right) \geq \frac{n}{2}\overline{N}\left(r,\frac{1}{f}\right), \quad N\left(r,\frac{1}{f-s_j}\right) \geq n\overline{N}\left(r,\frac{1}{f-s_j}\right), \quad j = 1,2.$$

Then

$$\Theta_f(s_j) = 1 - \limsup_{r \to \infty} \frac{\overline{N}\left(r, \frac{1}{f - s_j}\right)}{T(r, f)} \geq 1 - \frac{1}{n} \ (j = 1, 2), \quad \Theta_f(0) \geq 1 - \frac{2}{n},$$

and again by the second main theorem,

$$1 - \frac{2}{n} + 2\left(1 - \frac{1}{n}\right) \leq \Theta_f(0) + \sum_{j=1}^{2} \Theta_f(s_j) \leq 2,$$

which yields $n \leq 4$. This is impossible since $n \geq 10$. Thus we prove the claim.

Define

$$F = \frac{1}{F_{n,b}(f)}, \quad G = \frac{1}{F_{n,b}(g)}.$$

By Lemma 3.35, it follows that $G = AF + B$, and hence

$$-bBh_1 = h_3 - bB - A.$$

Thus it must be that $B = 0$, and therefore h_3 is a constant. If $h_3 \neq 1$, noting that

$$T(r, f) = T(r, g) + O(1),$$

then the second main theorem implies

$$
\begin{aligned}
nT(r, f) + \log r &= T(r, h_1) + \log r + O(1) \\
&\leq \overline{N}\left(r, \frac{1}{h_1}\right) + \overline{N}(r, h_1) + \overline{N}\left(r, \frac{1}{h_1 - 1 + h_3}\right) + O(1) \\
&\leq \overline{N}\left(r, \frac{1}{f}\right) + \overline{N}\left(r, \frac{1}{Q(f)}\right) + \overline{N}(r, f) + \overline{N}\left(r, \frac{1}{h_2}\right) + O(1) \\
&\leq 4T(r, f) + \overline{N}\left(r, \frac{1}{g}\right) + \overline{N}\left(r, \frac{1}{Q(g)}\right) + O(1) \\
&\leq 4T(r, f) + 3T(r, g) + O(1) \\
&\leq 7T(r, f) + O(1),
\end{aligned}
$$

which yields $n < 7$, a contradiction!

Therefore we must have $h_3 = 1$ and hence $f = g$ since $F_{n,b}$ is a unique polynomial for $\mathcal{M}(\kappa)$ by Theorem 3.23. $\qquad\square$

Corollary 3.37. *For any integer $n \geq 7$, then 2-tuple $(\infty, \dot{F}_{n,b})$ is a 2-urscm for $\mathcal{M}(\kappa)$.*

Proof. Take $f, g \in \mathcal{M}(\kappa) - \kappa$ such that $E_f(\dot{F}_{n,b}) = E_g(\dot{F}_{n,b})$ and $E_f(\infty) = E_g(\infty)$. Then $F_{n,b}(f)$ and $F_{n,b}(g)$ share 0 and ∞ CM. Thus $F_{n,b}(f) = h_3 F_{n,b}(g)$ for some constant h_3. The proof follows from the proof of Theorem 3.36. $\qquad\square$

Thus we obtain estimates $c_1(\mathcal{M}(\kappa)) \leq 10$ and $c_2(\mathcal{M}(\kappa)) \leq 8$.

Theorem 3.38. *The set $\dot{F}_{n,b}$ is an urscm for $\mathcal{A}(\kappa)$ if either $n \geq 6$ or $n \geq 4$, but $F_{n,b}$ is a unique polynomial for $\mathcal{A}(\kappa)$ satisfying*

$$b \neq -\frac{1}{2}. \tag{3.17}$$

Proof. Take $f, g \in A(\kappa) - \kappa$. Then the condition $E_f(\dot{F}_{n,b}) = E_g(\dot{F}_{n,b})$ implies that $\frac{F_{n,b}(f)}{F_{n,b}(g)}$ is an entire function on κ, which is never zero. This is a constant c. Hence

$$F_{n,b}(f) = cF_{n,b}(g).$$

Note that, by Theorem 2.11,

$$T(r, F_{n,b}(f)) = nT(r, f) + O(1).$$

We have

$$T(r, f) = T(r, g) + O(1).$$

Set

$$f_1 = -\frac{1}{b}f^{n-2}Q(f), \quad f_2 = \frac{c}{b}g^{n-2}Q(g),$$

where

$$Q(z) = \frac{(n-1)(n-2)}{2}z^2 - n(n-2)z + \frac{n(n-1)}{2}.$$

Then $f_1 + f_2 = 1 - c$. First of all, we consider the case $c \neq 1$. Note that

$$f_1' = -\frac{n(n-1)(n-2)}{2b}f^{n-3}f'(f-1)^2, \quad f_2' = \frac{n(n-1)(n-2)c}{2b}g^{n-3}g'(g-1)^2.$$

If the following systems

$$\begin{cases} f_1(z) - 1 + c = 0 \\ f(z) - 1 = 0 \end{cases} \tag{3.18}$$

and

$$\begin{cases} f_2(z) - 1 + c = 0 \\ g(z) - 1 = 0, \end{cases} \tag{3.19}$$

have solutions, solving (3.18) and (3.19) respectively, we will obtain

$$c = \frac{b+1}{b} \quad \text{and} \quad \frac{1}{c} = \frac{b+1}{b}.$$

Hence $c^2 = 1$ which means $c = 1$, or $c = -1$, that is, $b = -\frac{1}{2}$. Thus under the second class of conditions, we may assume that one of the systems (3.18) and (3.19), say (3.18), has no solutions. Thus we obtain

$$N\left(r, \frac{1}{f_1'}; 0, 1 - c\right) \geq N\left(r, \frac{1}{(f-1)^2}\right) = 2T(r, f) + O(1).$$

Therefore

$$\begin{aligned} nT(r, f) + \log r &= T(r, f_1) + \log r + O(1) \\ &\leq \overline{N}(r, f_1) + \overline{N}\left(r, \frac{1}{f_1}\right) + \overline{N}\left(r, \frac{1}{f_1 - 1 + c}\right) \\ &\quad -N\left(r, \frac{1}{f_1'}; 0, 1 - c\right) + O(1) \end{aligned}$$

$$\leq \ \overline{N}\left(r, \frac{1}{f}\right) + \overline{N}\left(r, \frac{1}{Q(f)}\right) + \overline{N}\left(r, \frac{1}{f_2}\right) - 2T(r, f) + O(1)$$
$$\leq \ T(r, f) + \overline{N}\left(r, \frac{1}{g}\right) + \overline{N}\left(r, \frac{1}{Q(g)}\right) + O(1)$$
$$\leq \ T(r, f) + 3T(r, g) + O(1)$$
$$= \ 4T(r, f) + O(1).$$

This is impossible since $n \geq 4$. If $n \geq 6$, we do not need the condition (3.17) and similarly obtain

$$nT(r, f) + \log r \leq 6T(r, f) + O(1).$$

This also is impossible under the first class of conditions. Thus, we have $c = 1$, and hence $f = g$ since $F_{n,b}$ is a unique polynomial for $\mathcal{A}(\kappa)$ by Theorem 3.23 or our assumption. $\qquad \square$

Theorem 3.39. *For $b \in \kappa_* - \{-1, -2\}$, the set $\dot{F}_{3,b}$ is an urscm for $\mathcal{A}(\kappa)$.*

Proof. Assume, to the contrary, that $\dot{F}_{3,b}$ is not an urscm for $\mathcal{A}(\kappa)$. By Theorem 3.28, $\dot{F}_{3,b}$ is not an ursim for $\mathrm{Aut}(\kappa)$, that is, there exists $\sigma \in \mathrm{Aut}(\kappa) - \{id\}$ such that $\sigma(\dot{F}_{3,b}) = \dot{F}_{3,b}$. Set $\dot{F}_{3,b} = \{r_1, r_2, r_3\}$ and write $\sigma(z) = Az + B$. Then we have

$$\begin{aligned} F_{3,b}(z) &= z^3 - 3z^2 + 3z + b = (z - r_1)(z - r_2)(z - r_3) \\ &= (z - \sigma(r_1))(z - \sigma(r_2))(z - \sigma(r_3)), \end{aligned}$$

which implies

$$\sum_{i=1}^{3} r_i = \sum_{i=1}^{3} \sigma(r_i) = 3, \quad r_1 r_2 r_3 = \sigma(r_1)\sigma(r_2)\sigma(r_3) = -b.$$

By simple computation, we find $A + B = 1$ and $A^3 = 1$. Thus $A \neq 1$, and hence Example 3.27 shows that the set $\dot{F}_{3,b}$ is of the following form

$$\dot{F}_{3,b} = \{1 + a, 1 + Aa, 1 + A^2 a\}.$$

Note that

$$\begin{aligned} 0 &= 1 + A + A^2, \\ 3 &= (1 + a)(1 + Aa) + (1 + a)(1 + A^2 a) + (1 + Aa)(1 + A^2 a), \\ -b &= (1 + a)(1 + Aa)(1 + A^2 a). \end{aligned}$$

We easily obtain $a = 1$, and hence $b = -2$. This is a contradiction. Hence $\dot{F}_{3,b}$ is an urscm for $\mathcal{A}(\kappa)$. $\qquad \square$

3.6   Some urscm for $\mathcal{M}(\kappa)$ and $\mathcal{A}(\kappa)$

We continue the discussion in § 3.5 and study the sets $\dot{Y}_{n,m}$. First we prove the main lemma in this section.

Lemma 3.40. *Take $m, n \in \mathbb{Z}^+$ and let $b \in \kappa_*$, $Q \in \kappa[z]$ with $\deg(Q) = n - m \geq 1$ such that the polynomial*

$$P(z) = z^m Q(z) + b$$

has only simple zeros. Let S_P be the set of zeros of P. Let f and g be non-constant meromorphic (resp., entire) functions on κ satisfying

$$E_f(S_P) = E_g(S_P),$$

and define

$$F = \frac{1}{P(f)}, \quad G = \frac{1}{P(g)}.$$

If $2m \geq n + 8$ (resp., $2m \geq n + 4$), then $G = F$ or $G + F = \frac{1}{b}$.

Proof. Note that F and G share ∞ CM. Thus, if (1) of Lemma 3.34 is true for $a_1 = 0$, $a_2 = 1/b$, we obtain

$$
\begin{aligned}
\frac{1}{2}\{T(r, F) + T(r, G)\} + \log r \ &\leq\ \sum_{j=1}^{2}\left\{ N_2\left(r, \frac{1}{F - a_j}\right) + N_2\left(r, \frac{1}{G - a_j}\right)\right\} + O(1) \\
&\leq\ 2\overline{N}(r, f) + 2\overline{N}\left(r, \frac{1}{f}\right) + N_2\left(r, \frac{1}{Q(f)}\right) \\
&\quad + 2\overline{N}(r, g) + 2\overline{N}\left(r, \frac{1}{g}\right) + N_2\left(r, \frac{1}{Q(g)}\right) + O(1) \\
&\leq\ (n - m + 2)\{T(r, f) + T(r, g)\} \\
&\quad + 2(\overline{N}(r, f) + \overline{N}(r, g)) + O(1) \\
&\leq\ \left(1 - \frac{m - 4}{n}\right)\{T(r, F) + T(r, G)\} + O(1),
\end{aligned}
$$

which is impossible since $2m \geq n + 8$. Therefore, (2) of Lemma 3.34 must be satisfied. The proof of the lemma is completed by the remark after Lemma 3.34.   □

Theorem 3.41 ([62]). *For any integer $n \geq 12$ with $(n, n - 2) = 1$, the set $\dot{Y}_{n,n-2}$ is an urscm for $\mathcal{M}(\kappa)$.*

Proof. Take $f, g \in \mathcal{M}(\kappa) - \kappa$ such that $E_f(\dot{Y}_{n,n-2}) = E_g(\dot{Y}_{n,n-2})$. Write $\dot{Y}_{n,n-2} = \{r_1, r_2, ..., r_n\}$. By two main theorems, we have the estimate

$$
\begin{aligned}
(n - 2)T(r, g) \ &\leq\ \sum_{k=1}^{n}\overline{N}\left(r, \frac{1}{g - r_k}\right) + O(1) \\
&=\ \sum_{k=1}^{n}\overline{N}\left(r, \frac{1}{f - r_k}\right) + O(1) \\
&\leq\ nT(r, f) + O(1).
\end{aligned}
$$

Similarly we can obtain the estimate

$$(n - 2)T(r, f) \leq nT(r, g) + O(1).$$

Define

$$h_1 = -\frac{1}{b} f^{n-2}(f^2 - a), \quad h_2 = \frac{h_3}{b} g^{n-2}(g^2 - a), \quad h_3 = \frac{f^n - af^{n-2} + b}{g^n - ag^{n-2} + b}.$$

Then we have

$$h_1 + h_2 + h_3 = 1.$$

Write $f = \frac{f_1}{f_2}$ and $g = \frac{g_1}{g_2}$, where pairs f_1, f_2 and g_1, g_2 are entire functions on κ without common factors, respectively. Then

$$h_3 = c \left(\frac{g_2}{f_2} \right)^n, \quad c = \frac{f_1^n - af_1^{n-2}f_2^2 + bf_2^n}{g_1^n - ag_1^{n-2}g_2^2 + bg_2^n}.$$

Note that c is an entire function on κ which is never zero and hence is constant. Thus we have

$$\overline{N}(r, h_3) \leq \overline{N}(r, f), \quad \overline{N}\left(r, \frac{1}{h_3}\right) \leq \overline{N}(r, g).$$

In the following, we will prove $h_3 \equiv 1$.

First we prove that h_1 can not be expressed linearly by $\{1, h_3\}$. Assume that we have a linear expression

$$h_1 = a_1 h_3 + a_2, \quad a_1, a_2 \in \kappa.$$

Since h_1 is not constant, then $a_1 \neq 0$, and h_3 is not constant. If $a_2 \neq 0$, then the second main theorem implies

$$
\begin{aligned}
nT(r, f) + \log r &= T(r, h_1) + \log r + O(1) \\
&\leq \overline{N}\left(r, \frac{1}{h_1}\right) + \overline{N}(r, h_1) + \overline{N}\left(r, \frac{1}{h_1 - a_2}\right) + O(1) \\
&\leq \overline{N}\left(r, \frac{1}{f}\right) + \overline{N}\left(r, \frac{1}{f^2 - a}\right) + \overline{N}(r, f) + \overline{N}\left(r, \frac{1}{h_3}\right) + O(1) \\
&\leq 4T(r, f) + \overline{N}(r, g) + O(1) \\
&\leq 4T(r, f) + T(r, g) + O(1) \\
&\leq \left(4 + \frac{n}{n-2}\right) T(r, f) + O(1),
\end{aligned}
$$

which yields $n \leq 4 + \frac{2}{n-2}$, a contradiction! If $a_2 = 0$, setting $z^2 - a = (z - s_1)(z - s_2)$, then by $h_1 = a_1 c \left(\frac{g_2}{f_2} \right)^n$ we see

$$N\left(r, \frac{1}{f}\right) \geq 2\overline{N}\left(r, \frac{1}{f}\right), \quad N\left(r, \frac{1}{f - s_j}\right) \geq n\overline{N}\left(r, \frac{1}{f - s_j}\right), \quad j = 1, 2.$$

Then

$$\Theta_f(s_j) = 1 - \limsup_{r \to \infty} \frac{\overline{N}\left(r, \frac{1}{f - s_j}\right)}{T(r, f)} \geq 1 - \frac{1}{n} \ (j = 1, 2), \quad \Theta_f(0) \geq \frac{1}{2},$$

and again by the second main theorem,

$$\frac{1}{2} + 2\left(1 - \frac{1}{n}\right) \leq \Theta_f(0) + \sum_{j=1}^{2} \Theta_f(s_j) \leq 2,$$

which implies $n \leq 4$. This is impossible since $n \geq 12$. Our claim is proved.

Define

$$F = \frac{1}{f^n - af^{n-2} + b}, \quad G = \frac{1}{g^n - ag^{n-2} + b}.$$

By Lemma 3.40, it follows that $1, F, G$ are linearly dependent when $n \geq 12$. Then there exists $(c_1, c_2, c_3) \in \kappa^3 - \{0\}$ such that

$$c_1 + c_2 F + c_3 G = 0,$$

and hence

$$-bc_1 h_1 + c_3 h_3 = -bc_1 - c_2.$$

Thus, it must be $c_1 = 0$, $c_3 \neq 0$, and therefore h_3 is a constant. If $h_3 \neq 1$, noting that

$$T(r, f) = T(r, g) + O(1),$$

then the second main theorem implies

$$\begin{aligned}
nT(r, f) + \log r &= T(r, h_1) + \log r + O(1) \\
&\leq \overline{N}\left(r, \frac{1}{h_1}\right) + \overline{N}(r, h_1) + \overline{N}\left(r, \frac{1}{h_1 - 1 + h_3}\right) + O(1) \\
&\leq \overline{N}\left(r, \frac{1}{f}\right) + \overline{N}\left(r, \frac{1}{f^2 - a}\right) + \overline{N}(r, f) + \overline{N}\left(r, \frac{1}{h_2}\right) + O(1) \\
&\leq 4T(r, f) + \overline{N}\left(r, \frac{1}{g}\right) + \overline{N}\left(r, \frac{1}{g^2 - a}\right) + O(1) \\
&\leq 4T(r, f) + 3T(r, g) + O(1) \\
&\leq 7T(r, f) + O(1),
\end{aligned}$$

which yields $n < 7$, a contradiction!

Therefore we must have $h_3 = 1$, and hence $f = g$ since $Y_{n,n-2}$ is a uniqueness polynomial for $\mathcal{M}(\kappa)$ by Theorem 3.21.   □

Corollary 3.42. *For any integer $n \geq 8$ with $(n, n - 2) = 1$, the set $\dot{Y}_{n,n-2}$ is an urscm for $\mathcal{A}(\kappa)$.*

Corollary 3.43 ([16]). *For any integer $n \geq 7$, then 2-tuple $(\infty, \dot{Y}_{n,n-2})$ is a 2-urscm for $\mathcal{M}(\kappa)$.*

Proof. Take $f, g \in \mathcal{M}(\kappa) - \kappa$ such that $E_f(\dot{Y}_{n,n-2}) = E_g(\dot{Y}_{n,n-2})$ and $E_f(\infty) = E_g(\infty)$. Then $Y_{n,n-2}(f)$ and $Y_{n,n-2}(g)$ share 0 and ∞ CM. Thus $Y_{n,n-2}(f) = h_3 Y_{n,n-2}(g)$ for some constant h_3. The proof follows from the proof of Theorem 3.41.   □

Further, Boutabaa and Escassut [16] show that for every point w in the projective completion of κ, for every $n \geq 5$, there exist sets S of n elements in κ such that 2-tuple (w, S) is a 2-urscm for $\mathcal{M}(\kappa)$. They [16] ask whether there exists a 2-urscm (w, S) for $\mathcal{M}(\kappa)$ with S a set of 3 or 4 points. Khoai [48] confirms existence of 2-urscm (w, S) $(w \in \kappa \cup \{\infty\})$ for $\mathcal{M}(\kappa)$ with S a set of 4 points.

Theorem 3.44. *Take an integer $n \geq 3$. If the constants $a, b \in \kappa_*$ in the polynomial $Y_{n,n-1}$ also satisfy*

$$\frac{a^n}{b} \neq 2 \frac{n^n}{(n-1)^{n-1}}, \tag{3.20}$$

then the set $\dot{Y}_{n,n-1}$ is an urscm for $\mathcal{A}(\kappa)$.

Proof. Take $f, g \in \mathcal{A}(\kappa) - \kappa$. Then the condition $E_f(\dot{Y}_{n,n-1}) = E_g(\dot{Y}_{n,n-1})$ implies that $\frac{f^n - af^{n-1} + b}{g^n - ag^{n-1} + b}$ is an entire function on κ, which is never zero. So this is a constant c. Hence

$$f^n - af^{n-1} + b = c(g^n - ag^{n-1} + b).$$

Note that, by Theorem 2.11

$$T(r, f^n - af^{n-1} + b) = nT(r, f) + O(1).$$

We have

$$T(r, f) = T(r, g) + O(1).$$

Set

$$f_1 = -\frac{1}{b}(f^n - af^{n-1}), \quad f_2 = \frac{c}{b}(g^n - ag^{n-1}).$$

Then $f_1 + f_2 = 1 - c$. First of all, we consider the case $c \neq 1$. Note that

$$f_1' = -\frac{n}{b} f^{n-2} f' \left(f - \frac{n-1}{n} a \right), \quad f_2' = \frac{nc}{b} g^{n-2} g' \left(g - \frac{n-1}{n} a \right).$$

Now we show that one of the following systems

$$\begin{cases} f_1(z) - 1 + c = 0 \\ f(z) - \frac{n-1}{n} a = 0 \end{cases} \tag{3.21}$$

or

$$\begin{cases} f_2(z) - 1 + c = 0 \\ g(z) - \frac{n-1}{n} a = 0 \end{cases} \tag{3.22}$$

has no solutions. Otherwise, solving (3.21) and (3.22) respectively, we will obtain

$$1 - c = A_n \text{ and } 1 - \frac{1}{c} = A_n,$$

where

$$A_n = \frac{a^n (n-1)^{n-1}}{bn^n}.$$

Hence $c^2 = 1$ which means $(1 - A_n)^2 = 1$, that is, $A_n = 2$. This is a contradiction. W. l. o. g., we may assume that the system (3.21) has no solutions. Thus we obtain

$$N\left(r, \frac{1}{f_1'}; 0, 1 - c\right) \geq N\left(r, \frac{1}{f - \frac{n-1}{n}a}\right) = T(r, f) + O(1).$$

Therefore

$$
\begin{aligned}
nT(r, f) &= T(r, f_1) + O(1) \\
&\leq \overline{N}(r, f_1) + \overline{N}\left(r, \frac{1}{f_1}\right) + \overline{N}\left(r, \frac{1}{f_1 - 1 + c}\right) \\
&\quad - N\left(r, \frac{1}{f_1'}; 0, 1 - c\right) - \log r + O(1) \\
&\leq \overline{N}\left(r, \frac{1}{f}\right) + \overline{N}\left(r, \frac{1}{f - a}\right) + \overline{N}\left(r, \frac{1}{f_2}\right) - T(r, f) - \log r + O(1) \\
&\leq T(r, f) + \overline{N}\left(r, \frac{1}{g}\right) + \overline{N}\left(r, \frac{1}{g - a}\right) - \log r + O(1) \\
&\leq T(r, f) + 2T(r, g) - \log r + O(1) \\
&= 3T(r, f) - \log r + O(1).
\end{aligned}
$$

This is impossible since $n \geq 3$. Thus, we have $c = 1$, and hence $f = g$ since $Y_{n,n-1}$ is a unique polynomial for $\mathcal{A}(\kappa)$ by Theorem 3.19.   □

If $n \geq 4$, the restriction (3.20) can be removed, which is proved independently by Boutabaa, A., Escassut, A. and Haddad, L. [19], Hu and Yang [62].

3.7   Some ursim for meromorphic functions

Let κ be an algebraically closed field of characteristic zero, complete for a non-trivial non-Archimedean absolute value $|\cdot|$. For the complex case, M. Reinders [107] remarked that if F and G share ∞ IM, the analogue of Lemma 3.34 also is true if the factor $\left(q - \frac{3}{2}\right)$ is replaced by $\left(q - \frac{5}{3}\right)$. However, A. Boutabaa and A. Escassut [14] pointed out that Reinders' remark is false and proved the following Lemma 3.45, where they missed the coefficient 2 at the front of $N_2(r, F) - \overline{N}(r, F)$ and $N_2(r, G) - \overline{N}(r, G)$.

Lemma 3.45. *Let F and G be non-constant meromorphic functions on κ sharing ∞ IM and let $a_1, ..., a_q$ be distinct elements of κ with $q \geq 2$. Then one of the following cases must occur*

1)

$$
\begin{aligned}
(3q - 5)\{T(r, F) + T(r, G)\} &\leq 2\sum_{j=1}^{q}\left\{N_2\left(r, \frac{1}{F - a_j}\right) + N_2\left(r, \frac{1}{G - a_j}\right)\right\} \\
&\quad + \sum_{j=1}^{q}\left\{\overline{N}\left(r, \frac{1}{F - a_j}\right) + \overline{N}\left(r, \frac{1}{G - a_j}\right)\right\}
\end{aligned}
$$

$$+2(N_2(r,F) - \overline{N}(r,F))$$
$$+2(N_2(r,G) - \overline{N}(r,G))$$
$$-N\left(r, \frac{1}{F'}; a_1, ..., a_q\right) - N\left(r, \frac{1}{G'}; a_1, ..., a_q\right)$$
$$-4\log r + O(1).$$

2) $G = AF + B$, where $A, B \in \kappa$ with $A \neq 0$, and

$$\#(\{a_1, ..., a_q\} \cap \{Aa_1 + B, ..., Aa_q + B\}) \geq 2.$$

Proof. Define

$$H = \frac{F''}{F'} - \frac{G''}{G'}.$$

First we suppose $H = 0$. Then $G = AF + B$ with $A, B \in \kappa$, $A \neq 0$. If (2) is not satisfied, then

$$\#\{a_1, ..., a_q, Aa_1 + B, ..., Aa_q + B\} \geq 2q - 1.$$

According to the proof of Lemma 3.34, we can prove

$$(2q - 3)(T(r, F) + T(r, G)) \leq 2 \sum_{j=1}^{q} \left\{ N_2\left(r, \frac{1}{F - a_j}\right) + N_2\left(r, \frac{1}{G - a_j}\right) \right\}$$
$$-2\log r + O(1).$$

The second main theorem applied to F and G gives

$$(q - 2)\{T(r, F) + T(r, G)\} \leq \sum_{j=1}^{q} \left\{ \overline{N}\left(r, \frac{1}{F - a_j}\right) + \overline{N}\left(r, \frac{1}{G - a_j}\right) \right\}$$
$$-N\left(r, \frac{1}{F'}; a_1, ..., a_q\right) - N\left(r, \frac{1}{G'}; a_1, ..., a_q\right)$$
$$-2\log r + O(1).$$

Summing two inequalities above, we obtain

$$(3q - 5)(T(r, F) + T(r, G)) \leq 2 \sum_{j=1}^{q} \left\{ N_2\left(r, \frac{1}{F - a_j}\right) + N_2\left(r, \frac{1}{G - a_j}\right) \right\}$$
$$+ \sum_{j=1}^{q} \left\{ \overline{N}\left(r, \frac{1}{F - a_j}\right) + \overline{N}\left(r, \frac{1}{G - a_j}\right) \right\}$$
$$-N\left(r, \frac{1}{F'}; a_1, ..., a_q\right) - N\left(r, \frac{1}{G'}; a_1, ..., a_q\right)$$
$$-4\log r + O(1),$$

and hence (1) of Lemma 3.45 follows.

Now we suppose $H \not\equiv 0$. Note that H has a zero at every point where F and G have a simple pole. It follows that

$$3(\overline{N}(r, F) + \overline{N}(r, G)) \leq 2 \left(N\left(r, \frac{1}{H}\right) + N(r, F) + N(r, G) \right).$$

By the first main theorem and the lemma of logarithmic derivatives, we see

$$3(\overline{N}(r,F) + \overline{N}(r,G)) \leq 2(N(r,H) + T(r,F) + T(r,G)) + O(1).$$

Since $\mu_H^\infty \leq 1$, then poles of H can only occur where F' or G' has a zero, or one of F and G has a multiple pole. Thus we have

$$
\begin{aligned}
N(r,H) \leq\ & \overline{N}\left(r,\frac{1}{F'}\right) + \overline{N}\left(r,\frac{1}{G'}\right) + N_2(r,F) \\
& -\overline{N}(r,F) + N_2(r,G) - \overline{N}(r,G) \\
\leq\ & \sum_{j=1}^{q}\left\{N_2\left(r,\frac{1}{F-a_j}\right) - \overline{N}\left(r,\frac{1}{F-a_j}\right)\right\} \\
& + \sum_{j=1}^{q}\left\{N_2\left(r,\frac{1}{G-a_j}\right) - \overline{N}\left(r,\frac{1}{G-a_j}\right)\right\} \\
& + N\left(r,\frac{1}{F'};a_1,...,a_q\right) + N\left(r,\frac{1}{G'};a_1,...,a_q\right) \\
& + N_2(r,F) - \overline{N}(r,F) \\
& + N_2(r,G) - \overline{N}(r,G).
\end{aligned}
$$

The second main theorem applied to F and G gives

$$
\begin{aligned}
(q-1)\{T(r,F) + T(r,G)\} \leq\ & \sum_{j=1}^{q}\left\{\overline{N}\left(r,\frac{1}{F-a_j}\right) + \overline{N}\left(r,\frac{1}{G-a_j}\right)\right\} \\
& + \overline{N}(r,F) + \overline{N}(r,G) \\
& - N\left(r,\frac{1}{F'};a_1,...,a_q\right) - N\left(r,\frac{1}{G'};a_1,...,a_q\right) \\
& - 2\log r + O(1).
\end{aligned}
$$

Hence we obtain

$$
\begin{aligned}
(3q-5)\{T(r,F) + T(r,G)\} \leq\ & 2\sum_{j=1}^{q}\left\{N_2\left(r,\frac{1}{F-a_j}\right) + N_2\left(r,\frac{1}{G-a_j}\right)\right\} \\
& + \sum_{j=1}^{q}\left\{\overline{N}\left(r,\frac{1}{F-a_j}\right) + \overline{N}\left(r,\frac{1}{G-a_j}\right)\right\} \\
& + 2(N_2(r,F) - \overline{N}(r,F)) \\
& + 2(N_2(r,G) - \overline{N}(r,G)) \\
& - N\left(r,\frac{1}{F'};a_1,...,a_q\right) - N\left(r,\frac{1}{G'};a_1,...,a_q\right) \\
& - 6\log r + O(1).
\end{aligned}
$$

Thus we complete the proof of the lemma.   □

Lemma 3.46. *Let f and g be non-constant meromorphic (resp., entire) functions on κ satisfying*

$$\overline{E}_f(\dot{F}_{n,b}) = \overline{E}_g(\dot{F}_{n,b}),$$

and define

$$F = \frac{1}{F_{n,b}(f)}, \quad G = \frac{1}{F_{n,b}(g)}.$$

If $n \geq 16$ (resp., $n \geq 9$), then $G = AF + B$, where $A, B \in \kappa$ with $A \neq 0$, and

$$\#(\{a_1, a_2, a_3\} \cap \{Aa_1 + B, Aa_2 + B, Aa_3 + B\}) \geq 2,$$

where $a_1 = 0$, $a_2 = \frac{1}{b}$, $a_3 = \frac{1}{b+1}$. In particular, we have

$$E_f(\dot{F}_{n,b}) = E_g(\dot{F}_{n,b}).$$

Proof. We will use the symbols and results in the proof of Lemma 3.35. Write $\dot{F}_{n,b} = \{c_1, ..., c_n\}$. Then we obtain

$$
\begin{aligned}
\overline{N}\left(r, \frac{1}{F - a_1}\right) &= \overline{N}(r, F_{n,b}(f)) = \overline{N}(r, f), \\
\overline{N}\left(r, \frac{1}{F - a_2}\right) &= \overline{N}\left(r, \frac{1}{F_{n,b}(f) - b}\right) \\
&\leq \overline{N}\left(r, \frac{1}{f}\right) + \overline{N}\left(r, \frac{1}{Q(f)}\right) \\
&\leq 3T(r, f) + O(1), \\
\overline{N}\left(r, \frac{1}{F - a_3}\right) &= \overline{N}\left(r, \frac{1}{F_{n,b}(f) - b - 1}\right) \\
&\leq \overline{N}\left(r, \frac{1}{f - 1}\right) + \overline{N}\left(r, \frac{1}{Q_1(f)}\right) \\
&\leq (n - 2)T(r, f) + O(1), \\
N_2(r, F) - \overline{N}(r, F) &= \sum_{j=1}^{n}\left\{N_2\left(r, \frac{1}{f - c_j}\right) - \overline{N}\left(r, \frac{1}{f - c_j}\right)\right\} \\
&\leq \sum_{j=1}^{n}\left\{T(r, f) - \overline{N}\left(r, \frac{1}{f - c_j}\right)\right\} + O(1) \\
&\leq \overline{N}(r, f) + T(r, f) - N\left(r, \frac{1}{f'}; c_1, ..., c_n\right) \\
&\quad - \log r + O(1).
\end{aligned}
$$

It follows that

$$\sum_{j=1}^{3} \overline{N}\left(r, \frac{1}{F - a_j}\right) + 2(N_2(r, F) - \overline{N}(r, F)) \leq (n + 3)T(r, f) + 3\overline{N}(r, f) + O(1),$$

and the same inequality holds with f and F replaced by g and G. Note that F and G share ∞ IM. Thus if (1) of Lemma 3.45 is true, we would get

$$
\begin{aligned}
4n\{T(r,f) + T(r,g)\} + 4\log r &\leq (3n+9)\{T(r,f) + T(r,g)\} \\
&\quad + 7\{\overline{N}(r,f) + \overline{N}(r,g)\} + O(1) \\
&\leq (3n+16)\{T(r,f) + T(r,g)\} + O(1),
\end{aligned}
$$

and hence $n < 16$ which is a contradiction to our assumptions. Therefore (2) of Lemma 3.45 must be satisfied. From $G = AF + B$, we see that F and G share ∞ CM, that is, $E_f(\dot{F}_{n,b}) = E_g(\dot{F}_{n,b})$. $\qquad\square$

Thus Lemma 3.46 and Theorem 3.36 yield the following Boutabaa-Escassut's result [14]:

Theorem 3.47. *For any integer $n \geq 16$, the set $\dot{F}_{n,b}$ is an ursim for $\mathcal{M}(\kappa)$.*

From the proof of Lemma 3.46 and Theorem 3.36, we also have

Theorem 3.48 ([14]). *For any integer $n \geq 9$, the set $\dot{F}_{n,b}$ is an ursim for $\mathcal{A}(\kappa)$.*

Thus Theorem 3.47 gives estimates $i_1(\mathcal{M}(\kappa)) \leq 16$ and $i_1(\mathcal{A}(\kappa)) \leq 9$.

Lemma 3.49. *Let f and g be non-constant meromorphic (resp., entire) functions on κ satisfying*

$$
\overline{E}_f(\dot{Y}_{n,n-2}) = \overline{E}_g(\dot{Y}_{n,n-2}),
$$

and define

$$
F = \frac{1}{Y_{n,n-2}(f)}, \quad G = \frac{1}{Y_{n,n-2}(g)}.
$$

If $n \geq 19$ (resp., $n \geq 12$), then $G = F$ or $G + F = \frac{1}{b}$. In particular, we have

$$
E_f(\dot{Y}_{n,n-2}) = E_g(\dot{Y}_{n,n-2}).
$$

Proof. Set $a_1 = 0$, $a_2 = 1/b$ and write $\dot{Y}_{n,n-2} = \{c_1, ..., c_n\}$. Note that

$$
F' = -\frac{nf'f^{n-3}}{(f^n - af^{n-2} + b)^2}\left(f^2 - \frac{n-2}{n}a\right).
$$

Then we obtain

$$
\begin{aligned}
N_2\left(r, \frac{1}{F - a_1}\right) &= N_2(r, Y_{n,n-2}(f)) = 2\overline{N}(r,f), \\
\overline{N}\left(r, \frac{1}{F - a_1}\right) &= \overline{N}(r, Y_{n,n-2}(f)) = \overline{N}(r,f), \\
N_2\left(r, \frac{1}{F - a_2}\right) &= N_2\left(r, \frac{1}{Y_{n,n-2}(f) - b}\right) \\
&\leq 2\overline{N}\left(r, \frac{1}{f}\right) + N_2\left(r, \frac{1}{f^2 - a}\right) \\
&\leq 4T(r,f) + O(1),
\end{aligned}
$$

$$\overline{N}\left(r, \frac{1}{F-a_2}\right) = \overline{N}\left(r, \frac{1}{Y_{n,n-2}(f)-b}\right)$$
$$\leq \overline{N}\left(r, \frac{1}{f}\right) + \overline{N}\left(r, \frac{1}{f^2-a}\right)$$
$$\leq 3T(r, f) + O(1),$$

$$N_2(r, F) - \overline{N}(r, F) = \sum_{j=1}^{n}\left\{N_2\left(r, \frac{1}{f-c_j}\right) - \overline{N}\left(r, \frac{1}{f-c_j}\right)\right\}$$
$$\leq \sum_{j=1}^{n}\left\{T(r, f) - \overline{N}\left(r, \frac{1}{f-c_j}\right)\right\} + O(1)$$
$$\leq \overline{N}(r, f) + T(r, f) - N\left(r, \frac{1}{f'}; c_1, ..., c_n\right)$$
$$- \log r + O(1),$$

$$N\left(r, \frac{1}{F'}; a_1, a_2\right) \geq N\left(r, \frac{1}{f^2 - \frac{n-2}{n}a}\right) \geq T(r, f) + O(1).$$

It follows that

$$2\sum_{j=1}^{2}N_2\left(r, \frac{1}{F-a_j}\right) + \sum_{j=1}^{2}\overline{N}\left(r, \frac{1}{F-a_j}\right)$$
$$+2(N_2(r, F) - \overline{N}(r, F)) - N\left(r, \frac{1}{F'}; a_1, a_2\right)$$
$$\leq 12T(r, f) + 7\overline{N}(r, f) + O(1),$$

and the same inequality holds with f and F replaced by g and G. Note that F and G share ∞ IM. Thus if (1) of Lemma 3.45 is true, we would get

$$n\{T(r, f) + T(r, g)\} + 4\log r \leq 12\{T(r, f) + T(r, g)\}$$
$$+7\{\overline{N}(r, f) + \overline{N}(r, g)\} + O(1)$$
$$\leq 19\{T(r, f) + T(r, g)\} + O(1),$$

and hence $n < 19$ which is a contradiction to our assumptions. Therefore (2) of Lemma 3.45 must be satisfied. Thus it follows that $G = F$ or $F + G = 1/b$, which means that F and G share ∞ CM, that is, $E_f(\dot{Y}_{n,n-2}) = E_g(\dot{Y}_{n,n-2})$. □

Thus Lemma 3.49 and Theorem 3.41 yield the following result:

Theorem 3.50. *For any integer* $n \geq 19$ *(resp.,* $n \geq 12$ *) with* $(n, n-2) = 1$*, the set* $\dot{Y}_{n,n-2}$ *is an ursim for* $\mathcal{M}(\kappa)$ *(resp., for* $\mathcal{A}(\kappa)$*).*

Lemma 3.51. *Let* f *and* g *be non-constant entire functions on* κ *satisfying*

$$\overline{E}_f(\dot{Y}_{n,n-1}) = \overline{E}_g(\dot{Y}_{n,n-1}),$$

and define

$$F = \frac{1}{Y_{n,n-1}(f)}, \quad G = \frac{1}{Y_{n,n-1}(g)}.$$

If $n \geq 9$, then $G = F$ or $G + F = \frac{1}{b}$. In particular, we have

$$E_f(\dot{Y}_{n,n-1}) = E_g(\dot{Y}_{n,n-1}).$$

Proof. Set $a_1 = 0$, $a_2 = 1/b$ and write $\dot{Y}_{n,n-1} = \{c_1, ..., c_n\}$. Note that

$$F' = -\frac{nf'f^{n-2}}{(f^n - af^{n-1} + b)^2}\left(f - \frac{n-1}{n}a\right).$$

Then we obtain

$$N_2\left(r, \frac{1}{F - a_1}\right) = N_2(r, Y_{n,n-1}(f)) = 0,$$

$$\overline{N}\left(r, \frac{1}{F - a_1}\right) = \overline{N}(r, Y_{n,n-1}(f)) = 0,$$

$$N_2\left(r, \frac{1}{F - a_2}\right) = N_2\left(r, \frac{1}{Y_{n,n-1}(f) - b}\right)$$

$$\leq 2\overline{N}\left(r, \frac{1}{f}\right) + N_2\left(r, \frac{1}{f - a}\right)$$

$$\leq 3T(r, f) + O(1),$$

$$\overline{N}\left(r, \frac{1}{F - a_2}\right) = \overline{N}\left(r, \frac{1}{Y_{n,n-1}(f) - b}\right)$$

$$\leq \overline{N}\left(r, \frac{1}{f}\right) + \overline{N}\left(r, \frac{1}{f - a}\right)$$

$$\leq 2T(r, f) + O(1),$$

$$N_2(r, F) - \overline{N}(r, F) = \sum_{j=1}^{n}\left\{N_2\left(r, \frac{1}{f - c_j}\right) - \overline{N}\left(r, \frac{1}{f - c_j}\right)\right\}$$

$$\leq \sum_{j=1}^{n}\left\{T(r, f) - \overline{N}\left(r, \frac{1}{f - c_j}\right)\right\} + O(1)$$

$$\leq \overline{N}(r, f) + T(r, f) - N\left(r, \frac{1}{f'}; c_1, ..., c_n\right)$$

$$- \log r + O(1),$$

$$N\left(r, \frac{1}{F'}; a_1, a_2\right) \geq N\left(r, \frac{1}{f - \frac{n-1}{n}a}\right) = T(r, f) + O(1).$$

It follows that

$$2\sum_{j=1}^{2}N_2\left(r, \frac{1}{F - a_j}\right) + \sum_{j=1}^{2}\overline{N}\left(r, \frac{1}{F - a_j}\right)$$

$$+ 2(N_2(r, F) - \overline{N}(r, F)) - N\left(r, \frac{1}{F'}; a_1, a_2\right)$$

$$\leq 9T(r, f) + O(1),$$

and the same inequality holds with f and F replaced by g and G. Note that F and G share ∞ IM. Thus if (1) of Lemma 3.45 is true, we would get

$$n\{T(r, f) + T(r, g)\} + 4\log r \leq 9\{T(r, f) + T(r, g)\} + O(1),$$

and hence $n < 9$, which is a contradiction to our assumptions. Therefore (2) of Lemma 3.45 must be satisfied. Thus it follows that $G = F$ or $F + G = 1/b$, which means that F and G share ∞ CM, that is, $E_f(\dot{Y}_{n,n-1}) = E_g(\dot{Y}_{n,n-1})$. $\square$

Thus Lemma 3.51 and the proof of Theorem 3.41 yield the following result:

Theorem 3.52. *For any integer $n \geq 9$, the set $\dot{Y}_{n,n-1}$ is an ursim for $\mathcal{A}(\kappa)$.*

3.8 Unique range sets for multiple values

In this section, we give uniqueness theorems dealing with multiple values of meromorphic functions. Let κ be an algebraically closed field of characteristic zero, complete for a non-trivial non-Archimedean absolute value $|\cdot|$. Let K denote the field κ or $\mathbb{C}$. Let f be a meromorphic function in K and take $a \in K \cup \{\infty\}$. For a positive integer k, define a set

$$E_f(a, k) = \left\{ (\mu_{f,k}^a(z), z) \mid z \in K \right\}.$$

Take a non-empty set $S \subset \bar{K}$ and write

$$E_f(S, k) = \bigcup_{a \in S} \{(\mu_{f,k}^a(z), z) \mid z \in K\} = \bigcup_{a \in S} E_f(a, k).$$

For more general settings, H. Fujimoto [34] gave some sufficient conditions for a finite subset S of $\mathbb{C}$ to be a unique range set for meromorphic (or entire) functions, which particularly implies the following fact:

Theorem 3.53. *Let f and g be non-constant meromorphic (resp., entire) functions on $\mathbb{C}$ satisfying*

$$E_f(\dot{F}_{n,b}, k) = E_g(\dot{F}_{n,b}, k) \ (2 \leq k \in \mathbb{Z}^+).$$

If $n > 10 + \frac{4}{k-1}$ (resp., $n > 6 + \frac{2}{k-1}$), then $f = g$.

Here we will use Frank-Reinders' technique to present a new approach.

Lemma 3.54. *Let F and G be non-constant meromorphic functions on κ satisfying*

$$E_F(\infty, k) = E_G(\infty, k) \ (2 \leq k \in \mathbb{Z}^+),$$

and let $a_1, ..., a_q$ be distinct elements of κ with $q \geq 2$. Then one of the following cases must occur

1)

$$\left(q - \frac{3}{2}\right) \{T(r, F) + T(r, G)\} \leq \sum_{j=1}^{q} \left\{ N_2\left(r, \frac{1}{F - a_j}\right) + N_2\left(r, \frac{1}{G - a_j}\right) \right\}$$
$$+ N_{k+1}(r, F) - N_k(r, F) + N_{k+1}(r, G) - N_k(r, G)$$
$$- \log r + O(1).$$

2) $G = AF + B$, where $A, B \in \kappa$ with $A \neq 0$, and

$$\#(\{a_1, ..., a_q\} \cap \{Aa_1 + B, ..., Aa_q + B\}) \geq 2.$$

The proof of Lemma 3.54 can be given completely similar to that of Lemma 3.34, where we only need to note that when $H \neq 0$, poles of H can only occur where F' or G' has a zero, or one of F and G has a pole with multiplicity $\geq k$. Thus we have

$$
\begin{aligned}
N(r, H) \leq\ & \overline{N}\left(r, \frac{1}{F'}\right) + \overline{N}\left(r, \frac{1}{G'}\right) \\
& + N_{k+1}(r, F) - N_k(r, F) + N_{k+1}(r, G) - N_k(r, G) \\
\leq\ & \sum_{j=1}^{q}\left\{ N_2\left(r, \frac{1}{F - a_j}\right) - \overline{N}\left(r, \frac{1}{F - a_j}\right)\right\} \\
& + \sum_{j=1}^{q}\left\{ N_2\left(r, \frac{1}{G - a_j}\right) - \overline{N}\left(r, \frac{1}{G - a_j}\right)\right\} \\
& + N\left(r, \frac{1}{F'}; a_1, ..., a_q\right) + N\left(r, \frac{1}{G'}; a_1, ..., a_q\right) \\
& + N_{k+1}(r, F) - N_k(r, F) + N_{k+1}(r, G) - N_k(r, G).
\end{aligned}
$$

Lemma 3.55. *Let f and g be non-constant meromorphic (resp., entire) functions on κ satisfying*

$$E_f(\dot{F}_{n,b}, k) = E_g(\dot{F}_{n,b}, k)\ (2 \leq k \in \mathbb{Z}^+),$$

and define

$$F = \frac{1}{F_{n,b}(f)},\ G = \frac{1}{F_{n,b}(g)}.$$

If $n \geq 10 + \frac{4}{k}$ (resp., $n \geq 6 + \frac{2}{k}$), then $G = AF + B$, where $A, B \in \kappa$ with $A \neq 0$, and

$$\#(\{a_1, a_2, a_3\} \cap \{Aa_1 + B, Aa_2 + B, Aa_3 + B\}) \geq 2,$$

where $a_1 = 0$, $a_2 = \frac{1}{b}$, $a_3 = \frac{1}{b+1}$. In particular, we have

$$E_f(\dot{F}_{n,b}) = E_g(\dot{F}_{n,b}).$$

The proof of Lemma 3.55 follows from that of Lemma 3.35 by noting that

$$
\begin{aligned}
N_{k+1}(r, F) - N_k(r, F) =\ & \sum_{j=1}^{n}\left\{ N_{k+1}\left(r, \frac{1}{f - c_j}\right) - N_k\left(r, \frac{1}{f - c_j}\right)\right\} \\
\leq\ & \frac{1}{k}\sum_{j=1}^{n}\left\{ N_{k+1}\left(r, \frac{1}{f - c_j}\right) - \overline{N}\left(r, \frac{1}{f - c_j}\right)\right\} \\
\leq\ & \frac{1}{k}\sum_{j=1}^{n}\left\{ T(r, f) - \overline{N}\left(r, \frac{1}{f - c_j}\right)\right\} + O(1) \\
\leq\ & \frac{1}{k}\overline{N}(r, f) + \frac{1}{k}T(r, f) - \frac{1}{k}N\left(r, \frac{1}{f'}; c_1, ..., c_n\right) \\
& - \frac{1}{k}\log r + O(1),
\end{aligned}
$$

where $\dot{F}_{n,b} = \{c_1, ..., c_n\}$. Thus we have the following result:

Theorem 3.56. *Let f and g be non-constant meromorphic (resp., entire) functions on κ satisfying*

$$E_f(\dot{F}_{n,b}, k) = E_g(\dot{F}_{n,b}, k) \ (2 \leq k \in \mathbf{Z}^+).$$

If $n \geq 10 + \frac{4}{k}$ (resp., $n \geq 6 + \frac{2}{k}$), then $f = g$.

Lemma 3.57. *Take $m, n \in \mathbf{Z}^+$ and let $b \in \kappa_*$, $Q \in \kappa[z]$ with $\deg(Q) = n - m \geq 1$ such that the polynomial*

$$P(z) = z^m Q(z) + b$$

has only simple zeros. Let S_P be the set of zeros of P. Let f and g be non-constant meromorphic (resp., entire) functions on κ satisfying

$$E_f(S_P, k) = E_g(S_P, k) \ (2 \leq k \in \mathbf{Z}^+),$$

and define

$$F = \frac{1}{P(f)}, \ G = \frac{1}{P(g)}.$$

If $2m \geq n + 8 + \frac{4}{k}$ (resp., $2m \geq n + 4 + \frac{2}{k}$), then $G = F$ or $G + F = \frac{1}{b}$.

Lemma 3.57 can be proved according to that of Lemma 3.40, and hence we obtain the following fact:

Theorem 3.58. *Let f and g be non-constant meromorphic (resp., entire) functions on κ satisfying*

$$E_f(\dot{Y}_{n,n-2}, k) = E_g(\dot{Y}_{n,n-2}, k) \ (2 \leq k \in \mathbf{Z}^+).$$

If $n \geq 12 + \frac{4}{k}$ (resp., $n \geq 8 + \frac{2}{k}$) with $(n, n - 2) = 1$, then $f = g$.

Remark. For the complex case, these results are true if the inequalities $* \geq *$ of integers are replaced by $* > *$. For example, we can prove the following result:

Theorem 3.59. *Let f and g be non-constant meromorphic (resp., entire) functions on $\mathbf{C}$ satisfying*

$$E_f(\dot{F}_{n,b}, k) = E_g(\dot{F}_{n,b}, k) \ (2 \leq k \in \mathbf{Z}^+).$$

If $n > 10 + \frac{4}{k}$ (resp., $n > 6 + \frac{2}{k}$), then $f = g$.

Obviously, Theorem 3.59 is better than Theorem 3.53 for the case $k = 2$, but the latter is the same as the former for larger k.

Chapter 4

Differential equations

In this chapter, we will give a survey of the non-Archimedean analogue of Malmquist-type theorems in ordinary differential equations based on the results of Yang-Hu [138].

4.1   Malmquist-type theorems

Let κ be an algebraically closed field of characteristic zero, complete for a non-trivial non-Archimedean absolute value $|\cdot|$. We talk of a non-Archimedean algebraic differential equation if it is of the form

$$\Omega\left(z, w, w', ..., w^{(n)}\right) = R(z, w), \tag{4.1}$$

where

$$\Omega\left(z, w, w', ..., w^{(n)}\right) = \sum_{i \in I} c_i(z) w^{i_0} (w')^{i_1} \cdots (w^{(n)})^{i_n} \tag{4.2}$$

and $i = (i_0, i_1, ..., i_n)$ are non-negative integer indices, I is a finite set, $c_i \in \mathcal{M}(\kappa)$, and $R(z, w)$ is a meromorphic function on κ^2, that is, when one variable is fixed $R(z, w)$ is a meromorphic function of another variable. Define the *degree* and *weight*, respectively, by

$$\deg(\Omega) = \max_{i \in I} \left\{ \sum_{\alpha=0}^{n} i_\alpha \right\},$$

$$\mathrm{wei}(\Omega) = \max_{i \in I} \left\{ \sum_{\alpha=0}^{n} (\alpha + 1) i_\alpha \right\},$$

and also write

$$\gamma(\Omega) = \max_{i \in I} \left\{ \sum_{\alpha=1}^{n} \alpha i_\alpha \right\}.$$

First of all, we show some properties of the differential operator Ω. Abbreviate

$$\Omega(z) = \Omega\left(z, w(z), w'(z), ..., w^{(n)}(z)\right)$$

for a meromorphic function $w = w(z)$ in $\mathcal{M}(\kappa)$. Then

$$N(r, w^{(\alpha)}) = N(r, w) + \alpha \overline{N}(r, w) \le (\alpha + 1) N(r, w).$$

115

Thus we have

$$N(r, \Omega) \leq \deg(\Omega) N(r, w) + \gamma(\Omega) \overline{N}(r, w) + \sum_{i \in I} N(r, c_i) \tag{4.3}$$

and, in particular,

$$N(r, \Omega) \leq \text{wei}(\Omega) N(r, w) + \sum_{i \in I} N(r, c_i). \tag{4.4}$$

Obviously,

$$m(r, \Omega) \leq \deg(\Omega) m(r, w) + \max_{i \in I} \left\{ m(r, c_i) + \sum_{\alpha=1}^{n} i_\alpha m\left(r, \frac{w^{(\alpha)}}{w}\right) \right\}. \tag{4.5}$$

By the lemma of logarithmic derivative, we obtain

$$T(r, \Omega) \leq \deg(\Omega) T(r, w) + \gamma(\Omega) \overline{N}(r, w) + \sum_{i \in I} T(r, c_i) + O(1) \tag{4.6}$$

and, in particular,

$$T(r, \Omega) \leq \text{wei}(\Omega) T(r, w) + \sum_{i \in I} T(r, c_i) + O(1). \tag{4.7}$$

Next, we will keep the notations introduced in §2.2, and consider the equation (4.1), where $R(z, w)$ is defined by (2.18). The following fact is a non-Archimedean analogue of Clunie theorem.

Lemma 4.1. *Let $w \in \mathcal{M}(\kappa)$ be a solution of (4.1), where $R(z, w)$ is defined by (2.18). If $q \geq k$, then*

$$m(r, \Omega) \leq \sum_{i \in I} m(r, c_i) + \sum_{j=0}^{k} m(r, a_j) + O\left(m\left(r, \frac{1}{b_q}\right) + \sum_{j=0}^{q} m(r, b_j)\right) \tag{4.8}$$

and

$$N(r, \Omega) \leq \sum_{i \in I} N(r, c_i) + \sum_{j=0}^{k} N(r, a_j) + O\left(\sum_{j=0}^{q} N\left(r, \frac{1}{b_j}\right)\right). \tag{4.9}$$

Proof. To prove (4.8), take $z \in \kappa$ with

$$w(z) \neq 0, \infty; \quad a_j(z) \neq 0, \infty \quad (0 \leq j \leq k);$$

$$c_i(z) \neq 0, \infty \quad (i \in I); \quad b_j(z) \neq 0, \infty \quad (0 \leq j \leq q).$$

Write

$$b(z) = \max_{0 \leq j < q} \left\{ 1, \left(\frac{|b_j(z)|}{|b_q(z)|}\right)^{\frac{1}{q-j}} \right\}.$$

If $|w(z)| > b(z)$, we have

$$|b_j(z)||w(z)|^j \leq |b_q(z)|b(z)^{q-j}|w(z)|^j < |b_q(z)||w(z)|^q,$$

and hence

$$|B(z, w(z))| = |b_q(z)||w(z)|^q.$$

Then

$$|\Omega(z)| = \frac{|A(z, w(z))|}{|B(z, w(z))|} \leq \frac{1}{|b_q(z)|} \max_{0 \leq j \leq k} |a_j(z)|.$$

If $|w(z)| \leq b(z)$,

$$|\Omega(z)| \leq b(z)^{\deg(\Omega)} \max_{i \in I} |c_i(z)| \left| \frac{w'(z)}{w(z)} \right|^{i_1} \cdots \left| \frac{w^{(n)}(z)}{w(z)} \right|^{i_n}.$$

Therefore, in any case, the inequality

$$\| \ \mu(r, \Omega) \ \leq \ \max_{0 \leq j \leq k, i \in I} \left\{ \frac{\mu(r, a_j)}{\mu(r, b_q)}, \mu(r, c_i) \mu\left(r, \frac{w'}{w}\right)^{i_1} \cdots \mu\left(r, \frac{w^{(n)}}{w}\right)^{i_n} \right.$$
$$\left. \cdot \max_{0 \leq j < q} \left\{ 1, \mu\left(r, \frac{b_j}{b_q}\right)^{\frac{\deg(\Omega)}{q-j}} \right\} \right\}$$

holds where $r = |z|$, which also holds for all $r > 0$ by continuity of the functions μ. Hence (4.8) follows from this inequality and the lemma of logarithmic derivative.

Next we prove (4.9). Take a point $z_0 \in \kappa$ with $w(z_0) = \infty$. According to the proof of Lemma 2.9, we have the inequalities

$$\mu_A^\infty(z_0) \leq k\mu_w^\infty(z_0) + \sum_{j=0}^{k} \mu_{a_j}^\infty(z_0)$$

and

$$\mu_B^\infty(z_0) \geq q\mu_w^\infty(z_0) - \sum_{j=0}^{q} \mu_{b_j}^0(z_0).$$

If $q\mu_w^\infty(z_0) - \sum_{j=0}^{q} \mu_{b_j}^0(z_0) > 0$, since $q \geq k$ we have

$$\mu_\Omega^\infty(z_0) \leq \mu_A^\infty(z_0) - \mu_B^\infty(z_0) \leq \sum_{j=0}^{k} \mu_{a_j}^\infty(z_0) + \sum_{j=0}^{q} \mu_{b_j}^0(z_0).$$

If $q\mu_w^\infty(z_0) - \sum_{j=0}^{q} \mu_{b_j}^0(z_0) \leq 0$, i.e.,

$$\mu_w^\infty(z_0) \leq \frac{1}{q} \sum_{j=0}^{q} \mu_{b_j}^0(z_0),$$

then

$$\mu_\Omega^\infty(z_0) \leq \text{wei}(\Omega)\mu_w^\infty(z_0) + \sum_{i \in I} \mu_{c_i}^\infty(z_0) \leq \frac{\text{wei}(\Omega)}{q} \sum_{j=0}^{q} \mu_{b_j}^0(z_0) + \sum_{i \in I} \mu_{c_i}^\infty(z_0).$$

Therefore,

$$\mu_\Omega^\infty \leq \sum_{j=0}^{k} \mu_{a_j}^\infty + \max\left\{1, \frac{\text{wei}(\Omega)}{q}\right\} \sum_{j=0}^{q} \mu_{b_j}^0 + \sum_{i \in I} \mu_{c_i}^\infty.$$

Hence (4.9) follows.   □

Definition 4.2. *A solution w of (4.1) with $R(z,w)$ defined by (2.18) is said to be admissible if $w \in \mathcal{M}(\kappa)$ satisfies (4.1) with*

$$\sum_{i \in I} T(r, c_i) + \sum_{j=0}^{k} T(r, a_j) + \sum_{j=0}^{q} T(r, b_j) = o(T(r, w)).$$

Theorem 4.3. *If R is of the form (2.18) and if (4.1) has an admissible non-constant solution w, then*

$$q = 0, \quad k \leq \min\{\text{wei}(\Omega), \deg(\Omega) + \gamma(\Omega)(1 - \Theta_w(\infty))\}.$$

Proof. By using the algorithm of division, we have

$$A(z, w) = A_1(z, w)B(z, w) + A_2(z, w)$$

with $\deg(A_2) < q$. Thus, the equation (4.1) can be rewritten as follows:

$$\Omega\left(z, w, w', ..., w^{(n)}\right) - A_1(z, w) = \frac{A_2(z, w)}{B(z, w)}.$$

Applying Lemma 4.1 to this equation, we obtain

$$T(r, \Omega - A_1) = o(T(r, w)).$$

Theorem 2.12 implies

$$T(r, \Omega - A_1) = T\left(r, \frac{A_2}{B}\right) = qT(r, w) + o(T(r, w)).$$

It follows that $q = 0$, and (4.1) assumes the following form

$$\Omega\left(z, w, w', ..., w^{(n)}\right) = A(z, w).$$

Thus, Theorem 2.11 implies

$$T(r, \Omega) = T(r, A) = kT(r, w) + o(T(r, w)).$$

Our result follows from this equation, (4.6), and (4.7).   □

For meromorphic functions on $\mathbb{C}$, this theorem is well known, called a Malmquist-type theorem, see Malmquist [88], Gackstatter-Laine [37], Laine [81], Toda [129] and Yosida [144] (Also see He-Xiao [54]). For several variables, see Hu-Yang [66] and [67]. The following result due to Boutabaa [11] follows from Theorem 4.3.

Corollary 4.4. *Let $\Omega\left(z, w, w', ..., w^{(n)}\right)$ be a differential polynomial with coefficients in $\kappa(z)$ and let $R(z, w) \in \kappa(z, w)$. If (4.1) has a non-Archimedean meromorphic solution $w = w(z) \in M(\kappa) - \kappa(z)$, then $R(z, w)$ is a polynomial in w of degree $\leq \mathrm{wei}(\Omega)$.*

Corollary 4.5. *Let $R(z, w)$ be a rational function of z and w. If there exists a transcendental meromorphic function $w = w(z)$ on κ satisfying*

$$\left(\frac{dw}{dz}\right)^n = R(z, w),$$

then $R(z, w)$ is a polynomial in w of degree $\leq 2n$.

If the equation (4.1) has constant coefficients, then Theorem 2.34 yields the following results:

Theorem 4.6 ([15]). *Let A and B be two relatively prime polynomials over κ, let t be the number of distinct zeros of B, and let $\Omega \in \kappa[x_0, x_1, ..., x_n] - \kappa[x_0, x_1, ..., x_{n-1}]$. Suppose that there exists a function $w \in M_{(\rho)}(\kappa)$ satisfying*

$$\Omega(w, w', ..., w^{(n)})B(w) = A(w).$$

1) Assume that w is unbounded ($\rho < \infty$). Then $(n+1)t \leq 2(n+1) + \deg(B)$. Moreover, if $\deg(A) > \deg(B)$, then $(n+1)t \leq \min\{n+1+\deg(A), 2(n+1)+\deg(B)\}$.
2) Assume $w \in M(\kappa) - \kappa$ ($\rho = \infty$). Then $(n+1)t < 2(n+1)+\deg(B)$. Moreover, if $\deg(A) > \deg(B)$, then $(n+1)t < \min\{n+1+\deg(A), 2(n+1)+\deg(B)\}$.

4.2 Generalized Malmquist-type theorems

In this section, we continue to study the differential equation (4.1) for more general $R(z, w)$.

Theorem 4.7. *Take $R \in M(\kappa)$. If the following differential equation*

$$\Omega\left(z, w, w', ..., w^{(n)}\right) = R(w)$$

has a non-constant solution $w \in M(\kappa)$ satisfying

$$\sum_{i \in I} T(r, c_i) = o(T(r, w)),$$

then R is a polynomial with

$$\deg(R) \leq \min\{\mathrm{wei}(\Omega), \deg(\Omega) + \gamma(\Omega)(1 - \Theta_w(\infty))\}.$$

Proof. Since

$$\limsup_{r \to \infty} \frac{T(r, R \circ w)}{T(r, w)} = \limsup_{r \to \infty} \frac{T(r, \Omega)}{T(r, w)} \leq \mathrm{wei}(\Omega),$$

Theorem 2.44 implies that R is a rational function. Thus Theorem 4.7 follows from Theorem 4.3. $\qquad \square$

Theorem 4.7 is also well known for meromorphic functions on $\mathbb{C}$ (see Rellich [108], Wittich [135], Laine [80], or see He-Xiao [54]). For the case of several variables, see Hu-Yang [65] and [66].

Theorem 4.8. *Let* $a_1, a_2, \ldots$ *be a sequence of distinct numbers in* κ *that tends to a finite limit value* a, *and let* $R(z, w)$ *be a meromorphic function on* κ^2. *If (4.1) has a non-constant meromorphic solution* w *on* κ *satisfying*

$$\sum_{i \in I} T(r, c_i) + T(r, R_j) = o(T(r, w)), \quad j = 1, 2, \ldots$$

where $R_j(z) = R(z, a_j)$, *then* $R(z, w)$ *is a polynomial in* w *with*

$$\deg(R) \leq \min\{\text{wei}(\Omega), \deg(\Omega) + \gamma(\Omega)(1 - \Theta_w(\infty))\}.$$

Proof. Write

$$\varphi[a_1] = \frac{\Omega - R_1}{w - a_1},$$

$$\varphi[a_1, a_2] = \frac{\varphi[a_1] - \varphi[a_2]}{a_1 - a_2} = \frac{\Omega}{(w - a_1)(w - a_2)}$$
$$- \frac{R_1}{(a_1 - a_2)(w - a_1)} + \frac{R_2}{(a_1 - a_2)(w - a_2)},$$

and inductively define

$$\varphi[a_1, \ldots, a_l] = \frac{\varphi[a_1, \ldots, a_{l-1}] - \varphi[a_1, \ldots, a_{l-2}, a_l]}{a_{l-1} - a_l}$$

$$= \frac{\Omega}{(w - a_1) \cdots (w - a_l)} + \sum_{j=1}^{l} \frac{\hat{a}_{lj} R_j}{w - a_j}$$

$$= \frac{\Omega - Q_l(z, w)}{(w - a_1) \cdots (w - a_l)} \quad (l \geq 3),$$

where $\hat{a}_{lj}$ are constants depending on $\{a_1, \ldots, a_l\}$, and $Q_l(z, w)$ is a polynomial in w of degree $< l$ such that its coefficients are linear combinations in $R_j (1 \leq j \leq l)$. Abbreviate

$$\nu = \text{wei}(\Omega); \quad \varphi_l = \varphi[a_{l(\nu+1)+1}, \ldots, a_{(l+1)(\nu+1)}], \quad l = 0, 1, \ldots .$$

We claim that $\varphi_l \equiv 0$ for some $l \geq 0$.

Assume, on the contrary, that $\varphi_l \not\equiv 0$ for all $l \geq 0$. By using the Jensen formula, we have

$$T(r, w) = T(r, w - a_{\nu+1}) + O(1)$$
$$\leq T(r, (w - a_{\nu+1})\varphi_0) + T\left(r, \frac{1}{\varphi_0}\right) + O(1)$$
$$= T(r, (w - a_{\nu+1})\varphi_0) + T(r, \varphi_0) + O(1). \tag{4.10}$$

On the other hand, noting that

$$\left| \frac{w(z)}{w(z) - a_j} \right| \leq \hat{a} \max\left\{ 1, \frac{1}{|w(z) - a_j|} \right\}, \quad j = 1, \ldots, \nu + 1,$$

where

$$\hat{a} = \max_{1 \le j \le \nu+1}\{1 + |a_j|\},$$

and by the lemma of logarithmic derivative, we have

$$
\begin{aligned}
m(r, \varphi_0) &\le m\left(r, \frac{\Omega}{(w - a_1)\cdots(w - a_{\nu+1})}\right) + m\left(r, \sum_{j=1}^{\nu+1}\frac{\hat{a}_{\nu+1,j}R_j}{w - a_j}\right)\\
&\le 2\sum_{j=1}^{\nu+1} m\left(r, \frac{1}{w - a_j}\right) + \sum_{i \in I} m(r, c_i) + \sum_{j=1}^{\nu+1} m(r, R_j) + O(1), \quad (4.11)
\end{aligned}
$$

and

$$
\begin{aligned}
m(r, (w - a_{\nu+1})\varphi_0) &\le 2\sum_{j=1}^{\nu} m\left(r, \frac{1}{w - a_j}\right)\\
&\quad + \sum_{i \in I} m(r, c_i) + \sum_{j=1}^{\nu+1} m(r, R_j) + O(1). \quad (4.12)
\end{aligned}
$$

Further, we estimate the poles of φ_0. Take $z_0 \in \kappa$. Since w is the solution of (4.1), then

$$\operatorname{supp}\mu_w^{a_1} \subset \operatorname{supp}\mu_{\Omega - R_1}^0.$$

Thus, when $\mu_w^{a_1}(z_0) > 0$,

$$\mu_{\varphi[a_1]}^\infty(z_0) \le \mu_w^{a_1}(z_0) - 1.$$

If $\mu_w^{a_j}(z_0) > 0$ for some j with $1 \le j \le \nu + 1$, but $c_i(z_0) \ne \infty (i \in I)$, $R_l(z_0) \ne \infty (1 \le l \le \nu + 1)$, then, by induction, we have

$$\mu_{\varphi_0}^\infty(z_0) \le \mu_w^{a_j}(z_0) - 1.$$

If z_0 is a pole of w, that is, $\mu_w^\infty(z_0) > 0$, then

$$
\begin{aligned}
\mu_{\varphi_0}^\infty(z_0) &\le \max\{0, \max\{\mu_\Omega^\infty(z_0), \mu_{Q_{\nu+1}}^\infty(z_0)\} - (\nu + 1)\mu_w^\infty(z_0)\}\\
&\le \sum_{i \in I}\mu_{c_i}^\infty(z_0) + \sum_{j=1}^{\nu+1}\mu_{R_j}^\infty(z_0).
\end{aligned}
$$

Therefore, we have the estimate

$$
\begin{aligned}
N(r, \varphi_0) &\le \sum_{j=1}^{\nu+1}\left\{N\left(r, \frac{1}{w - a_j}\right) - \overline{N}\left(r, \frac{1}{w - a_j}\right)\right\}\\
&\quad + \sum_{i \in I} N(r, c_i) + \sum_{j=1}^{\nu+1} N(r, R_j). \quad (4.13)
\end{aligned}
$$

Similarly, we can obtain

$$N(r, (w - a_{\nu+1})\varphi_0) \leq \sum_{j=1}^{\nu+1} \left\{ N\left(r, \frac{1}{w - a_j}\right) - \overline{N}\left(r, \frac{1}{w - a_j}\right) \right\}$$

$$+ \sum_{i \in I} N(r, c_i) + \sum_{j=1}^{\nu+1} N(r, R_j). \tag{4.14}$$

Therefore, by (4.10)-(4.14), we obtain

$$T(r, w) \leq 2 \sum_{j=1}^{\nu+1} \left\{ N\left(r, \frac{1}{w - a_j}\right) - \overline{N}\left(r, \frac{1}{w - a_j}\right) \right\}$$

$$+ 4 \sum_{j=1}^{\nu+1} m\left(r, \frac{1}{w - a_j}\right) + o(T(r, w)).$$

In a similar fashion, we have, for all $l \geq 0$,

$$T(r, w) \leq 2 \sum_{j=l(\nu+1)+1}^{(l+1)(\nu+1)} \left\{ N\left(r, \frac{1}{w - a_j}\right) - \overline{N}\left(r, \frac{1}{w - a_j}\right) \right\}$$

$$+ 4 \sum_{j=l(\nu+1)+1}^{(l+1)(\nu+1)} m\left(r, \frac{1}{w - a_j}\right) + o(T(r, w)).$$

Therefore,

$$lT(r, w) \leq 2 \sum_{j=1}^{l(\nu+1)} \left\{ N\left(r, \frac{1}{w - a_j}\right) - \overline{N}\left(r, \frac{1}{w - a_j}\right) \right\}$$

$$+ 4 \sum_{j=1}^{l(\nu+1)} m\left(r, \frac{1}{w - a_j}\right) + o(T(r, w)).$$

By using the second main theorem, we finally obtain

$$(l - 8)T(r, w) \leq o(T(r, w)).$$

This is impossible when $l > 8$. Thus the assertion is proved.

Hence $\varphi_l \equiv 0$ for some $l \geq 0$, say, $l = 0$, that is, w satisfies the following equation

$$\Omega\left(z, w, w', ..., w^{(n)}\right) = Q_{\nu+1}(z, w).$$

Write

$$H(z, w) = R(z, w) - Q_{\nu+1}(z, w), \quad H_j(z) = H(z, a_j).$$

If $H_j \not\equiv 0$, then

$$\overline{N}\left(r, \frac{1}{w - a_j}\right) \leq N\left(r, \frac{1}{H_j}\right) \leq T(r, H_j) + O(1)$$

$$\leq T(r, R_j) + \sum_{l=1}^{\nu+1} T(r, R_l) + O(1) = o(T(r, w)).$$

By the second main theorem, there are two values a_j at most such that the above inequality holds. Hence $H_j \equiv 0$, that is,

$$R(z, a_j) = Q_{\nu+1}(z, a_j), \quad z \in \kappa,$$

holds except for two values a_j at most. Then $R(z, w) \equiv Q_{\nu+1}(z, w)$ by the standard argument in proving the identity of two analytic functions in complex analysis. Thus Theorem 4.8 follows from Theorem 4.3. □

Theorem 4.8 is proved by Steinmetz [125] for meromorphic functions on $\mathbb{C}$ (or see He-Xiao [54]). For the case of several variables, see Hu-Yang [66].

4.3 Further results on Malmquist-type theorems

This is a continuation of § 4.1. In this section, we introduce the differential operator

$$\Phi\left(z, w, w', ..., w^{(n)}\right) = \sum_{i \in J} d_i w^{i_0} (w')^{i_1} \cdots (w^{(n)})^{i_n}, \tag{4.15}$$

where $\#J < \infty, d_i \in \mathcal{M}(\kappa)$, and consider the following differential equation:

$$\Omega\left(z, w, w', ..., w^{(n)}\right) = R(z, w)\Phi\left(z, w, w', ..., w^{(n)}\right). \tag{4.16}$$

First of all, we prove the following Clunie-type result.

Lemma 4.9. *Let* $R(z, w)$ *be defined by (2.18) and let* w *be a solution of (4.16). If* $q \geq k$, *then*

$$T\left(r, \frac{\Omega}{\Phi}\right) \leq T(r, \Phi) + \sum_{i \in I} T(r, c_i) + \sum_{j=0}^{k} T(r, a_j) + O\left(\sum_{j=0}^{q} T(r, b_j)\right).$$

If $q \geq k + \deg(\Phi)$, *then*

$$m(r, \Omega) \leq \sum_{i \in I} m(r, c_i) + \sum_{i \in J} m(r, d_i) + \sum_{j=0}^{k} m(r, a_j)$$

$$+ O\left(\sum_{j=0}^{q} m(r, b_j) + m\left(r, \frac{1}{b_q}\right)\right).$$

Proof. When $q \geq k$, according to the proof of Lemma 4.1, we have

$$m\left(r, \frac{\Omega}{\Phi}\right) \leq m\left(r, \frac{1}{\Phi}\right) + \sum_{i \in I} m(r, c_i) + \sum_{j=0}^{k} m(r, a_j)$$

$$+ O\left(\sum_{j=0}^{q} m(r, b_j) + m\left(r, \frac{1}{b_q}\right)\right) \tag{4.17}$$

and

$$N\left(r, \frac{\Omega}{\Phi}\right) \leq N\left(r, \frac{1}{\Phi}\right) + \sum_{i \in I} N(r, c_i) + \sum_{j=0}^{k} N(r, a_j)$$

$$+ O\left(\sum_{j=0}^{q} N\left(r, \frac{1}{b_j}\right)\right), \tag{4.18}$$

and so the first inequality follows from (4.17) and (4.18). The second inequality can be obtained similarly.   □

Theorem 4.10. *If there exists a non-constant solution w of (4.16) with $R(z, w)$ defined by (2.18) such that*

$$\sum_{i \in I} T(r, c_i) + \sum_{i \in J} T(r, d_i) + \sum_{j=0}^{k} T(r, a_j) + \sum_{j=0}^{q} T(r, b_j) = o(T(r, w)),$$

then

$$q \leq \min\{\text{wei}(\Phi), \deg(\Phi) + \gamma(\Phi)(1 - \Theta_w(\infty))\},$$
$$k \leq \min\{\text{wei}(\Omega), \deg(\Omega) + \gamma(\Omega)(1 - \Theta_w(\infty))\}.$$

Proof. By using the algorithm of division, we have

$$A(z, w) = A_1(z, w)B(z, w) + A_2(z, w)$$

with $\deg(A_2) < q$, and hence the equation (4.16) can be rewritten as follows

$$\Omega\left(z, w, w', ..., w^{(n)}\right) - A_1(z, w)\Phi\left(z, w, w', ..., w^{(n)}\right) = \frac{A_2(z, w)}{B(z, w)}\Phi\left(z, w, w', ..., w^{(n)}\right).$$

By Lemma 4.9, we obtain

$$T\left(r, \frac{\Omega - A_1\Phi}{\Phi}\right) \leq T(r, \Phi) + o(T(r, w)).$$

By Theorem 2.12,

$$T\left(r, \frac{\Omega - A_1\Phi}{\Phi}\right) = T\left(r, \frac{A_2}{B}\right) = qT(r, w) + o(T(r, w)).$$

Thus, we have

$$qT(r, w) \leq T(r, \Phi) + o(T(r, w)).$$

Combining this inequality with (4.6) and (4.7), we obtain the bound of q.

By rewriting (4.16) as follows

$$\frac{\Phi}{\Omega} = \frac{1}{R} = \frac{B}{A},$$

and by the conclusion above, the asserted upper bound for k can be obtained. □

Similarly, we can prove the following result:

Theorem 4.11. *Take $R \in \mathcal{M}(\kappa)$. If the following differential equation*

$$\Omega\left(z, w, w', ..., w^{(n)}\right) = R(w)\Phi\left(z, w, w', ..., w^{(n)}\right),$$

has a non-constant solution $w \in \mathcal{M}(\kappa)$ satisfying

$$\sum_{i \in I} T(r, c_i) + \sum_{i \in J} T(r, d_i) = o(T(r, w)),$$

then $R = \frac{A}{B}$ is a rational function with

$$\deg(B) \leq \min\{\text{wei}(\Phi), \deg(\Phi) + \gamma(\Phi)(1 - \Theta_w(\infty))\},$$

$$\deg(A) \leq \min\{\text{wei}(\Omega), \deg(\Omega) + \gamma(\Omega)(1 - \Theta_w(\infty))\}.$$

Theorem 4.10 is proved by Tu [131] for meromorphic functions on $\mathbb{C}$. For several variables, see Hu-Yang [66].

4.4 Admissible solutions of some differential equations

Let κ be an algebraically closed field of characteristic zero, complete for a non-trivial non-Archimedean absolute value $|\cdot|$. In this section, we study the following differential equations

$$\Omega\left(z, w, w', ..., w^{(n)}\right) = \sum_{j=0}^{k} a_j(z)w^j, \tag{4.19}$$

for some special Ω. For $w \in \mathcal{M}(\kappa)$, here and in the sequel, $\Omega\left(z, w, w', ..., w^{(n)}\right)$ is called a *differential polynomial of w* if

$$T(r, c_i) = o(T(r, w)) \quad (i \in I).$$

Lemma 4.12. *If $w_0, w_1 \in \mathcal{M}(\kappa)$ are linearly independent, then*

$$\begin{aligned}
T(r, w_0) \leq \ & m(r, w_0 + w_1) + N(r, w_0) + \bar{N}(r, w_0) \\
& + \bar{N}\left(r, \frac{1}{w_0}\right) + \bar{N}(r, w_1) + \bar{N}\left(r, \frac{1}{w_1}\right) + O(1).
\end{aligned}$$

Proof. Setting $w = w_0 + w_1$, then

$$w' = \frac{w_0'}{w_0}w_0 + \frac{w_1'}{w_1}w_1,$$

which yields

$$w_0 = w\left(\frac{w_1'}{w_1} - \frac{w'}{w}\right)\left(\frac{w_1'}{w_1} - \frac{w_0'}{w_0}\right)^{-1}.$$

By using the first main theorem and the lemma of logarithmic derivative,

$$
\begin{aligned}
m(r, w_0) &\leq m(r, w) + m\left(r, \frac{w_1'}{w_1} - \frac{w'}{w}\right) + m\left(r, \left(\frac{w_1'}{w_1} - \frac{w_0'}{w_0}\right)^{-1}\right) \\
&\leq m(r, w) + m\left(r, \frac{w_1'}{w_1}\right) + m\left(r, \frac{w'}{w}\right) \\
&\quad + m\left(r, \frac{w_1'}{w_1} - \frac{w_0'}{w_0}\right) + N\left(r, \frac{w_1'}{w_1} - \frac{w_0'}{w_0}\right) + O(1) \\
&\leq m(r, w) + \bar{N}(r, w_0) + \bar{N}(r, w_1) \\
&\quad + \bar{N}\left(r, \frac{1}{w_0}\right) + \bar{N}\left(r, \frac{1}{w_1}\right) + O(1).
\end{aligned}
$$

Thus, the lemma follows.   □

Theorem 4.13. *Assume*

$$\Omega\left(z, w, w', ..., w^{(n)}\right) = \left\{P\left(z, w, w', ..., w^{(n)}\right) + Q\left(z, w, w', ..., w^{(n)}\right)\right\}^l, \qquad (4.20)$$

where P is a differential monomial of w and Q is a differential polynomial of w with

$$\deg(P) \geq \deg(Q), \quad \gamma(P) > \gamma(Q).$$

If $k < l$, and if (4.19) has an admissible non-constant meromorphic solution, then (4.19) assumes the following form

$$\Omega\left(z, w, w', ..., w^{(n)}\right) = a_k(z)(w + b(z))^k, \quad b(z) = \frac{a_{k-1}(z)}{ka_k(z)}. \qquad (4.21)$$

Proof. The case $k = 1$ is obvious. Next we consider the case: $1 < k < l$. Assume, on the contrary, that

$$\sum_{j=0}^{k} a_j w^j - a_k(w + b)^k = \sum_{j=0}^{m} A_j w^j \not\equiv 0, \quad b = \frac{a_{k-1}}{ka_k},$$

where $0 \leq m \leq k - 2$, $A_m \not\equiv 0$ and A_j are rational functions of $\{a_j\}$. Then meromorphic functions $w_0 = -a_k(w + b)^k$ and $w_1 = \Omega = (P + Q)^l$ are linearly independent. In fact, if

$$\alpha w_0 + \beta w_1 \equiv 0, \quad \{\alpha, \beta\} \subset \kappa,$$

then

$$\sum_{j=0}^{k} \beta a_j w^j = \alpha a_k (w + b)^k = \alpha a_k w^k + \alpha a_{k-1} w^{k-1} + \cdots + \alpha a_k b^k.$$

Let $\mathcal{M}_w(\kappa)$ be the set of meromorphic functions f on κ satisfying $T(r, f) = o(T(r, w))$. Then $1, w, w^2, ..., w^k$ are linear independent over $\mathcal{M}_w(\kappa)$. Thus we obtain $\alpha = \beta$ from the equation above. Since

$$w_0 + w_1 = \sum_{j=0}^{m} A_j w^j \not\equiv 0,$$

then $\alpha = \beta = 0$ follows.

By the assumptions of Theorem 4.13, we obtain

$$
\begin{aligned}
N(r, \Omega) &= lN(r, P + Q) \\
&= l\{\deg(P)N(r, w) + \gamma(P)\bar{N}(r, w) + o(T(r, w))\} \\
&= N(r, \sum_{j=0}^{k} a_j w^j) = kN(r, w) + o(T(r, w)),
\end{aligned}
$$

and hence

$$N(r, w) = o(T(r, w))$$

since $\deg(P) > 0, l > k$. Similarly, we can prove

$$T(r, P + Q) = \frac{k}{l} T(r, w) + o(T(r, w)).$$

Now we have

$$T(r, w_0) = kT(r, w) + o(T(r, w)),$$

$$T(r, w_1) = lT(r, P + Q) = kT(r, w) + o(T(r, w)),$$

$$N(r, w_0) = o(T(r, w)), \quad N(r, w_1) = o(T(r, w)),$$

$$\bar{N}\left(r, \frac{1}{w_0}\right) = \bar{N}\left(r, \frac{1}{w + b}\right) + o(T(r, w)) \leq T(r, w) + o(T(r, w)),$$

$$
\begin{aligned}
\bar{N}\left(r, \frac{1}{w_1}\right) &= \bar{N}\left(r, \frac{1}{P + Q}\right) \leq T(r, P + Q) \\
&= \frac{k}{l} T(r, w) + o(T(r, w)),
\end{aligned}
$$

$$m(r, w_0 + w_1) = m \, m(r, w) + o(T(r, w)) \leq mT(r, w) + o(T(r, w)).$$

From this and Lemma 4.12, we obtain

$$kT(r, w) \leq mT(r, w) + T(r, w) + \frac{k}{l} T(r, w) + o(T(r, w)),$$

which implies $k \leq m + 1 + \frac{k}{l}$. This is impossible since $k \geq m + 2 > m + 1 + \frac{k}{l}$. $\square$

Corollary 4.14. *If*

$$(w')^n = \sum_{j=0}^{k} a_j(z)w^j \quad (k < n) \tag{4.22}$$

has an admissible non-constant meromorphic solution, then (4.22) assumes the following form

$$(w')^n = a_k(z)(w + b(z))^k, \quad b(z) = \frac{a_{k-1}(z)}{ka_k(z)}. \tag{4.23}$$

Corollary 4.15. *If $n > k$ and if $n - k$ is not a factor of n, then (4.22) with constant coefficients a_j has no admissible non-constant meromorphic solutions.*

The corollary can be proved easily by using Corollary 4.14 and by comparing the multiplicity of poles and $(-b)$-valued points of w in (4.23).

Conjecture 4.16. *The equation (4.22) has no admissible transcendental meromorphic solutions.*

For the complex case, this is the conjecture due to Gackstatter and Laine [37]. For the case of constant coefficients, Boutabaa and Escassut [15] prove it as follows:

Theorem 4.17. *If the equation (4.22) with constant coefficients a_j has a non-constant meromorphic solution w, then $n - k$ divides n, and w assumes the following form*

$$w(z) = a(z - c)^{\frac{n}{n-k}} - b \ (b, c \in \kappa), \tag{4.24}$$

where a satisfies $a^{n-k} \left(\frac{n}{n-k} \right)^n = a_k$.

Proof. By Corollary 4.14 and Corollary 4.15, the equation (4.22) assumes the form (4.23) with $n - k$ a factor of n, where a_k and b are constant. By comparing the multiplicity of poles and $(-b)$-valued points of w in (4.23), it follow that w is entire such that each zero of $w + b$ has the multiplicity $l = \frac{n}{n-k}$. Thus there is an entire function f satisfying $w + b = f^l$. By using (4.23), we obtain $l^n(f')^n = a_k$, that is, f has the following form $f(z) = A(z - c)$ with $l^n A^n = a_k$. Hence the theorem is proved by setting $a = A^l$. $\qquad \square$

Lemma 4.18. *Assume*

$$\Omega\left(z, w, w', ..., w^{(n)}\right) = B(z, w)P\left(z, w, w', ..., w^{(n)}\right) + Q\left(z, w, w', ..., w^{(n)}\right) \not\equiv 0, \tag{4.25}$$

where $P(\not\equiv 0)$ and $Q(\not\equiv 0)$ are differential polynomial of w, and where $B(z,w)$ is defined by (2.17). If

$$q = \deg(B) > \min\{\text{wei}(Q), \deg(Q) + \gamma(Q)(1 - \Theta_w(\infty))\},$$

then

$$(q - \deg(Q))T(r, w) \leq (\text{wei}(Q) - \deg(Q) + 1)\bar{N}(r, w)$$
$$+ \bar{N}\left(r, \frac{1}{\Omega}\right) + \bar{N}\left(r, \frac{1}{B}\right) + o(T(r, w)).$$

Proof. By Theorem 4.10, $\frac{BP}{Q}$ is not constant, and so is $\frac{\Omega}{Q}$, i.e., Ω and Q are linearly independent. Hence

$$Q^* = \left(\frac{Q'}{Q} - \frac{\Omega'}{\Omega}\right) Q \not\equiv 0.$$

Noting that

$$\frac{\Omega'}{\Omega} BP + \frac{\Omega'}{\Omega} Q = \Omega' = B'P + BP' + Q',$$

we have $BP^* = Q^*$, where

$$P^* = \left(\frac{\Omega'}{\Omega} - \frac{B'}{B} - \frac{P'}{P}\right) P.$$

Lemma 4.9 implies

$$m(r, P^*) = o(T(r, w)).$$

Now

$$m(r, Q^*) \le \deg(Q)m(r, w) + o(T(r, w))$$

and by the first main theorem, we have

$$
\begin{aligned}
m\left(r, \frac{1}{P^*}\right) &= m(r, P^*) + N(r, P^*) - N\left(r, \frac{1}{P^*}\right) + O(1) \\
&= N(r, P^*) - N\left(r, \frac{1}{P^*}\right) + o(T(r, w)).
\end{aligned}
$$

Hence

$$
\begin{aligned}
qm(r, w) &= m(r, B) + o(T(r, w)) \qquad\qquad (4.26)\\
&\le m(r, Q^*) + m\left(r, \frac{1}{P^*}\right) + o(T(r, w)) \\
&\le \deg(Q)m(r, w) + N(r, P^*) - N\left(r, \frac{1}{P^*}\right) + o(T(r, w)). \qquad (4.27)
\end{aligned}
$$

Next we estimate the zeros and poles of P^*. Take $z_0 \in \kappa$. If $\mu_B^0(z_0) > 0$, but z_0 is not a pole or a zero of the coefficients of B, P and Q, then we have

$$w(z_0) \ne \infty, \quad \mu_{P^*}^\infty(z_0) \le 1.$$

If $\mu_w^\infty(z_0) > 0$, but z_0 is not a pole or a zero of the coefficients of B, P and Q, then the estimate

$$\mu_{Q^*}^\infty(z_0) \le \deg(Q)\mu_w^\infty(z_0) + \text{wei}(Q) - \deg(Q) + 1$$

holds, and further,

$$\mu_{P^*}^\infty(z_0) = \mu_{Q^*/B}^\infty(z_0) \le \deg(Q)\mu_w^\infty(z_0) + \text{wei}(Q) - \deg(Q) + 1 - q\mu_w^\infty(z_0)$$

holds if $\mu_{P^*}^\infty(z_0) > 0$, or

$$\mu_{1/P^*}^\infty(z_0) = \mu_{B/Q^*}^\infty(z_0) \ge q\mu_w^\infty(z_0) - \{\deg(Q)\mu_w^\infty(z_0) + \text{wei}(Q) - \deg(Q) + 1\}$$

holds when $\mu_{P_*}^\infty(z_0) = 0$. Therefore

$$N(r, P^*) - N\left(r, \frac{1}{P^*}\right) \leq \bar{N}\left(r, \frac{1}{\Omega}\right) + \bar{N}\left(r, \frac{1}{B}\right) - (q - \deg(Q))N(r, w)$$
$$+ (\text{wei}(Q) - \deg(Q) + 1)\bar{N}(r, w) + o(T(r, w)), \quad (4.28)$$

and hence the lemma follows from (4.27) and (4.28). □

Theorem 4.19. *Assume*

$$\Omega\left(z, w, w', ..., w^{(n)}\right) = \left\{w^q P\left(z, w, w', ..., w^{(n)}\right) + Q\left(z, w, w', ..., w^{(n)}\right)\right\}^l, \quad (4.29)$$

where $P(\not\equiv 0)$ and $Q(\not\equiv 0)$ are differential polynomials of w with

$$q > \max\{\deg(Q) + 2, \text{wei}(Q)\}.$$

If $k < l$, then (4.19) has no admissible non-constant meromorphic solutions.

Proof. Assume, to the contrary, that (4.19) has an admissible transcendental meromorphic solution w. Then we obtain

$$N(r, \Omega) = N\left(r, \sum_{j=0}^{k} a_j w^j\right) = kN(r, w) + o(T(r, w)),$$

$$N(r, \Omega) = lN(r, w^q P + Q) \geq lqN(r, w) + o(T(r, w)),$$

and so

$$N(r, w) = o(T(r, w)).$$

Following the proof of Theorem 4.13, we can prove that (4.19) assumes the form (4.21). Then Lemma 4.18 gives

$$\begin{aligned}
(q - \deg(Q))T(r, w) &\leq \bar{N}\left(r, \frac{1}{w^q P + Q}\right) + \bar{N}\left(r, \frac{1}{w}\right) + o(T(r, w)) \\
&= \bar{N}\left(r, \frac{1}{w + b}\right) + \bar{N}\left(r, \frac{1}{w}\right) + o(T(r, w)) \\
&\leq 2T(r, w) + o(T(r, w)),
\end{aligned}$$

which implies $q - \deg(Q) \leq 2$. This is impossible since $q > \deg(Q) + 2$. □

Similarly, we can prove

Theorem 4.20. *Assume that Ω is defined by (4.29) with $q > \text{wei}(Q) + 3$. Then (4.21) has no admissible non-constant meromorphic solutions for any positive integers k and l.*

For work related to the topics in this section in one complex variable case, see Hu [57], and Hu-Yang [73].

4.5 Differential equations of constant coefficients

We continue to study the differential equation (4.19), but in this section, we consider only some of its special cases when the coefficients are constants.

Theorem 4.21. *The following non-Archimedean differential equation*

$$w^{(n)} + a_n w^{(n-1)} + \cdots + a_2 w' + a_1 w + a_0 = 0 \tag{4.30}$$

with $a_j \in \kappa$ has no transcendental meromorphic solutions.

Proof. Assume, on the contrary, that (4.30) has a transcendental meromorphic solution $w = w(z)$. Substituting it into (4.30) and comparing the order of poles of w, it follows that w has no poles, that is, w must be entire. Next, we distinguish two cases.

Case 1. Assume that $a_0 = 0$. Then the equation (4.30) becomes

$$w^{(n)} + a_n w^{(n-1)} + \cdots + a_j w^{(j-1)} = 0,$$

where $a_j \neq 0$ for a minimal integer $j \geq 1$. Hence

$$
\begin{aligned}
|a_j| &= \left| \frac{w^{(n)}(z)}{w^{(j-1)}(z)} + a_n \frac{w^{(n-1)}(z)}{w^{(j-1)}(z)} + \cdots + a_{j+1} \frac{w^{(j)}(z)}{w^{(j-1)}(z)} \right| \\
&\leq \max \left\{ \frac{1}{|z|^{n-j+1}}, \frac{|a_n|}{|z|^{n-j}}, \ldots, \frac{|a_{j+1}|}{|z|} \right\} \to 0,
\end{aligned}
$$

as $|z| \to \infty$, and so $a_j = 0$. This is a contradiction.

Case 2. Assume that $a_0 \neq 0$. Then the equation (4.30) becomes

$$w^{(n)} + a_n w^{(n-1)} + \cdots + a_j w^{(j-1)} = -a_0,$$

where $a_j \neq 0$ for a minimal integer $j \geq 1$. Thus, we have

$$w^{(n+1)} + a_n w^{(n)} + \cdots + a_j w^{(j)} = 0.$$

According to the proof of Case 1, we shall arrive at a contradiction. Therefore, (4.30) has no transcendental meromorphic solutions. □

Further, one can easily check that every solution of (4.30) is a polynomial of degree $j - 1$, where j is the smallest positive integer such that $a_j \neq 0$. We will need the following:

Lemma 4.22. *Let $R_i(z, w_1, ..., w_n)$ $(1 \leq i \leq n)$ be n holomorphic functions in the polydisc*

$$\Delta_\rho^{n+1}(x_0) = \{(z, w_1, ..., w_n) \in \kappa^{n+1} \mid |z - z_0| < r, \ |w_i - w_{i,0}| < \rho_i, 1 \leq i \leq n\},$$

that is, when n variables are fixed each of $R_i(z, w_1, ..., w_n)$ is a holomorphic function of the rest variable, where $x_0 = (z_0, w_{1,0}, ..., w_{n,0})$, $\rho = (r, \rho_1, ..., \rho_n)$. If the Cauchy's problem

$$\frac{dw_i}{dz} = R_i(z, w_1, ..., w_n), \quad w_i(z_0) = w_{i,0}, \quad 1 \leq i \leq n \tag{4.31}$$

has holomorphic solutions, then it is unique.

Proof. Setting $Z = z - z_0$, $W_i = w_i - w_{i,0}$, then (4.31) is converted into the following Cauchy's problem

$$\frac{dW_i}{dZ} = f_i(Z, W_1, ..., W_n), \quad W_i(0) = 0, \quad 1 \le i \le n, \qquad (4.32)$$

where $f_i(Z, W_1, ..., W_n)$ are holomorphic in $\Delta_\rho^{n+1}(0)$. Write

$$f_i(Z, W_1, ..., W_n) = \sum_{j \ge 0, m_l \ge 0} a^i_{j,m_1,...,m_n} Z^j W_1^{m_1} \cdots W_n^{m_n}.$$

Let $W_i = W_i(Z)$ $(1 \le i \le n)$ be a holomorphic solution of (4.32). Putting

$$W_i(Z) = \sum_{j=0}^{\infty} a^i_j Z^j,$$

then for $i = 1, ..., n$, we have

$$a^i_0 = W_i(0) = 0,$$

$$a^i_1 = \frac{dW_i}{dZ}(0) = f_i(0, 0, ..., 0) = a^i_{0,0,...,0},$$

$$a^i_2 = \frac{1}{2!} \frac{d^2 W_i}{dZ^2}(0) = \frac{1}{2}\left(\frac{\partial f_i}{\partial Z} + \sum_{l=1}^{n} \frac{\partial f_i}{\partial W_l} \frac{dW_l}{dZ} \right)_0$$

$$= \frac{1}{2}(a^i_{1,0,...,0} + a^i_{0,1,0,...0} a^1_{0,0,...,0} + \cdots + a^i_{0,0,0,...1} a^n_{0,0,...,0}), \quad \cdots,$$

$$a^i_m = \frac{1}{m!} \frac{d^m W_i}{dZ^m}(0) = P^i_m(..., a^l_{j,m_1,...,m_n}, ...), \quad \cdots,$$

where P^i_m is a polynomial of $a^l_{j,m_1,...,m_n}$ $(j+m_1+\cdots+m_n \le m-1, 1 \le l \le n)$ with rational coefficients. Thus a^i_m $(m \ge 0)$ are uniquely determined by the functions $f_i(Z, W_1, ..., W_n)$ $(1 \le i \le n)$, and hence the solution of (4.31) is unique. $\qquad\square$

Corollary 4.23. *Let* $R(z, w_1, ..., w_n)$ *be a holomorphic function in the polydisc*

$$\Delta_\rho^{n+1}(x_0) = \{(z, w_1, ..., w_n) \in \kappa^{n+1} \mid |z - z_0| < r, \ |w_{i+1} - w_0^{(i)}| < \rho_i, 0 \le i \le n-1\},$$

where $x_0 = (z_0, w_0^{(0)}, ..., w_0^{(n-1)})$, $\rho = (r, \rho_1, ..., \rho_n)$. *If the Cauchy's problem*

$$\frac{d^n w}{dz^n} = R(z, w, w'..., w^{(n-1)}), \quad w(z_0) = w_0^{(0)}, \quad w^{(i)}(z_0) = w_0^{(i)}, \quad 1 \le i \le n-1 \quad (4.33)$$

has holomorphic solutions, then it is unique.

Proof. Set

$$\frac{dw}{dz} = w_1, \quad \frac{dw_1}{dz} = w_2, ..., \quad \frac{dw_{n-1}}{dz} = R(z, w, w_1..., w_{n-1}).$$

Then Corollary 4.23 follows from Lemma 4.22. $\qquad\square$

Theorem 4.24 ([15]). *Take $k \in \mathbb{Z}^+$, $a_j \in \kappa$ $(0 \leq j \leq k)$ with $a_k \neq 0$. If the following differential equation*

$$\frac{dw}{dz} = a_k w^k + a_{k-1} w^{k-1} + \cdots + a_1 w + a_0 \tag{4.34}$$

has a non-constant meromorphic solution $w = w(z)$, then $k = 2$, the equation (4.34) becomes the form $w' = a_2(w - c)^2$, and w is of the following form

$$w(z) = c - \frac{1}{a_2(z - b)} \quad (b, c \in \kappa).$$

Proof. Assume, on the contrary, that (4.34) has a non-constant meromorphic solution $w = w(z)$. By Theorem 4.3, we have $k \leq 2$. We can rule out the case $k \leq 1$ from Theorem 4.21. Hence we can derive $k = 2$. Rewriting (4.34) as follows

$$\frac{dw}{dz} = a_2(w - c_1)(w - c_2) \tag{4.35}$$

for $c_j \in \kappa$ $(j = 1, 2)$, then (4.34) or (4.35) has constant solutions $w = c_j$ $(j = 1, 2)$. By Lemma 4.22 or by comparing the multiplicity of c_j-valued points of w, then w has Picard exceptional values c_j $(j = 1, 2)$. If $c_1 \neq c_2$, this is a contradiction since non-constant meromorphic functions have at most one Picard exceptional value. Assume $c_1 = c_2 = c$, that is, w satisfies

$$\frac{dw}{dz} = a_2(w - c)^2.$$

Since each non-constant function in $\mathcal{A}(\kappa)$ has no finite Picard exceptional values, then w has at least one pole. We can write

$$w = c + \frac{1}{h} \quad (h \in \mathcal{A}(\kappa) - \kappa).$$

Then we have $h' = -a_2$, that is, $h(z) = -a_2(z - b)$ for some $b \in \kappa$. The theorem is proved. □

Generally, Boutabaa and Escassut [15] prove the following result:

Theorem 4.25. *Take $n, k \in \mathbb{Z}^+$, $a_j \in \kappa$ $(0 \leq j \leq k)$ with $a_k \neq 0$. If the following differential equation*

$$\left(\frac{dw}{dz}\right)^n = a_k w^k + a_{k-1} w^{k-1} + \cdots + a_1 w + a_0 \tag{4.36}$$

has a non-constant meromorphic solution w, then $|n - k|$ is a factor of n, and the equation (4.36) and the solution w assume respectively the following form

$$\left(\frac{dw}{dz}\right)^n = a_k(w - c)^k, \quad c = -\frac{a_{k-1}}{k a_k} \tag{4.37}$$

and

$$w(z) = a(z - b)^{\frac{n}{n-k}} + c \quad (b, c \in \kappa), \tag{4.38}$$

where a satisfies $a^{n-k} \left(\frac{n}{n-k}\right)^n = a_k$.

Proof. According to Theorem 4.3, Theorem 4.24 and Theorem 4.17, we may assume that $2n \geq k \geq n \geq 2$. Write

$$A(z) = a_k w^k + a_{k-1} w^{k-1} + \cdots + a_1 w + a_0 = a_k \prod_{i=1}^{s} (z - c_i)^{k_i},$$

where $c_1, ..., c_s$ are distinct. We distinguish two cases that w have (or not) poles. First of all, we assume that w have not poles. Then $k_i < n$, $n - k_i$ divides n and each zero of $w - c_i$ has the multiplicity $\frac{n}{n-k_i}$. Hence for $i = 1, ..., s$,

$$\overline{N}\left(r, \frac{1}{w - c_i}\right) = \frac{n - k_i}{n} N\left(r, \frac{1}{w - c_i}\right) \leq \frac{n - k_i}{n} T(r, w) + O(1).$$

If $s \geq 2$, then the second main theorem implies

$$(s - 1)T(r, w) \; < \; \overline{N}(r, w) + \sum_{i=1}^{s} \overline{N}\left(r, \frac{1}{w - c_i}\right) - \log r + O(1)$$
$$\leq \; \left(s - \frac{k}{n}\right) T(r, w) - \log r + O(1),$$

which is impossible. Thus it has to be $s = 1$ and $k_1 = k < n$, which contradicts with our assumption. Hence the equation (4.36) has no non-constant entire function solutions if $k \geq n \geq 2$.

Next we assume that w have poles. For this case, it follow from (4.36) that $k > n$, $k - n$ divides n and each pole of w has the multiplicity $\frac{n}{k-n}$. Hence

$$\overline{N}(r, w) = \frac{k - n}{n} N(r, w) \leq \frac{k - n}{n} T(r, w).$$

If $s \geq 2$, then the second main theorem implies

$$(s - 1)T(r, w) \; < \; \overline{N}(r, w) + \sum_{i=1}^{s} \overline{N}\left(r, \frac{1}{w - c_i}\right) - \log r + O(1)$$
$$\leq \; \frac{k - n}{n} T(r, w) + \left(s - \frac{k}{n}\right) T(r, w) - \log r + O(1),$$

which is impossible. Thus we have $s = 1$ and $k_1 = k > n$, that is, (4.36) has the form (4.37). Since $k > n$, the equation (4.37) implies that c is a Picard exceptional value of w. Hence w is of the following form

$$w = c + \frac{1}{h} \; (h \in \mathcal{A}(\kappa) - \kappa).$$

Then h satisfies

$$(h')^n = (-1)^n a_k h^{2n-k}.$$

Note that each zero of h has the multiplicity $l = \frac{n}{k-n}$. We can write $h = f^l$ for some $f \in \mathcal{A}(\kappa) - \kappa$. Therefore we obtain

$$(f')^n = \frac{(-1)^n a_k}{l^n},$$

that is, f has the following form $f(z) = A(z-b)$ with $l^n A^n = (-1)^n a_k$. Hence the theorem is proved by setting $a = A^{-l}$. □

We make some remarks about Lemma 4.22. Assume that the functions $R_i(z, w_1, ..., w_n)$ in Lemma 4.22 are of the following form:

$$R_i(z, w_1, ..., w_n) = \sum_{j=1}^{n} a_{ij}(z) w_j.$$

Then the equations (4.31) can be rewritten as follows

$$\frac{dw}{dz} = A(z)w, \tag{4.39}$$

where $A(z) = (a_{ij}(z))$, and w is the transpose of the matrix $(w_1, ..., w_n)$.

Lemma 4.26. *The equation (4.39) has at most n linearly independent solutions.*

Proof. Let $w^1, ..., w^{n+1}$ be solutions of (4.39) satisfying original values

$$w^i(z_0) = w_0^i, \quad i = 1, ..., n+1.$$

If $w_0^1, ..., w_0^n$ are linearly dependent, then there are constants $(c_1, ..., c_n) \in \kappa^n - \{0\}$ such that

$$c_1 w_0^1 + \cdots + c_n w_0^n = 0.$$

Write

$$w(z) = c_1 w^1(z) + \cdots + c_n w^n(z).$$

Then $w(z_0) = 0$. By Lemma 4.22, we obtain $w \equiv 0$, that is, $w^1, ..., w^n$ are linearly dependent, and so are $w^1, ..., w^{n+1}$. Suppose that $w_0^1, ..., w_0^n$ are linearly independent. Then there are constants $(c_1, ..., c_n) \in \kappa^n$ such that

$$w_0^{n+1} = c_1 w_0^1 + \cdots + c_n w_0^n.$$

Again by Lemma 4.22, we have

$$w^{n+1} = c_1 w^1 + \cdots + c_n w^n,$$

that is, $w^1, ..., w^{n+1}$ are linearly dependent. □

Corollary 4.27. *The following equation*

$$w^{(n)} + a_n(z)w^{(n-1)} + \cdots + a_2(z)w' + a_1(z)w = 0 \tag{4.40}$$

with $a_j \in \mathcal{A}(\kappa)$ has at most n linearly independent solutions.

We conjectured the problem: The following non-Archimedean differential equation

$$w^{(n)} + a_n(z)w^{(n-1)} + \cdots + a_2(z)w' + a_1(z)w + a_0(z) = 0 \tag{4.41}$$

with $a_j \in \mathcal{M}(\kappa)$ has no admissible transcendental meromorphic solutions. In [12], Boutabaa studied meromorphic solutions of (4.41)and proved the following result:

Theorem 4.28. *Let $\overline{\mathbb{Q}}$ be the algebraic closure of $\mathbb{Q}$ in the field $\mathbb{C}_p$. Suppose that the equation (4.41) is such that $a_1(z), ..., a_n(z) \in \overline{\mathbb{Q}}(z)$, $a_0(z) \equiv 0$ and let $w(z) \in \mathcal{M}(\mathbb{C}_p)$ be a solution of (4.41). Then $w(z) \in \mathbb{C}_p(z)$.*

If $a_1(z), ..., a_n(z)$ are not all in $\overline{\mathbb{Q}}(z)$, Boutabaa [13] shows that the conjecture may be false. Here we introduce the counter-example due to Boutabaa [13]. Consider the application $f : \mathbb{Z}_+ \longrightarrow \mathbb{Z}_+$ defined by

$$f(0) = 0, \quad f(i+1) = p^{f(i)+i} \ (i \in \mathbb{Z}_+).$$

Then one has

$$\frac{f(i+1)}{p^{f(i)}} = p^i \ (i \in \mathbb{Z}_+).$$

Define two transcendental p-adic integers (cf. [27] or [30])

$$a = -\sum_{i=0}^{\infty} p^{f(2i)}, \quad b = -\sum_{i=0}^{\infty} p^{f(2i+1)}.$$

Take $c \in \mathbb{C}_p$ such that $v_p(c) \leq \frac{1}{1-p}$. Then the hyper-geometric function

$$w(z) = F(a, b, c; z) = \sum_{n=0}^{\infty} \frac{(a)_n (b)_n}{n!(c)_n} z^n$$

is an entire function on $\mathbb{C}_p$, and is a solution of the Gaussian differential equation

$$z(1-z)\frac{d^2 w}{dz^2} + (c - (a+b+1)z)\frac{dw}{dz} - abw = 0, \tag{4.42}$$

where, by definition,

$$(\alpha)_n = \alpha(\alpha+1) \cdots (\alpha+n-1), \quad \alpha \in \mathbb{C}_p.$$

It is well known and easily checked (cf. [30] or [106]) that the function $w(z) = F(a, b, c; z)$ is a formal solution of the equation (4.42). It remains to prove $w \in \mathcal{A}(\mathbb{C}_p)$. The radius ρ of convergence of w is defined by

$$\frac{1}{\rho} = \limsup_{n \to +\infty} \sqrt[n]{\frac{|(a)_n (b)_n|}{n!|(c)_n|}}.$$

Hence

$$\frac{\log \rho}{\log p} = \liminf_{n \to +\infty} \frac{v_p((a)_n) + v_p((b)_n) - v_p(n!) - v_p((c)_n)}{n}.$$

By using the facts

$$v_p(n!) \leq \frac{n}{p-1}, \quad v_p((c)_n) = n v_p(c) \leq -\frac{n}{p-1},$$

one obtains $v_p(n!) + v_p((c)_n) \leq 0$, and hence

$$\frac{\log \rho}{\log p} \geq \liminf_{n \to +\infty} \frac{v_p((a)_n) + v_p((b)_n)}{n}.$$

For n sufficiently large, let m be the greatest integer such that

$$p^{f(m)} \leq n < p^{f(m+1)}.$$

Then one has the following inequality:

$$\frac{v_p((a)_n) + v_p((b)_n)}{n} \geq \begin{cases} p^{m-1} & : \quad p^{f(m)} \leq n < p^{f(m)+1} \\ p^{m+1} & : \quad p^{f(m)+1} \leq n < p^{f(m+1)} \end{cases}. \tag{4.43}$$

We first consider the case: $p^{f(m)} \leq n < p^{f(m)+1}$. Then one has

$$\frac{1}{n} > \frac{1}{p} \cdot \frac{1}{p^{f(m)}}. \tag{4.44}$$

If $m = 2l$, then

$$0 < \sum_{i=0}^{l-1} p^{f(2i+1)} < p^{f(2l)} = p^{f(m)} \leq n$$

and note that

$$b + \sum_{i=0}^{l-1} p^{f(2i+1)} = -\sum_{i=l}^{\infty} p^{f(2i+1)}.$$

Thus it follows

$$v_p((b)_n) = \sum_{k=0}^{n-1} v_p(b+k) \geq v_p \left(b + \sum_{i=0}^{l-1} p^{f(2i+1)} \right) = f(2l+1) = f(m+1).$$

If $m = 2l + 1$, then

$$0 < \sum_{i=0}^{l} p^{f(2i)} < p^{f(2l+1)} = p^{f(m)} \leq n$$

and note that

$$a + \sum_{i=0}^{l} p^{f(2i)} = -\sum_{i=l+1}^{\infty} p^{f(2i)}.$$

Hence

$$v_p((a)_n) = \sum_{k=0}^{n-1} v_p(a+k) \geq v_p \left(a + \sum_{i=0}^{l} p^{f(2i)} \right) = f(2l+2) = f(m+1).$$

So one has

$$v_p((a)_n) + v_p((b)_n) \geq f(m+1). \tag{4.45}$$

From (4.44) and (4.45), it follows that

$$\frac{v_p((a)_n) + v_p((b)_n)}{n} \geq \frac{1}{p} \cdot \frac{f(m+1)}{p^{f(m)}} = p^{m-1} \tag{4.46}$$

which is the first part of the inequality (4.43). Next we study the case: $p^{f(m)+1} \leq n < p^{f(m+1)}$. Then one has

$$\frac{1}{n} > \frac{1}{p^{f(m+1)}}.$$ (4.47)

If $m = 2l$, then

$$0 < \sum_{i=0}^{l} p^{f(2i)} < p^{f(2l)+1} = p^{f(m)+1} \leq n$$

and note that

$$a + \sum_{i=0}^{l} p^{f(2i)} = - \sum_{i=l+1}^{\infty} p^{f(2i)}.$$

Thus it follows

$$v_p((a)_n) = \sum_{k=0}^{n-1} v_p(a + k) \geq v_p \left(a + \sum_{i=0}^{l} p^{f(2i)} \right) = f(2l + 2) = f(m + 2).$$

If $m = 2l + 1$, then

$$0 < \sum_{i=0}^{l} p^{f(2i+1)} < p^{f(2l+1)+1} = p^{f(m)+1} \leq n$$

and note that

$$b + \sum_{i=0}^{l} p^{f(2i+1)} = - \sum_{i=l+1}^{\infty} p^{f(2i+1)}.$$

Hence

$$v_p((b)_n) = \sum_{k=0}^{n-1} v_p(b + k) \geq v_p \left(b + \sum_{i=0}^{l} p^{f(2i+1)} \right) = f(2l + 3) = f(m + 2).$$

So one has

$$v_p((a)_n) + v_p((b)_n) \geq f(m + 2).$$ (4.48)

From (4.47) and (4.48), it follows that

$$\frac{v_p((a)_n) + v_p((b)_n)}{n} \geq \frac{f(m + 2)}{p^{f(m+1)}} = p^{m+1}$$ (4.49)

which is the second part of the inequality (4.43). From (4.43) one obtains that

$$\lim_{n \to +\infty} \frac{v_p((a)_n) + v_p((b)_n)}{n} = +\infty.$$

Therefore $\rho = +\infty$, that is, $w \in A(\mathbb{C}_p)$.

Chapter 5

Dynamics

In this chapter, we will study the Fatou-Julia-type theory of entire (or rational) functions defined on a non-Archimedean field.

5.1 Attractors and repellers

First of all, we introduce some basic notations. If f is a mapping of a set M into itself, a subset E of M is:

    (a) *forward invariant* if $f(E) = E$;
    (b) *backward invariant* if $f^{-1}(E) = E$;
    (c) *completely invariant* if $f^{-1}(E) = E = f(E)$.

If f is injective, forward invariance implies backward invariance and completely invariant. Generally, we have the following relations:

Lemma 5.1 ([68]). *1) If E is backward invariant, then*

$$f(E) = E \cap f(M) \subset E.$$

2) If $f^{-1}(E) \subset E, f(E) \subset E$, then E is backward invariant.

We will denote by $C(M, N)$ the set of continuous mappings between topological spaces M and N. Take $f \in C(M, M)$ and define the *iterate of f* by

$$f^0 = id, \quad f^n = f^{n-1} \circ f = f \circ f^{n-1} \ (n > 0),$$

where id is the identity mapping in M. Denote the *forward orbit* of $x \in M$ by

$$O^+(x) = \bigcup_{n \geq 0} \{f^n(x)\}.$$

The elements in $O^+(x)$ are called *successors* of x. Define the *backward orbit* of x by

$$O^-(x) = \bigcup_{n \geq 0} \{f^{-n}(x)\}.$$

139

The elements in $O^-(x)$ are called *predecessors* of x. Define the *total orbit*:

$$O(x) = O^+(x) \cup O^-(x).$$

Generally, for a subset $E \subset M$, we can also define the forward, backward and total orbits of E respectively by

$$O^+(E) = \bigcup_{n \geq 0} f^n(E), \quad O^-(E) = \bigcup_{n \geq 0} f^{-n}(E), \quad O(E) = O^+(E) \cup O^-(E).$$

Obviously, we have

$$O^+(E) = \bigcup_{x \in E} O^+(x), \quad O^-(E) = \bigcup_{x \in E} O^-(x).$$

For any x and y in M, we define the relation $\sim$ on M by $x \sim y$ if and only if there exist non-negative integers m and n such that

$$f^m(x) = f^n(y),$$

that is, x and y have a common successor. Obviously, the relation $\sim$ is symmetric and reflexive, and it is also transitive since

$$f^m(x) = f^n(y), f^k(y) = f^l(z) \implies f^{m+k}(x) = f^{n+l}(z),$$

thus $\sim$ is an equivalence relation on M. We denote the equivalence class containing x by $[x]$, and we call this the *(grand) orbit* of x. Since $\sim$ is an equivalence relation, the two orbits are either identical or disjoint. Obviously, an orbit consists precisely of all successors and all predecessors of all successors of any one of its elements, that is,

$$[x] = O^+(x) \cup \left\{ \bigcup_{n \geq 0} O^-(f^n(x)) \right\} = \bigcup_{n \geq 0} O(f^n(x)).$$

Theorem 5.2 (cf. [68]). *The (grand) orbits are precisely the minimal sets that are backward invariant.*

Corollary 5.3. *A subset E of M is backward invariant if and only if it is a union of equivalence classes $[x]$. If this is the case, then its complement $M - E$ must also be a union of equivalence classes and, therefore, also backward invariant.*

Corollary 5.4. *Assume that $f : M \longrightarrow M$ is a continuous open mapping and suppose that E is backward invariant. Then so are the interior E^o, the boundary ∂E, and the closure $\overline{E}$ of E.*

For a subset E of M, define

$$[E] = \bigcup_{x \in E} [x].$$

Then $[E]$ is backward invariant. Obviously, if E is backward invariant, then $[E] = O(E) = E$. The backward invariant set $[E]$ is the minimal element in all backward invariant sets containing E.

A *fixed point* of the self-mapping f on M is a point $x \in M$ such that $f(x) = x$. Denote the set of fixed points of f by $\mathrm{Fix}(f)$. A *k-cycle* (or *cycle of order k*) of f is a k-tuple of pairwise different elements $x_0, ..., x_{k-1}$ of M such that

$$f(x_i) = x_{i+1} \ (0 \le i < k - 1); \ f(x_{k-1}) = x_0.$$

For a k-cycle $\{x_0, ..., x_{k-1}\}$, obviously x_i is of $f^k(x_i) = x_i$, i.e.,

$$\{x_0, ..., x_{k-1}\} \subset \mathrm{Fix}(f^k).$$

Also each x_i is said to be a *fixed point of exact order k* of f. A 1-cycle of f is just a fixed point of f. Set

$$\mathrm{Per}(f) = \bigcup_{k=1}^{\infty} \mathrm{Fix}(f^k).$$

The points in $\mathrm{Per}(f)$ are called *periodic points* of the mapping f. Obviously, $x \in \mathrm{Per}(f)$ if and only if $f^k(x) = x$ for some $k \in \mathbb{Z}^+$. The minimal such k is called the *period* of the point x. If k is the period of x, then $k \ge 1$, and $\{x, f(x), ..., f^{k-1}(x)\}$ is a k-cycle of f. Hence $\mathrm{Per}(f)$ just is the union of cycles of f.

For any point $x \in M$, define the *ω-limit set* $L^+(x)$ of x as

$$L^+(x) = \bigcap_{k \ge 0} \overline{\bigcup_{n \ge k} f^n(x)}.$$

It can be characterized by saying that $y \in L^+(x)$ if and only if there is a sequence $n_j \to +\infty$ such that $f^{n_j}(x) \to y$ as $j \to +\infty$. For a subset A of M, we denote by $\mathrm{Att}(A)$ its *basin of attraction*, that is the set

$$\mathrm{Att}(A) = \{x \in M \mid L^+(x) \subset A\}.$$

Definition 5.5. *A set A is asymptotically stable if there exists a neighborhood U of A satisfying $U \subset \mathrm{Att}(A)$.*

It is easy to prove that if A is asymptotically stable, $\mathrm{Att}(A)$ is open. A fixed point x_0 of f is called an *attractor* of f if there exists a neighborhood U of x_0 such that

$$\lim_{n \to +\infty} f^n(x) = x_0 \quad \text{for all} \quad x \in U,$$

that is, a fixed point x_0 of f is an attractor if and only if the set $\{x_0\}$ is asymptotically stable.

Definition 5.6 (cf. [68]). *A set A is called repulsive*
1) if for every neighborhood V of A, $V - [A]$ is non-empty;
2) if there exists a neighborhood U of A such that for each $x \in U - [A]$ there is an $n_0 > 0$ with $f^n(x) \notin U$ for all $n \ge n_0$.

In particular, a repulsive fixed point of f is called a *repeller* of f.

Theorem 5.7 (cf. [68]). *If one point of a k-cycle ($k \geq 1$) of a continuous self-mapping f of a topological space M is an attractor (repeller) of f^k, then every point of the cycle is an attractor (repeller) of f^k. Hence the cycle is asymptotically stable (repulsive).*

Theorem 5.8 (cf. [68]). *If a k-cycle ($k \geq 1$) of a continuous self-mapping f on a topological space M is repulsive, then every point of the cycle is a repeller of f^k.*

Definition 5.9. *A k-cycle ($k \geq 1$) of a continuous self-mapping f on a topological space M is said to be an attractor (repeller) of f if one point of the cycle is an attractor (repeller) of f^k.*

Theorem 5.7 and Theorem 5.8 show that a k-cycle ($k \geq 1$) of f is repulsive if and only if it is a repeller, that is, one point of the cycle is a repeller of f^k. Also, if a cycle is an attractor, it is asymptotically stable, but if a cycle is asymptotically stable, is the cycle an attractor (see [68])?

Let κ be an algebraically closed field of characteristic zero, complete for a non-trivial non-Archimedean absolute value $|\cdot|$. Let K denote the field κ or $\mathbb{C}$ and denote $K \cup \{\infty\}$ by $\bar{K}$. Note that according to the definition of complex manifolds over $\mathbb{C}$, we can define *non-Archimedean manifolds* of arbitrary dimensions over κ so that C^k-*differentiable* and *holomorphic* mappings can be similarly defined on non-Archimedean manifolds. Thus, an entire function f on K can be identified with a holomorphic mapping $f : K \longrightarrow K$, a rational function R on K can be identified with a holomorphic mapping $f : \bar{K} \longrightarrow \bar{K}$, and a meromorphic function f on K can be identified with a holomorphic mapping $f : K \longrightarrow \bar{K}$.

Now let f be a differentiable self-mapping on a manifold M over K and consider a k-cycle

$$f : x_0 \mapsto x_1 \mapsto \cdots \mapsto x_{k-1} \mapsto x_k = x_0.$$

If M is K, then the derivative

$$\lambda = (f^k)'(x_i) = f'(x_1) \cdot f'(x_2) \cdots f'(x_k)$$

is well defined, called the *multiplier* or the *eigenvalue* of this cycle. More generally, for self-mappings of an arbitrary Riemann surface over K the multiplier of a cycle can be defined using a local coordinate chart around any point of the cycle. By definition, the cycle is either *attracting* or *repelling* or *indifferent* (= *neutral*) according to if its multiplier satisfies $|\lambda| < 1$ or $|\lambda| > 1$ or $|\lambda| = 1$. The cycle is called *superattracting* if $\lambda = 0$

Theorem 5.10. *Take an open set $D \subset \bar{\kappa}$. Suppose that $f : D \longrightarrow D$ is a holomorphic mapping. Then a fixed point z_0 of f is an attractor if and only if z_0 is attracting.*

Proof. W. l. o. g., we may assume $z_0 = 0$. Set $\lambda = f'(0)$. First we assume $|\lambda| < 1$. Write

$$f(z) = \lambda z + \sum_{j=2}^{\infty} a_j z^j.$$

Take $r \in \mathbb{R}^+$ such that $\kappa(0;r) \subset D$ and

$$s = \max_{j \geq 2} |a_j| r^{j-1} < 1, \tag{5.1}$$

and choose a constant β so that

$$\max\{|\lambda|, s\} < \beta < 1.$$

Note that

$$|f(z)| \leq \max_{j \geq 2}\{|\lambda||z|, |a_j||z|^j\} \leq \max_{j \geq 2}\{\beta|z|, |a_j||z|^j\}.$$

Thus $z \in \kappa(0;r)$ implies

$$|f(z)| \leq \beta|z| < r. \tag{5.2}$$

If $|z| < r$, then iterations of (5.2) yield

$$\|f^n(z)\| \leq \beta^n |z|, \tag{5.3}$$

for all $z \in \kappa(0;r)$, which means $\kappa(0;r) \subset \mathrm{Att}(0)$ so that the point $z = 0$ is an attractor.

Next we assume that $z_0 (= 0)$ is an attractor so that we can choose a neighborhood U of z_0 such that $U \subset \mathrm{Att}(0)$. Take a small positive number r such that the disc $\kappa(0;r) \subset D \cap U$. Then

$$\begin{aligned} |\lambda|^n &= \left| \frac{df^n}{dz}(0) \right| \leq \max_{\zeta \in \kappa(0;r)} \left| \frac{df^n}{dz}(\zeta) \right| \\ &\leq \frac{1}{r} \max_{\zeta \in \kappa(0;r)} |f^n(\zeta)| \to 0 \end{aligned}$$

as $n \to \infty$. Therefore, $|\lambda| < 1$. $\qquad \square$

Corollary 5.11. *Take an open set $D \subset \bar{\kappa}$. Suppose that $f : D \longrightarrow D$ is a holomorphic mapping. Then a cycle A of f is an attractor if and only if A is attracting.*

Theorem 5.12. *Take an open set $D \subset \bar{\kappa}$. Suppose that $f : D \longrightarrow D$ is a holomorphic mapping. Then a fixed point z_0 of f is a repeller if and only if z_0 is repelling.*

Proof. W. l. o. g., we may assume $z_0 = 0$. Set $\lambda = f'(0)$. Assume $|\lambda| > 1$. By the inverse function theorem (Theorem 1.41), the inverse f^{-1} exists in a neighborhood of 0. Note that $(f^{-1})'(0) = 1/\lambda$. According to the proof above and by Theorem 1.41, there is some $r > 0$ such that f^{-1} exists on $\kappa(0;r)$, and such that

$$|f^{-n}(z)| \leq \frac{|z|}{|\lambda|^n} \text{ for all } z \in \kappa(0;r), \ n > 0.$$

Set $U = \kappa(0;r)$ and note that $[0]$ contains only one point 0 since f^{-1} is injective on $\kappa(0;r)$. For any $\xi \in U - [0]$, then $f^n(\xi) \notin U$ for all $n > \log(r/|\xi|)/\log|\lambda|$. In fact, if $\xi_n = f^n(\xi) \in U$ for some $n > \log(r/|\xi|)/\log|\lambda|$, then

$$|\xi| = \|f^{-n}(\xi_n)\| \leq \frac{|\xi_n|}{|\lambda|^n} < \frac{r}{|\lambda|^n},$$

which is a contradiction! Thus 0 is a repeller.

Next we assume that $z_0 = 0$ is a repeller so that we can choose a neighborhood U of z_0 such that for any $x \in U - [0]$, there is an $n_0 > 0$ with $f^n(x) \notin U$ for all $n \geq n_0$. Assume, on the contrary, that $|\lambda| \leq 1$. If $|\lambda| < 1$, then 0 is an attractor of f which is obviously not a repeller. Thus, it has to be $|\lambda| = 1$. Write

$$f(z) = \lambda z + \sum_{j=2}^{\infty} a_j z^j.$$

Thus we can take $r \in \mathbb{R}^+$ such that $\kappa[0;r] \subset U$ and such that $|a_j| r^{j-1} < 1$ for all $j \geq 2$. Hence

$$|f(z)| = |z|, \quad z \in \kappa[0;r],$$

and therefore

$$|f^n(z)| = |z|, \quad z \in \kappa[0;r],\ n \in \mathbb{Z}^+.$$

This is contrary to assumption about the repeller.   □

Corollary 5.13. *Take an open set $D \subset \bar{\kappa}$. Suppose that $f : D \longrightarrow D$ is a holomorphic mapping. Then a cycle A of f is a repeller if and only if A is repelling.*

In the proofs of Theorem 5.10 and Theorem 5.12, we see that if z_0 is a fixed point of f with $f'(z_0) \neq 0$, then there is a $r \in \mathbb{R}^+$ such that

$$|f(z) - z_0| = |f'(z_0)||z - z_0|, \quad z \in \kappa[z_0;r]. \tag{5.4}$$

By Theorem 1.40, the number r can be determined by

$$\max_{j \geq 2} \left| \frac{1}{j!} \frac{d^j f}{dz^j}(z_0) \right| r^{j-1} < |f'(z_0)|; \quad \kappa[z_0;r] \subset D. \tag{5.5}$$

Thus if z_0 is an attractor of f, then

$$\kappa[z_0;r] \subset \text{Att}(z_0).$$

Theorem 5.12 yields the following result:

Corollary 5.14. *Take an open set $D \subset \bar{\kappa}$. Suppose that $f : D \longrightarrow D$ is a holomorphic mapping. Then a fixed point z_0 of f is a repeller if and only if there is a neighborhood U of z_0 such that*

$$|f(z) - z_0| > |z - z_0|, \quad z \in U - \{z_0\}. \tag{5.6}$$

The sufficient conditions of Theorem 5.10 and Theorem 5.12 are also proved in [75], where Khrennikov used the condition (5.6) to define repellers. More general cases are discussed in [59], [68] and [69], where the necessary conditions of Theorem 5.12 are conjectured for holomorphic mappings on complex manifolds.

Assume that z_0 is an indifferent fixed point of f. Then (5.4) gives

$$|f(z) - z_0| = |z - z_0|, \quad z \in \kappa(z_0;r), \tag{5.7}$$

and iterations of the equalities yield

$$|f^n(z) - z_0| = |z - z_0|, \quad z \in \kappa(z_0; r), \tag{5.8}$$

that is, $O^+(z) \subset \kappa\langle z_0; |z - z_0|\rangle$ if $z \in \kappa(z_0; r)$. The ball $\kappa(z_0; r)$ (contained in D) is said to be a *Siegel disc*. The union of all Siegel discs with their centers at z_0 is said a *maximal Siegel disc*, which is denoted by $\mathrm{Sie}(z_0)$.

Proposition 5.15. *If z_0 is an indifferent fixed point of f, then z_0 is a center of a Siegel disc satisfying*

$$\kappa[z_0; r] \subset \mathrm{Sie}(z_0),$$

where r is determined by (5.5).

In the same way we define a Siegel disc with center at a periodic point $x_0 \in D$ of f with the corresponding k-cycle $A = \{x_0, x_1, ..., x_{k-1}\}$, that is, a ball $\kappa(x_0; r)$ (contained in D) is said to be a *Siegel disc* if one takes an initial point z on one of the spheres $\kappa\langle z_0; r'\rangle$, $r' < r$, all iterated points on f^k will also be on it. We also denote by $\mathrm{Sie}(x_0)$ the union of all Siegel discs with their centers at x_0, that is, the *maximal Siegel disc*. We will write

$$\mathrm{Sie}(A) = \bigcup_{j=0}^{k-1} \mathrm{Sie}(x_j).$$

Proposition 5.16. *If $A = \{x_0, x_1, ..., x_{k-1}\}$ is an indifferent k-cycle of f, then each x_i is a center of a Siegel disc satisfying*

$$\kappa[x_i; r_i] \subset \mathrm{Sie}(x_i), \quad i = 0, ..., k - 1,$$

where r_i is determined by

$$\max_{j \geq 2} \left| \frac{1}{j!} \frac{d^j f^k}{dz^j}(x_i) \right| r_i^{j-1} < 1; \quad \kappa[x_i; r_i] \subset D. \tag{5.9}$$

5.2 Riemann-Hurwitz relation

Let κ be an algebraically closed field of characteristic zero, complete for a non-trivial non-Archimedean absolute value $|\cdot|$. First of all, we introduce some notations. We say that two rational mappings f and g are *conjugacy* if there is some $\sigma \in \mathrm{Aut}(\bar{\kappa})$ with

$$g = \sigma \circ f \circ \sigma^{-1}.$$

Conjugacy is clearly an equivalence relation, and equivalence classes are the *conjugacy classes* of rational mappings. If $g = \sigma \circ f \circ \sigma^{-1}$, the following facts are clear

$$\deg(g) = \deg(f), \quad g^n = \sigma \circ f^n \circ \sigma^{-1}.$$

Let us characterize the polynomials within the class of rational mappings. Note that a non-constant rational mapping f is a polynomial if and only if f has a pole at ∞ and no poles in κ, that is, $f^{-1}(\infty) = \infty$. More generally, it is easy to prove the following

Theorem 5.17. *A non-constant rational mapping f is conjugate to a polynomial if and only if there is some $a \in \bar{\kappa}$ with $f^{-1}(a) = a$.*

Theorem 5.18. *A rational mapping of degree $d \geq 1$ has precisely $d + 1$ fixed points in $\bar{\kappa}$.*

Proof. Note that the number of fixed points is invariant under Möbius transformations. W. l. o. g., we may assume that f does not fix ∞. Write $f = P/Q$ with P and Q coprime and take $\zeta \in \text{Fix}(f)$, so ζ is finite. Since $Q(\zeta) \neq 0$, the number of zeros of $f(z) - z$ at ζ is exactly the same as the number of zeros of $P(z) - zQ(z)$ at ζ. Hence the number of fixed points of f is exactly the number of solutions of $P(z) = zQ(z)$ in $\mathbb{C}$. Since f does not fix ∞, one has

$$\deg(P) \leq \deg(Q) = \deg(R) = d$$

and, hence,

$$\deg(P(z) - zQ(z)) = d + 1.$$

This also completes the proof. $\square$

Consider any function f that is non-constant and holomorphic near the point z_0 in κ. Then there exists a unique positive integer m such that the limit

$$\lim_{z \to z_0} \frac{f(z) - f(z_0)}{(z - z_0)^m}$$

is a finite non-zero constant. We denote this integer m by $\mu_f(z_0)$ and call it *mapping degree* or *valency* of f at z_0. Note that $\mu_f(z_0) = 1$ if and only if $f'(z_0) \neq 0$, i.e., if and only if f is injective in some neighborhood of z_0 by Theorem 1.40. The valency function satisfies the *chain rule*:

$$\mu_{f \circ g}(z_0) = \mu_f(g(z_0))\mu_g(z_0), \tag{5.10}$$

where z_0, $g(z_0)$ and $f(g(z_0))$ are all in κ. The proof is easy (see [5]).

The equality enables us to extend the definition of $\mu_f(z_0)$ to the case where $z_0 = \infty$ or $f(z_0) = \infty$ (or both). We select σ, $\xi \in \text{Aut}(\bar{\kappa})$ with $\sigma(z_0)$, $\xi(f(z_0)) \in \kappa$, and then define

$$\mu_f(z_0) = \mu_{\xi \circ f \circ \sigma^{-1}}(\sigma(z_0)).$$

It is easy to check that $\mu_f(z_0)$ is well defined, that is, $\mu_f(z_0)$ is independent of the choice of σ and ξ. According to the definition, a rational mapping f satisfies

$$\sum_{z \in f^{-1}(a)} \mu_f(z) = \sum_{z \in f^{-1}(a)} \mu_f^a(z) = \deg(f)$$

for any $a \in \bar{\kappa}$.

A point z is a *critical point* of a rational mapping f if f fails to be injective in any neighborhood of z. If f is not constant, critical points of f are precisely the points at which $\mu_f(z) > 1$. A value a is a *critical value* for f if $f^{-1}(a)$ contains some critical point. Now we prove the *Riemann-Hurwitz relation*:

Theorem 5.19. *For any non-constant rational mapping f,*

$$\sum_{z \in \bar{\kappa}} (\mu_f(z) - 1) = 2 \deg(f) - 2. \tag{5.11}$$

Proof. Here we will follow the elegant elementary proof by Beardon [5]. Note that both sides of (5.11) are invariant under conjugation. Thus it is sufficient to prove (5.11) for any conjugate of f. Take a point ζ such that $\zeta \notin \mathrm{Fix}(f)$, $\mu_f(\zeta) = 1$, and such that ζ is not a critical value of f, and take $\sigma \in \mathrm{Aut}(\bar{\kappa})$ satisfying

$$\sigma(\zeta) = \infty, \quad \sigma(f(\zeta)) = 1.$$

Then $g = \sigma \circ f \circ \sigma^{-1}$ satisfies

$$g(\infty) = 1, \quad \mu_g(\infty) = 1,$$

and g has distinct simple poles $z_1, ..., z_d$ in κ, where $d = \deg(f) = \deg(g)$. Since $\mu_g(z_j) = 1$ $(j = 1, ..., d)$, it is sufficient to prove the following formula

$$\sum_{z \in \kappa - g^{-1}(\infty)} (\mu_g(z) - 1) = 2d - 2.$$

Obviously, we have

$$\sum_{z \in \kappa - g^{-1}(\infty)} (\mu_g(z) - 1) = \sum_{z \in \kappa - g^{-1}(\infty)} \mu_{g'}^0(z).$$

Write $g = P/Q$ such that P and Q are coprime polynomials on κ, and note that the numerator and denominator of

$$g'(z) = \frac{P'(z)Q(z) - P(z)Q'(z)}{Q(z)^2}$$

also are coprime. Then

$$\sum_{z \in \kappa - g^{-1}(\infty)} \mu_{g'}^0(z) = \sum_{z \in \kappa} \mu_{P'Q-PQ'}^0(z) = \deg(P'Q - PQ').$$

Since $g(\infty) = 1$, one has $\deg(P) = \deg(Q) = d$, and hence $Q(z)^2/z^{2d}$ tends to a finite non-zero limit as $z \to \infty$. Also $\mu_g(\infty) = 1$ means that g is injective in some neighborhood of ∞ and so

$$g\left(\frac{1}{z}\right) = 1 + cz + \cdots, \quad c \neq 0,$$

near the origin. Differentiating both sides of this, and replacing z by $1/z$, one finds that $z^2 g'(z)$ tends to a finite non-zero limit as $z \to \infty$, and finally

$$\deg(P'Q - PQ') = 2d - 2$$

as required. □

Corollary 5.20. *A rational mapping of degree $d \geq 2$ has at most $2d - 2$ critical points in $\bar{\kappa}$. A polynomial of degree $d \geq 2$ has at most $d - 1$ critical points in κ.*

Theorem 5.21. *Let f be a rational mapping of degree at least two and suppose that a finite set E is completely invariant under f. Then E has at most two elements.*

Proof. Assume that E has m elements. Since E is a completely invariant finite set, then f must act as a permutation of E, and thus f^q is the identity mapping of E into itself for some suitable integer q. Set $\deg(f^q) = d$. It follows that for every $a \in E$, the equation $f^q(z) = a$ has d roots which are all at a, and so by applying the Riemann-Hurwitz relation to f^q, one has

$$m(d - 1) \leq 2d - 2,$$

and hence $m \leq 2$ since $d \geq 2$.   □

Definition 5.22. *A point z is said to be exceptional for a rational mapping f when $[z]$ is finite, and the set of such points is denoted by $\mathrm{Exc}(f)$.*

Theorem 5.23. *A rational mapping f of degree at least two has at most two exceptional points. If $\mathrm{Exc}(f) = \{\zeta\}$, then f is conjugate to a polynomial with ζ corresponding to ∞. If $\mathrm{Exc}(f) = \{\zeta_1, \zeta_2\}$ with $\zeta_1 \neq \zeta_2$, then f is conjugate to some mapping $z \mapsto z^d$ such that ζ_1 and ζ_2 correspond to 0 and ∞.*

Proof. Here we follow Beardon's observation [5]. Note that $[z] \subset \mathrm{Exc}(f)$ if $z \in \mathrm{Exc}(f)$. By Corollary 5.3, $\mathrm{Exc}(f)$ is completely invariant under f and hence, by Theorem 5.21, f has at most two exceptional points. After a suitable conjugation, there are the following four possibilities:

(i) $\mathrm{Exc}(f) = \emptyset$;
(ii) $\mathrm{Exc}(f) = \{\infty\} = [\infty]$;
(iii) $\mathrm{Exc}(f) = \{0, \infty\}$, $[0] = \{0\}$, $[\infty] = \{\infty\}$;
(iv) $\mathrm{Exc}(f) = \{0, \infty\} = [0] = [\infty]$.

There is nothing to say about (i). If (ii) or (iii) holds, by Theorem 5.17, f is a polynomial. For the case (iii), obviously $f(z) = az^d$ for some $d \in \mathbb{Z}^+$, $a \in \kappa_*$. If (iv) holds, it must be $f(0) = \infty$, $f(\infty) = 0$, and all zeros and poles of f are in $\{0, \infty\}$ so that it is of the form az^d for some negative integer d. The proof is thus completed.   □

According to Theorem 4.1.4 of Beardon in [5], it is easy to prove the following characterization of exceptional points:

Theorem 5.24. *The backward orbit $O^-(z)$ of z is finite if and only if z is exceptional.*

5.3 Fixed points of entire functions

Let κ be an algebraically closed field of characteristic zero, complete for a non-trivial non-Archimedean absolute value $|\cdot|$. Let f be an entire function on κ. Next we prove the non-Archimedean analogue of Rosenbloom-Baker's theorem (cf. [109],[3]).

Theorem 5.25. *If f is a transcendental entire function on κ, then f possesses infinitely many fixed points of exact order n, except for at most one value of n.*

Proof. We will follow Hayman's proof [52]. Suppose that f has only a finite number of fixed points of the exact order k, say, $a_1, a_2, ..., a_q$, and assume $n > k$. If the equation $f^n(z) - f^{n-k}(z) = 0$ has a root z_0, then z_0 satisfies

$$f^k(f^{n-k}(z_0)) = f^n(z_0) = f^{n-k}(z_0),$$

so that $w = f^{n-k}(z_0) \in \mathrm{Fix}(f^k)$. Thus, either $w = a_j$ for some j, or w is a fixed point of exact order l less that k, so that z_0 is a root of the equation $f^{n-k+l}(z) - f^{n-k}(z) = 0$. Hence

$$\overline{N}\left(r, \frac{1}{f^n - f^{n-k}}\right) \leq \sum_{l=1}^{k-1} \overline{N}\left(r, \frac{1}{f^{n-k+l} - f^{n-k}}\right) + \sum_{j=1}^{q} \overline{N}\left(r, \frac{1}{f^{n-k} - a_j}\right)$$

$$= O\left(\sum_{l=1}^{k-1} T(r, f^l)\right) = o(T(r, f^n))$$

by Theorem 2.44 since $f^n = f^l \circ f^{n-l}$. We now apply Theorem 2.21 to f^n with $a_1(z) = z$, $a_2(z) = f^{n-k}(z)$, and $a_3(z) = \infty$, and obtain

$$T(r, f^n) \leq \overline{N}\left(r, \frac{1}{f^n(z) - z}\right) + o(T(r, f^n)).$$

Since the contribution of fixed points of exact order less than n to $\overline{N}\left(r, \frac{1}{f^n(z)-z}\right)$ is at most

$$\sum_{l=1}^{n-1} \overline{N}\left(r, \frac{1}{f^l(z) - z}\right) = O\left(\sum_{l=1}^{n-1} T(r, f^l)\right) = o(T(r, f^n)),$$

then f has infinitely many fixed points of exact order n, and so there can be at most one value of k for which f has only finitely many fixed points of exact order k. $\square$

For the complex case, the exceptional case occurs only for $n = 1$. For example, the function $f(z) = e^z + z$ has no fixed points of exact order 1, but for the non-Archimedean case, each transcendental entire function has infinitely many fixed points of exact order 1. Thus we have the following result:

Theorem 5.26. *If f is a transcendental entire function on κ, then f possesses infinitely many fixed points of exact order n for all $n \geq 1$.*

Proof. Note that

$$N\left(r, \frac{1}{f^n(z) - z}\right) = T(r, f^n) + O(\log r).$$

If $n = 1$, it is obvious since $\log r = o(T(r, f))$. Assume $n \geq 2$. Since the contribution of fixed points of exact order less than n to $N\left(r, \frac{1}{f^n(z)-z}\right)$ is at most

$$\sum_{l=1}^{n-1} N\left(r, \frac{1}{f^l(z) - z}\right) = \sum_{l=1}^{n-1} T(r, f^l) + O(\log r) = o(T(r, f^n)),$$

thus f has infinitely many fixed points of exact order n. $\square$

Following Baker [3] or Hayman [52], we can prove the following:

Theorem 5.27. *If f is a polynomial of degree at least 2 on κ, then f has at least one fixed point of exact order n, except for at most one value of n.*

Proof. Write $d = \deg(f) \geq 2$. Then $\deg(f^n) = d^n$, and hence

$$\mathrm{Fix}(f^n) \neq \emptyset, \quad n = 1, 2, \ldots .$$

Assume, on the contrary, that f has no fixed points of exact orders n, k, where $n > k \geq 2$. Define

$$\Psi(z) = \frac{f^n(z) - z}{f^{n-k}(z) - z}.$$

Let z_0 be a zero of Ψ, that is, $f^n(z_0) = z_0$. Then z_0 is a fixed point of exact order j of f for some $j < n$ so that $\Lambda = \{z_0, f^1(z_0), \ldots, f^{j-1}(z_0)\}$ is a j-cycle with $\Lambda \subset \mathrm{Fix}(f^n)$. Thus $j|n$. If $n = 3$, it must be $j = 1$, so that there are at most d distinct zeros of Ψ. If $n = 4$, then $j = 1$ or 2, so that Ψ has at most d^2 distinct zeros by noting that $\mathrm{Fix}(f) \subset \mathrm{Fix}(f^2)$. If $n > 4$, we must have $j \leq n - 3$, and so Ψ has at most

$$\sum_{j=1}^{n-3} d^j < d^{n-2}$$

distinct zeros. Thus, in all cases, Ψ has at most d^{n-2} distinct zeros.

Let x_0 be a root of $\Psi(z) - 1 = 0$, that is,

$$f^{n-k}(x_0) = f^n(x_0) = f^k(f^{n-k}(x_0)).$$

Then $f^{n-k}(x_0)$ is a fixed point of f^k and so it is a fixed point of exact order j of f for some divisor j of k with $1 \leq j < k$. Hence

$$f^{n-k+j}(x_0) = f^j(f^{n-k}(x_0)) = f^{n-k}(x_0).$$

Since $\deg(f^{n-k+j} - f^{n-k}) = d^{n-k+j}$, then the number of distinct roots of $\Psi(z) - 1 = 0$ is at most

$$\sum_j d^{n-k+j} \begin{cases} \leq \sum_{j=1}^{k-2} d^{n-k+j} \leq d^{n-1} & : \quad k \geq 3 \\ = d^{n-1} & : \quad k = 2. \end{cases}$$

Let N be the number of zeros of $\Psi'(z)$ with due count of multiplicity. Then the total number of solutions of the equations $\Psi(z) = 0, 1$ is at most $N + d^{n-2} + d^{n-1}$. Note that Ψ has a pole of order $d^n - d^{n-k}$ at ∞. Then Ψ has $d^n - d^{n-k} + q$ poles, where q is the number of finite poles of Ψ (counting multiplicity), and so Ψ has $2(d^n - d^{n-k} + q)$ zeros and 1-points altogether in $\kappa \cup \{\infty\}$, counting multiplicity. Since Ψ' has a pole of order $d^n - d^{n-k} - 1$ at ∞ and at most $2q$ finite poles, then $N \leq d^n - d^{n-k} + 2q - 1$, and hence

$$2(d^n - d^{n-k} + q) \leq N + d^{n-2} + d^{n-1} \leq d^n - d^{n-k} + d^{n-2} + d^{n-1} + 2q - 1,$$

which means

$$d^n \leq d^{n-k} + d^{n-2} + d^{n-1} - 1 \leq 2d^{n-1} - 1 \leq d^n - 1,$$

giving a contradiction. $\square$

Example 5.28 (cf. [52]). *Take $f(z) = z^2 - z$. By soving the equations*

$$f(z) = z, \quad f^2(z) = z,$$

that is,

$$z^2 - z = z, \quad (z^2 - z)^2 - (z^2 - z) = z,$$

we obtain $\text{Fix}(f) = \text{Fix}(f^2) = \{0, 2\}$. *Thus f has no fixed points of exact order 2.*

In the one complex variable case, Example 5.28 is the only example of this type (up to conjugates).

Theorem 5.29 (cf. [5]). *Let f be a polynomial of degree at least 2 on $\mathbb{C}$ and suppose that f has no fixed points of exact order n. Then $n = 2$ and f is conjugate to $z \mapsto z^2 - z$.*

Conjecture 5.30. *Theorem 5.29 is true for polynomials of degree at least 2 on κ.*

There is also a corresponding result for rational functions on $\mathbb{C}$, namely

Theorem 5.31. *Let f be a rational function of degree d on $\mathbb{C}$, where $d \geq 2$, and suppose that f has no fixed points of exact order n. Then (d, n) is one of the pairs*

$$(2, 2), \quad (2, 3), \quad (3, 2), \quad (4, 2);$$

moreover, each such pair does arise from some f in this way.

This result is also due to Baker [3], and we conjecture that it is even true for rational functions on κ.

5.4 Normal families

Let κ be an algebraically closed field of characteristic zero, complete for a non-trivial non-Archimedean absolute value $|\cdot|$. First we prove the following non-Archimedean analogue of Schwarz's lemma in complex analysis:

Theorem 5.32. *Take $f \in A_{(1}^*(\kappa) - \kappa$ and assume that*

$$f(0) = 0; \quad |f(z)| \leq 1, \quad z \in \kappa(0; 1). \tag{5.12}$$

Then either

$$|f(z)| < |z|, \quad z \in \kappa(0; 1) - \{0\},$$

or

$$|f(z)| = |z|, \quad z \in \kappa(0; 1).$$

Proof. Write

$$f(z) = \sum_{n=1}^{\infty} a_n z^n.$$

Note that Theorems 1.22 and (5.12) imply

$$\mu(r, f) = \max_{n \geq 1} |a_n| r^n \leq 1, \quad 0 < r < 1,$$

which means
$$\max_{n\geq 1}|a_n| \leq 1.$$

If $|a_1| < 1$, then

$$|f(z)| \leq \max_{n\geq 1}|a_n||z|^n = \left\{\max_{n\geq 1}|a_n||z|^{n-1}\right\}|z| < |z|, \quad z \in \kappa(0;1) - \{0\}.$$

If $|a_1| = 1$, then
$$\max_{n\geq 2}|a_n||z|^{n-1} < 1 = |a_1|,$$

and hence

$$|f(z)| = |a_1 z| = |z|.$$

$\square$

Corollary 5.33. *Under the assumptions of Theorem 5.32, $|f'(0)| \leq 1$.*

In the unit disc $\mathbb{C}(0;1)$ of the complex field $\mathbb{C}$, it is well-known that the *hyperbolic metric*

$$\rho(z,w) = \frac{1}{2}\log\frac{|1-\bar{z}w|_\infty + |z-w|_\infty}{|1-\bar{z}w|_\infty - |z-w|_\infty}$$

is complete. For non-Archimedean absolute value $|\cdot|$ on κ, the metric is of the following form

$$\rho(z,w) = \frac{1}{2}\log\frac{1+|z-w|}{1-|z-w|}. \tag{5.13}$$

The non-Archimedean analogue of Schwarz's lemma given by G. Pick in 1916 holds:

Theorem 5.34. *Suppose that $f \in \mathcal{A}_{(1}^*(\kappa)$ and that $|f(z)| < 1$ for $z \in \kappa(0;1)$. Then either*

$$\rho(f(z),f(w)) < d(z,w)$$

for every z and w, $z \neq w$, or
$$\rho(f(z),f(w)) = d(z,w)$$

for all z and w.

The assertions can be proved by using the technique in the proofs of Theorem 1.40 and Theorem 5.32. Let us check the *chordal metric*

$$\chi(z,w) = \begin{cases} \frac{|z-w|_\infty}{(1+|z|_\infty^2)^{1/2}(1+|w|_\infty^2)^{1/2}} & : \quad z,w \in \mathbb{C} \\ \frac{1}{(1+|z|_\infty^2)^{1/2}} & : \quad w = \infty \end{cases}$$

defined on $\bar{\mathbb{C}}$. If the absolute value $|\cdot|_\infty$ is replaced by the non-Archimedean absolute value $|\cdot|$ on κ, then χ can not serve as an ultrametric on $\bar{\kappa}$. However, it is easy to check that

$$\chi(z,w) = \begin{cases} \frac{|z-w|}{|z|^\vee|w|^\vee} & : \quad z,w \in \kappa \\ \frac{1}{|z|^\vee} & : \quad w = \infty \end{cases} \tag{5.14}$$

is an ultrametric on $\bar{\kappa}$, where, by definition,

$$x^\vee = \max\{1, x\}.$$

If we set

$$\frac{1}{0} = \infty, \quad \frac{1}{\infty} = 0,$$

then we have

$$\chi\left(\frac{1}{z}, \frac{1}{w}\right) = \chi(z, w), \quad z, w \in \bar{\kappa}.$$

Take an open set $D \subset \kappa$. A function $f(z)$ defined in D is *spherically continuous* at $z = z_0 \in D$, provided that for any $\varepsilon > 0$ there exists a $\delta > 0$ such that

$$\chi(f(z), f(z_0)) < \varepsilon, \quad z \in D \cap \kappa(z_0; \delta).$$

Also f is spherically continuous in D if it is spherically continuous at all points of D.

Lemma 5.35. *A local meromorphic function f is spherically continuous in its domain of definition.*

Proof. If a point z_0 is not a pole of f, then f is continuous in the ordinary sense, and from the inequality

$$\chi(f(z), f(z_0)) < |f(z) - f(z_0)|, \tag{5.15}$$

it follows that f is also spherically continuous. At a pole of f, the same argument applies to $1/f$, by noting the equality

$$\chi\left(\frac{1}{f(z)}, \frac{1}{f(w)}\right) = \chi(f(z), f(w)).$$

$\square$

Definition 5.36. *A family $\mathcal{F}$ of functions of an open set $D \subset \kappa$ into $\bar{\kappa}$ is called spherically equicontinuous or a spherically equicontinuous family at $z_0 \in D$ if and only if for every positive ε there exists a positive δ such that for all z in D, and for all f in $\mathcal{F}$,*

$$|z - z_0| < \delta \implies \chi(f(z), f(z_0)) < \varepsilon.$$

The family $\mathcal{F}$ is said to be spherically equicontinuous on D if and only if $\mathcal{F}$ is spherically equicontinuous at each point of D.

The inequality (5.15) implies that a family $\mathcal{F}$ of functions, equicontinuous in the ordinary sense on an open set D, is also spherically equicontinuous. A sequence of functions f_n defined on a set D *converges spherically uniformly* on D if, given any $\varepsilon > 0$, there exists an N such that

$$\chi(f_m(z), f_n(z)) < \varepsilon, \quad m, n > N,$$

for every $z \in D$, and is said to be *locally spherically uniformly convergent* on D if each point $z_0 \in D$ has a disc $\kappa[z_0; r] \subset D$ on which it converges spherically uniformly, and it is said to be *uniformly bounded* in D if there exists a $B \in \mathbb{R}^+$ such that

$$|f_n(z)| < B, \quad n \geq 1, \ z \in D.$$

Lemma 5.37. *If a sequence of spherically continuous functions* $\{f_n\}$ *converges spherically uniformly on a closed set* D, *then it converges to a function* f *which is spherically continuous on* D, *and the functions* f_n *are spherically equicontinuous on* D.

Proof. Obviously, f is uniquely defined. Note that f may take the value ∞ everywhere in D. From the following inequality,

$$\chi(f(z), f(z_0)) \leq \max \left\{ \chi(f(z), f_n(z)), \chi(f_n(z), f_n(z_0)), \chi(f_n(z_0), f(z_0)) \right\},$$

it is easy to prove that f is spherically continuous on D.

To prove equicontinuity, it is suffices to note the following inequality

$$\chi(f_n(z), f_n(z_0)) \leq \max \left\{ \chi(f_n(z), f(z)), \chi(f(z), f(z_0)), \chi(f(z_0), f_n(z_0)) \right\}.$$

$\square$

Definition 5.38. *A family* $\mathcal{F}$ *of local meromorphic functions defined on an open set* D *is called normal in* D *if every sequence* $\{f_n\}$ *in* $\mathcal{F}$ *contains a subsequence which is locally spherically uniformly convergent on* D.

Theorem 5.39. *Let* $\{f_n\}$ *be a sequence of local meromorphic functions defined on an open set* D. *If* $\{f_n\}$ *converges locally spherically uniformly on* D, *then it converges to a local meromorphic function* f *(including the infinite constant).*

Proof. Here we follow the proof of Hille [55]. By Lemma 5.37, f is spherically continuous in D, and the functions f_n are spherically equicontinuous on D. In particular, for each $z_0 \in D$, they have this property in some disc Δ containing z_0. We distinguish three possibilities:

1) $f(z_0) \in \kappa$. By equicontinuity, there exists a neighborhood $U \subset \Delta$ of z_0 and a $B > 1$ such that $|f(z)| < B$ in U. Further, the locally spherically uniform convergence implies the existence of an integer n_0 such that

$$|f_n(z)| < 2B, \quad n > n_0, \ z \in U,$$

which means that f_n has no poles in U for any $n > n_0$, that is, they are holomorphic in U. Note that

$$|f_n(z) - f(z)| < 2B^2 \chi(f_n(z), f(z)), \quad n > n_0, \ z \in U.$$

It follows that the sequence $\{f_n\}$ converges uniformly to f on U. Thus f is holomorphic in U.

2) $f(z_0) = \infty$, but there is a neighborhood $U \subset \Delta$ of z_0 in which $f(z) \neq \infty$ except at $z = z_0$. We may assume U to be small enough so that $f(z) \neq 0$ in U. Then $1/f(z)$ is well defined in U and takes the value 0 at $z = z_0$. Note that

$$\chi \left(\frac{1}{f(z)}, \frac{1}{f_n(z)} \right) = \chi(f(z), f_n(z)).$$

Then there exists an integer n_0 such that the sequence $\{1/f_n \mid n > n_0\}$ consists of holomorphic functions in U which converges uniformly to the limit $1/f$ on U. Thus $1/f$ is holomorphic in U and has a zero at $z = z_0$. It follows that f is meromorphic in U and has a pole at $z = z_0$.

3) $f(z_0) = \infty$, and f takes ∞ in every neighborhood of z_0 except at $z = z_0$. Now note that

$$\chi(f(z), f(z_0)) = \frac{1}{|f(z)|^\vee}.$$

By spherical continuity, we must have

$$|f(z)| > B, \quad |z - z_0| < \delta(B)$$

for any $B > 1$, and the spherically uniform convergence shows that

$$|f_n(z)| > \frac{1}{2}B, \quad n > n_0, \ |z - z_0| < \delta(B).$$

Thus, the functions $1/f_n$ $(n > n_0)$ are holomorphic in $\kappa(z_0; \delta(B))$ and uniformly small. Again we have uniform convergence of $1/f_n(z)$ to $1/f(z)$ in this neighborhood, and hence $1/f(z)$ is holomorphic. Since $z = z_0$ is a limit point of zeros of $1/f(z)$, then $f(z)$ must be identically ∞ in the neighborhood under consideration.

Thus, we see that for every point $z_0 \in D$, there is a neighborhood in which either $f(z)$ or $1/f(z)$ is holomorphic, and hence f is meromorphic in the neighborhood. Therefore f is locally meromorphic in D. $\qquad\square$

Theorem 5.40. *If a family $\mathcal{F}$ of local meromorphic functions defined on an open set D is normal, then $\mathcal{F}$ is spherically equicontinuous on D.*

Proof. Suppose that the family $\mathcal{F}$ is normal, and assume, on the contrary, that $\mathcal{F}$ is not spherically equicontinuous on D. The latter means that for a particular number $\varepsilon > 0$, we can find a point z_0 and a sequence of points $\{z_n\}$ in D, together with sequence $\{f_n\}$ in $\mathcal{F}$ such that $z_n \to z_0$ but

$$\chi(f_n(z_n), f_n(z_0)) > \varepsilon, \quad n \in \mathbb{Z}^+. \tag{5.16}$$

On the other hand, the normality of $\mathcal{F}$ implies that the sequence $\{f_n(z)\}$ contains a subsequence $\{f_{n_k}(z)\}$ that converges locally spherically uniformly on D. In particular, the point z_0 has a disc $\kappa(z_0; r) \subset D$ on which $\{f_{n_k}(z)\}$ converges spherically uniformly. W. l. o. g., we may assume $\{z_n\} \subset \kappa(z_0; r)$. However, by Lemma 5.37, $\{f_{n_k}(z)\}$ is spherically equicontinuous at $z = z_0$, and for sufficiently large k, we obtain a contradiction of (5.16). Thus $\mathcal{F}$ must be spherically equicontinuous on D. $\qquad\square$

The converse theorem is perhaps not true since we will need local compactness of the spaces to prove the normality of $\mathcal{F}$ from equicontinuity. Thus, the dynamics deduced from normality and equicontinuity have different consequences.

5.5   Montel's theorems

First of all, we will need a type of non-Archimedean integration introduced by Shnirelman in 1938 [123]. To define it, we note that $|\cdot|$ is equivalent to a p-adic absolute value $|\cdot|_p$ for some prime p.

Definition 5.41. *Take* $r \in |\kappa_*|$, $a, \Gamma \in \kappa$ *with* $|\Gamma| = r$, *and let* f *be an* κ-*valued function defined on* $\kappa\langle a; r \rangle$. *The Shnirelman integral is defined as the following limit if it exists:*

$$\int_{a,\Gamma} f(x)dx := \lim_{p \nmid n, n \to \infty} \frac{1}{n} \sum_{\xi^n=1} f(a + \xi\Gamma).$$

By the definition, it is easy to show that if f and g are integrable on $\kappa\langle a; r \rangle$, then $\alpha f + \beta g$ is integrable for any $\alpha, \beta \in \kappa$ such that

$$\int_{a,\Gamma} (\alpha f(x) + \beta g(x))dx = \alpha \int_{a,\Gamma} f(x)dx + \beta \int_{a,\Gamma} g(x)dx. \qquad (5.17)$$

Here we recall some basic properties of the Shnirelman integrals (see [76]):

(a) If $\int_{a,\Gamma} f(x)dx$ exists, then

$$\left| \int_{a,\Gamma} f(x)dx \right| \le \max_{x \in \kappa\langle a;r \rangle} |f(x)|.$$

In fact, noting that $|n| = 1$ if $p \nmid n$, we have

$$
\begin{aligned}
\left| \int_{a,\Gamma} f(x)dx \right| &\le \max_{p \nmid n, n \ge 1} \left| \frac{1}{n} \sum_{\xi^n=1} f(a + \xi\Gamma) \right| \\
&\le \max_{p \nmid n, n \ge 1} \max_{\xi^n=1} |f(a + \xi\Gamma)| \\
&\le \max_{x \in \kappa\langle a;r \rangle} |f(x)|.
\end{aligned}
$$

Here we used the following simple fact: if $a_n \to a$ ($a_n, a \in \kappa$) as $n \to \infty$, then $|a| \le \max\{|a_n|\}$. It is trivial when $a = 0$. If $a \ne 0$, then $|a_n - a| < |a|$ if n is large sufficiently, and so $|a| = |a - a_n + a_n| \le \max\{|a_n - a|, |a_n|\} = \max\{|a_n|\}$ as required.

(b) $\int_{a,\Gamma}$ commutes with limits of functions that are uniform limits on $\kappa\langle a; r \rangle$, that is, if f_n ($n = 1, 2, ...$) are integrable and converges uniformly to an integrable function f on $\kappa\langle a; r \rangle$, then

$$\lim_{n \to \infty} \int_{a,\Gamma} f_n(x)dx = \int_{a,\Gamma} f(x)dx.$$

This fact follows from the uniform convergence of $\{f_n\}$ and the inequality

$$\left| \int_{a,\Gamma} f_n(x)dx - \int_{a,\Gamma} f(x)dx \right| = \left| \int_{a,\Gamma} (f_n(x) - f(x))dx \right| \le \max_{x \in \kappa\langle a;r \rangle} |f_n(x) - f(x)|.$$

(c) If $r_1 \le r \le r_2$ and if f is given by a convergent Laurent series

$$f(z) = \sum_{m=-\infty}^{\infty} a_m(z - a)^m, \quad r_1 \le |z - a| \le r_2,$$

then $\int_{a,\Gamma} f(x)dx$ exists and equals a_0. More generally,

$$\int_{a,\Gamma} \frac{f(x)}{(x - a)^m}dx = a_m.$$

To prove property (c), we will utilize the fact that if $m \neq 0$, then

$$\sum_{\xi^n=1} \xi^m = 0, \quad n > |m|. \tag{5.18}$$

To show (5.18), we first consider a general polynomial

$$f(x) = (x - \xi_1)(x - \xi_2) \cdots (x - \xi_n)$$

and set

$$s_k = \xi_1^k + \xi_2^k + \cdots + \xi_n^k, \quad k = 0, 1, 2, \cdots.$$

Then for each integer $k \geq 0$, there exists a polynomial g_k of degree $< n$ such that

$$x^{k+1} f'(x) = (s_0 x^k + s_1 x^{k-1} + \cdots + s_{k-1} x + s_k) f(x) + g_k(x). \tag{5.19}$$

In fact, noting that

$$
\begin{aligned}
x^{k+1} f'(x) &= x^{k+1} \sum_{j=1}^{n} \prod_{i \neq j} (x - \xi_i) = x^k \sum_{j=1}^{n} (x - \xi_j + \xi_j) \prod_{i \neq j} (x - \xi_i) \\
&= s_0 x^k f(x) + x^k \sum_{j=1}^{n} \xi_j \prod_{i \neq j} (x - \xi_i),
\end{aligned}
$$

then the claim (5.19) is deduced inductively. In particular, taking $f(x) = x^n - 1$, then (5.19) imply that $s_1 = \cdots s_{n-1} = 0$. This fact yields (5.18) since

$$x^n - 1 = \prod_{\xi^n=1} \left(x - \frac{1}{\xi} \right).$$

Now the property (c) follows from the observation:

$$
\begin{aligned}
\lim_{p \wedge n, n \to \infty} \frac{1}{n} \sum_{\xi^n=1} f(a + \xi \Gamma) &= \lim_{p \wedge n, n \to \infty} \frac{1}{n} \sum_{\xi^n=1} \sum_{m=-\infty}^{\infty} a_m \xi^m \Gamma^m \\
&= \lim_{p \wedge n, n \to \infty} \sum_{m=-\infty}^{\infty} \frac{a_m \Gamma^m}{n} \sum_{\xi^n=1} \xi^m \\
&= \lim_{p \wedge n, n \to \infty} \left\{ a_0 + \sum_{|m| \geq n} \frac{a_m \Gamma^m}{n} \sum_{\xi^n=1} \xi^m \right\} \\
&= a_0.
\end{aligned}
$$

(d) If $\Gamma \in \kappa$ with $|\Gamma| = r > 0$, and f is given by a convergent power series

$$f(z) = \sum_{m=0}^{\infty} a_m (z - a)^m, \quad |z - a| \leq r,$$

then for fixed $z \in \kappa$, the following non-Archimedean *Cauchy integral formula*

$$\int_{a,\Gamma} \frac{f(x)(x-a)}{x-z} dx = \begin{cases} f(z) & : \quad |z-a| < r \\ 0 & : \quad |z-a| > r, \end{cases} \tag{5.20}$$

holds. More generally,

$$\int_{a,\Gamma} \frac{f(x)(x-a)}{(x-z)^{n+1}} dx = \begin{cases} \frac{1}{n!} f^{(n)}(z) & : \quad |z-a| < r \\ 0 & : \quad |z-a| > r. \end{cases}$$

Note that

$$\frac{1}{1-x} = \sum_{n=0}^{\infty} x^n$$

holds when $|x| < 1$. Thus if $|z-a| < r$,

$$\begin{aligned}
\int_{a,\Gamma} \frac{f(x)(x-a)}{x-z} dx &= \int_{a,\Gamma} \left(\sum_{m=0}^{\infty} \frac{a_m(x-a)^m}{1 - \frac{z-a}{x-a}} \right) dx \\
&= \sum_{m=0}^{\infty} \sum_{n=0}^{\infty} a_m(z-a)^n \int_{a,\Gamma} (x-a)^{m-n} dx \\
&= \sum_{m=0}^{\infty} a_m(z-a)^m = f(z).
\end{aligned}$$

Similarly, we can prove the second part of (5.20).

Now we give the non-Archimedean analogue of Montel's theorem.

Theorem 5.42. *Let $\mathcal{F}$ be a family of holomorphic functions in an open set D of κ. If $\mathcal{F}$ is locally uniformly bounded in D, that is, each point $z_0 \in D$ has a disc $\kappa[z_0; r] \subset D$ on which it is uniformly bounded, then $\mathcal{F}$ is equicontinuous in D.*

Proof. First, equicontinuity is a local property so we need only prove that for every $a \in D$, the family $\mathcal{F}$ is equicontinuous at a. Since $\mathcal{F}$ is locally uniformly bounded in D, then there are $r \in |\kappa_*|$, $B \in \mathbb{R}^+$ such that

$$|f(z)| \leq B, \quad z \in \kappa[a; r] \subset D, \quad f \in \mathcal{F}.$$

Since each function can be expressed as a power series on a maximal analytic disc centered at a, it can as well be on $\kappa[a; r]$. Take $\Gamma \in \kappa$ with $|\Gamma| = r$. If $f \in \mathcal{F}$, and if $z \in \kappa(a; r)$, then Cauchy's integral formula implies

$$f(z) = \int_{a,\Gamma} \frac{f(x)(x-a)}{x-z} dx.$$

Then for $z_1, z_2 \in \kappa(a; r)$, we have

$$f(z_1) - f(z_2) = (z_1 - z_2) \int_{a,\Gamma} \frac{f(x)(x-a)}{(x-z_1)(x-z_2)} dx,$$

and, hence,

$$|f(z_1) - f(z_2)| \leq \max_{x \in \kappa(a;r)} \frac{|f(x)(x-a)|}{|(x-z_1)(x-z_2)|} \leq \frac{B}{r}|z_1 - z_2|,$$

and, therefore, $\mathcal{F}$ is equicontinuous at a. □

Theorem 5.43. Let $\mathcal{F}$ be a family of holomorphic functions in an open set D of $\bar{\kappa}$. If each function in $\mathcal{F}$ does not take the value 0, then $\mathcal{F}$ is spherically equicontinuous in D.

Proof. Take a point $z_0 \in D$. It is suffices to prove that $\mathcal{F}$ is equicontinuous at z_0. W. l. o. g., we may assume $z_0 = 0$. Note that each function f in $\mathcal{F}$ can be expressed as a power series on the maximal analytic disc centered at 0, say $\kappa(0; \rho)$. By Weierstrass preparation theorem (Theorem 1.21), the central index of f satisfies $\nu(r, f) = 0$ for $0 \leq r < \rho$, that is, $\mu(r, f)$ is a constant. Thus we have

$$|f(z)| = |f(0)|, \quad z \in \kappa(0; \rho), \quad f \in \mathcal{F}.$$

Set

$$\mathcal{F}_1 = \{f \in \mathcal{F} \mid |f(0)| \leq 1\}, \quad \mathcal{F}_2 = \{f \in \mathcal{F} \mid |f(0)| > 1\}.$$

Then $\mathcal{F}_1$ and $\{1/f \mid f \in \mathcal{F}_2\}$ all are uniformly bounded in $\kappa(0; \rho)$. By Theorem 5.42, they are equicontinuous in $\kappa(0; \rho)$. Note that

$$\chi(f(z), f(w)) = \chi\left(\frac{1}{f(z)}, \frac{1}{f(w)}\right) \leq \left|\frac{1}{f(z)} - \frac{1}{f(w)}\right|.$$

Then $\mathcal{F}_2$ is spherically equicontinuous in $\kappa(0; \rho)$. We finally obtain that $\mathcal{F}$ is spherically equicontinuous in $\kappa(0; \rho)$. □

Corollary 5.44. Let $\mathcal{F}$ be a family of holomorphic functions in an open set D of $\bar{\kappa}$. If each function in $\mathcal{F}$ does not take a value $a \in \kappa$, then $\mathcal{F}$ is spherically equicontinuous in D.

Proof. The conclusion follows from Theorem 5.43 and the following inequality

$$\chi(f(z), f(w)) \leq (|a|^\vee)^2 \chi(f(z) - a, f(w) - a).$$

□

Corollary 5.45. Let $\mathcal{F}$ be a family of meromorphic functions in an open set D of $\bar{\kappa}$. If each function in $\mathcal{F}$ does not take two distinct values a and b in $\bar{\kappa}$, then $\mathcal{F}$ is spherically equicontinuous in D.

Proof. If one of a and b, say b, is ∞, Corollary 5.45 follows from Corollary 5.44. When a and b both are finite, it is sufficient to consider the family

$$\mathcal{G} = \left\{\frac{1}{f-a} \mid f \in \mathcal{F}\right\}.$$

Obviously, each function in $\mathcal{G}$ is holomorphic and does not take the value $1/(b-a)$. Hence $\mathcal{G}$ is spherically equicontinuous in D. Thus Corollary 5.45 follows from the following inequality

$$\chi(f(z), f(w)) \leq (|a|^\vee)^2 \chi(f(z) - a, f(w) - a) = (|a|^\vee)^2 \chi\left(\frac{1}{f(z) - a}, \frac{1}{f(w) - a}\right).$$

□

Theorem 5.46. *A rational mapping f satisfies some Lipschitz condition*

$$\chi(f(z), f(w)) \le \lambda\chi(z, w)$$

on $\bar{\kappa}$, and hence is uniformly spherically continuous on $\bar{\kappa}$.

Proof. Here we will follow Beardon's idea [5]. When f maps a point z to $f(z)$, the change of scale (measured in the metric χ) at z is the ratio

$$R_f(z) = \left(\frac{|z|^\vee}{|f(z)|^\vee}\right)^2 |f'(z)|, \tag{5.21}$$

so it is sufficient to prove that this function R_f is bounded above on $\bar{\kappa}$. To do this, it is only necessary to check that R_f is bounded above near ∞ and near every pole of f, and this follows from standard, elementary arguments because f behaves like some term $(z - z_0)^m$ near each of these points z_0. Obviously, the supremum of (5.21) gives a sharp value of the constant λ in Theorem 5.46. $\square$

For a Möbius transformation

$$f(z) = \frac{az + b}{cz + d}, \quad ad - bc = 1,$$

we estimate the bound λ in Theorem 5.46. Now the function R_f in (5.21) is as follows:

$$R_f(z) = \left(\frac{|z|^\vee}{\max\{|az + b|, |cz + d|\}}\right)^2,$$

which obviously satisfies the following inequality:

$$R_f(z) \ge \|f\|^{-2},$$

where

$$\|f\| = \max\{|a|, |b|, |c|, |d|\}.$$

Note the equality $\|f^{-1}\| = \|f\|$, and the chain rule

$$R_{f^{-1}}(f(z))R_f(z) = 1.$$

We obtain

$$R_{f^{-1}}(f(z)) = \frac{1}{R_f(z)} \le \|f\|^2.$$

This is true for any f. Thus replacing f by f^{-1}, one has

$$R_f(f^{-1}(z)) \le \|f\|^2,$$

and hence for all $w \in \bar{\kappa}$,

$$R_f(w) \le \|f\|^2.$$

Thus the Möbius transformation f satisfies the Lipschitz condition

$$\chi(f(z), f(w)) \le \|f\|^2\chi(z, w). \tag{5.22}$$

Further we assume that there is a positive number α such that

$$\chi(f(0), f(1)) \geq \alpha, \quad \chi(f(1), f(\infty)) \geq \alpha, \quad \chi(f(\infty), f(0)) \geq \alpha. \qquad (5.23)$$

Then

$$\chi(f(0), f(1))\chi(f(1), f(\infty))\chi(f(\infty), f(0)) \geq \alpha^3.$$

We evaluate the left-hand side of the inequality and obtain

$$(\max\{|a|, |c|\} \max\{|b|, |d|\} \max\{|a + b|, |c + d|\})^2 \leq \frac{1}{\alpha^3}.$$

Note that

$$1 = ad - bc = d(a + b) - b(c + d),$$

and, hence,

$$1 \leq \max\{|d(a + b)|, |b(c + d)|\} \leq \max\{|b|, |d|\} \max\{|a + b|, |c + d|\}.$$

Thus we deduce that

$$\max\{|a|, |c|\} \leq \left(\frac{1}{\alpha}\right)^{\frac{3}{2}},$$

and a similar argument shows that

$$\max\{|b|, |d|\} \leq \left(\frac{1}{\alpha}\right)^{\frac{3}{2}}.$$

Therefore, we obtain

$$\|f\|^2 \leq \frac{1}{\alpha^3},$$

and hence

$$\chi(f(z), f(w)) \leq \frac{1}{\alpha^3}\chi(z, w). \qquad (5.24)$$

Here the proof follows Beardon's idea [5]. Now we prove the following variation of Corollary 5.45:

Theorem 5.47. *Let $\mathcal{F}$ be a family of meromorphic functions in an open set D of $\bar{\kappa}$. Let φ_1, φ_2 and φ_3 be meromorphic functions in D such that*

$$\inf_{z,w \in D} \chi(\varphi_i(z), \varphi_j(w)) \geq \alpha > 0, \quad i \neq j.$$

If each function $f \in \mathcal{F}$ satisfies

$$f(z) \neq \varphi_i(z), \quad i = 1, 2, 3, \quad z \in D,$$

then $\mathcal{F}$ is spherically equicontinuous in D.

Proof. Note that if g is a Möbius transformation, then

$$\|g\|^{-2}\chi(g \circ f(z), g \circ f(w)) \le \chi(f(z), f(w)) \le \|g\|^2 \chi(g \circ f(z), g \circ f(w)).$$

Thus $\mathcal{F}$ is spherically equicontinuous in D if and only if $g \circ \mathcal{F}$ does for some Möbius transformation g. Hence w. l. o. g., we may assume $\infty \in \varphi_3(D)$. Now we consider the family

$$\mathcal{G} = \left\{ F \mid F = \frac{f - \varphi_1}{f - \varphi_2}, \ f \in \mathcal{F} \right\}.$$

Obviously, each function in $\mathcal{G}$ is holomorphic and does not take the value 0. Hence $\mathcal{G}$ is spherically equicontinuous in D. It is easy to prove that the following family

$$\mathcal{G}_1 = \left\{ h \mid h = \frac{1}{F-1}, \ f \in \mathcal{G} \right\}$$

is spherically equicontinuous in D. Next we consider the family

$$\mathcal{G}_2 = \left\{ \Psi h \mid \Psi = \varphi_2 - \varphi_1, \ h \in \mathcal{G}_1 \right\}.$$

Fix a point $z_0 \in D$. Note that

$$\chi(\Psi(z)h(z), \Psi(z_0)h(z_0)) \le \max\{A(z)\chi(h(z), h(z_0)), B(z)\chi(\Psi(z), \Psi(z_0))\},$$

where

$$A(z) = \frac{|\Psi(z)||h(z)|^{\vee}|h(z_0)|^{\vee}}{|\Psi(z)h(z)|^{\vee}|\Psi(z_0)h(z_0)|^{\vee}}, \quad B(z) = \frac{|h(z_0)||\Psi(z)|^{\vee}|\Psi(z_0)|^{\vee}}{|\Psi(z)h(z)|^{\vee}|\Psi(z_0)h(z_0)|^{\vee}}.$$

Since Ψ is analytic at z_0, then Ψ is bounded in a neighborhood of z_0, say $\kappa(z_0; r)$, and hence A and B are uniformly bounded in $\kappa(z_0; r)$. Thus the continuity of Ψ at z_0 and the equicontinuity of $\mathcal{G}_1$ imply that $\mathcal{G}_2$ is spherically equicontinuous at z_0, and is so in D. Note that $F = (f - \varphi_1)/(f - \varphi_2) \in \mathcal{G}$ means that

$$f = \frac{\varphi_2 F - \varphi_1}{F - 1} = \varphi_2 + \frac{\varphi_2 - \varphi_1}{F - 1},$$

that is,

$$\mathcal{F} = \{\varphi_2 + \xi \mid \xi \in \mathcal{G}_2\}.$$

Now we have

$$\begin{aligned}
\chi(f(z), f(z_0)) &= \chi(\varphi_2(z) + \xi(z), \varphi_2(z_0) + \xi(z_0)) \\
&\le \max\{C(z)C(z_0)\chi(\varphi_2(z), \varphi_2(z_0)), H(z)H(z_0)\chi(\xi(z), \xi(z_0))\},
\end{aligned}$$

where

$$C(z) = \frac{|\varphi_2(z)|^{\vee}}{|\varphi_2(z) + \xi(z)|^{\vee}} \le |\varphi_2(z)|^{\vee}, \quad H(z) = \frac{|\xi(z)|^{\vee}}{|\varphi_2(z) + \xi(z)|^{\vee}}.$$

Note that

$$H(z) \le \frac{[\max\{|\varphi_2(z) + \xi(z)|, |\varphi_2(z)|\}]^{\vee}}{|\varphi_2(z) + \xi(z)|^{\vee}} \le |\varphi_2(z)|^{\vee}.$$

Since φ_2 is analytic at z_0, then φ_2 is bounded in a neighborhood of z_0, say $\kappa(z_0; r)$, and hence C and H are uniformly bounded in $\kappa(z_0; r)$. Therefore, $\mathcal{F}$ is spherically equicontinuous at z_0, and is so in D. □

By imitating the proof of Theorem 5.47, we can prove the following result:

Theorem 5.48. *Let $\mathcal{F}$ be a family of meromorphic functions in an open set D of $\bar{\kappa}$. Let φ_1 and φ_2 be analytic functions in D such that*

$$\inf_{z,w \in D} \chi(\varphi_1(z), \varphi_2(w)) \geq \alpha > 0.$$

If each function $f \in \mathcal{F}$ satisfies

$$f(z) \neq \varphi_i(z), \quad i = 1, 2, \quad z \in D,$$

then $\mathcal{F}$ is spherically equicontinuous in D.

L. C. Hsia [56] also proved independently the results in Corollary 5.45 and Theorem 5.48.

5.6 Fatou-Julia theory

Let κ be an algebraically closed field of characteristic zero, complete for a non-trivial non-Archimedean absolute value $|\cdot|$. Let D denote κ or $\bar{\kappa}$. In this section, we will consider a non-constant holomorphic mapping $f : D \longrightarrow D$. If $D = \kappa$, then f is an entire function (including polynomial). If $D = \bar{\kappa}$, then f is a rational function, and will be called a *rational mapping*.

Definition 5.49. *A family $\mathcal{F}$ of local meromorphic functions defined on an open set D is called normal at $z_0 \in D$ if there exists a disc $\kappa[z_0; r] \subset D$ such that $\mathcal{F}$ is normal on $\kappa[z_0; r]$.*

Obviously, the family $\mathcal{F}$ is normal on D if and only if it is normal at each point of D. Taking the collection $\{U_\alpha\}$ to be the class of all open subsets of D on which $\mathcal{F}$ is normal, this leads to the following general principle.

Theorem 5.50. *Let $\mathcal{F}$ be a family of local meromorphic functions defined on an open set D. Then there is a maximal open subset $F(\mathcal{F})$ of D on which $\mathcal{F}$ is normal. In particular, if $f : D \longrightarrow D$ is a holomorphic mapping, then there is a maximal open subset $F(f)$ of D on which the family of iterates $\{f^n\}$ is normal.*

The sets $F(\mathcal{F})$ and $F(f)$ in Theorem 5.50 are usually called *Fatou sets* of $\mathcal{F}$ and f respectively. *Julia sets* of $\mathcal{F}$ and f are defined respectively by

$$J(\mathcal{F}) = D - F(\mathcal{F}), \quad J(f) = D - F(f),$$

which are closed subsets of D. If $\mathcal{F}$ is finite, we define $J(\mathcal{F}) = \emptyset$.

Similarly, taking the collection $\{V_\beta\}$ to be the class of all open subsets of D on which $\mathcal{F}$ is spherically equicontinuous , this leads to the following general principle.

Theorem 5.51. *Let $\mathcal{F}$ be a family of local meromorphic functions defined on an open set D. Then there is a maximal open subset $F_{equ}(\mathcal{F})$ of D on which $\mathcal{F}$ is spherically equicontinuous. In particular, if $f : D \longrightarrow D$ is a holomorphic mapping, then there is a maximal open subset $F_{equ}(f) = F_{equ}(f, \chi)$ of D on which the family of iterates $\{f^n\}$ is spherically equicontinuous.*

Define the closed sets

$$J_{equ}(\mathcal{F}) = D - F_{equ}(\mathcal{F}), \quad J_{equ}(f) = J_{equ}(f, \chi) = D - F_{equ}(f, \chi).$$

By Theorem 5.40, we have

$$F(f) \subset F_{equ}(f), \quad J_{equ}(f) \subset J(f).$$

The following result is basic:

Theorem 5.52. *The sets $F = F(f)$ and $J = J(f)$ are backward invariant, that is,*

$$f^{-1}(F) = F, \quad f^{-1}(J) = J. \tag{5.25}$$

Proof. Let $\Delta \subset D$ be any disc and let D' be any component of $f^{-1}(\Delta)$. Since F and J divide D, then the assertion follows from the trivial identity

$$f^n|_{D'} = f^{n-1}|_D \circ f|_{D'}$$

and by distinguishing two cases:

(a) $\Delta \subseteq F$. Then $\{f^n\}$ is normal in D', i.e., $D' \subset F$. Since Δ and D' are arbitrary, this shows that $f^{-1}(F) \subset F$.

(b) $\Delta \cap J \neq \emptyset$. This means that the sequence $\{f^n\}$ is not normal in D', and hence $D' \cap J \neq \emptyset$. If we let Δ shrink to some point $z_0 \in J$ then $f^{-1}(z_0) \subset J$ and so $f^{-1}(J) \subset J$ because z_0 is arbitrary. $\qquad\square$

Note that f is surjective, that is, $f(D) = D$. By Lemma 5.1 , (5.25) implies

$$f(F) = F, \quad f(J) = J. \tag{5.26}$$

Hence F and J are completely invariant. Also it is easy to prove that the sets $F_{equ}(f)$ and $J_{equ}(f)$ are completely invariant.

Theorem 5.53. *For each positive integer $m \geq 2$,*

$$F(f) \subset F(f^m), \quad J(f^m) \subset J(f). \tag{5.27}$$

Furthermore, if $D = \bar{\kappa}$, then

$$F(f) = F(f^m), \quad J(f^m) = J(f). \tag{5.28}$$

Proof. It suffices to prove the assertion for, say, the Fatou sets. Since the family $\{f^{mn}\}$ is contained in the family $\{f^n\}$, we thus obtain (5.27). Assume that $D = \bar{\kappa}$. Given any disc $\Delta \subset \bar{\kappa}$ we set

$$\mathcal{F} = \{f^n|_\Delta \mid n \geq 0\}, \quad \mathcal{F}_j = \{f^j \circ f^{mn}|_\Delta \mid n \geq 0\}.$$

Then obviously,

$$\mathcal{F} = \mathcal{F}_0 \cup ... \cup \mathcal{F}_{m-1},$$

and since f^j is uniformly spherically continuous on $\bar{\kappa}$ by Theorem 5.46, $\mathcal{F}$ is normal if and only if $\mathcal{F}_0$ is normal. □

Theorem 5.53 is true for sets $F_{equ}(f)$ and $J_{equ}(f)$.

Theorem 5.54. *The Julia set $J(f)$ contains all repellers.*

Proof. Assume that a fixed point ξ of f is a repeller. By definition, there exists a neighborhood U of ξ such that for every $z_j \in U - [\xi](j = 1, 2, ...)$ there is an $n_j \in \mathbb{Z}^+$ with $f^n(z_j) \notin U$ for all $n \geq n_j$. Take the sequence $\{z_j\} \subset U - [\xi]$ such that $z_j \to \xi$ as $j \to \infty$. Assume $\xi \in F(f)$. Then we can find a disc $\Delta \subset U$ centered at ξ and subsequence $\{f^{n_k}\}$ of $\{f^n\}$ which converges spherically uniformly to a function ϕ on Δ. Obviously, $\phi(\xi) = \xi$ and ϕ is spherically continuous on Δ. Write $\Delta = \kappa[\xi; r]$ for some $r > 0$. Take some j with $z_j \in \kappa[\xi; r]$ and $\chi(\phi(z_j), \xi) < r$. Then there is k_0 such that $\chi(f^{n_k}(z_j), \phi(z_j)) < r$ for all $k \geq k_0$ so that

$$\chi(f^{n_k}(z_j), \xi) \leq \max\{\chi(f^{n_k}(z_j), \phi(z_j)), \chi(\phi(z_j), \xi)\} < r,$$

which implies $f^{n_k}(z_j) \in \Delta$ for all $k \geq k_0$, but $f^{n_k}(z_j) \notin U$ if $n_k \geq n_j$. This is a contradiction. □

Thus if $D = \bar{\kappa}$, Theorem 5.54 and Theorem 5.53 show that the Julia set $J(f)$ contains the closure of its set of repellers since $J(f)$ is closed. According to the proof of Theorem 5.10, we have in fact proved the following result:

Theorem 5.55. *The Fatou set $F(f)$ contains all attractors.*

If z_0 is an attracting fixed point, by the proof of Theorem 5.10, in particular, by (5.3), there is a $r \in \mathbb{R}^+$ such that $\kappa(z_0; r) \subset F(f)$. Since $F(f)$ is completely invariant, we have

$$\text{Att}(z_0) = O^-(\kappa(z_0; r)) \subset F(f),$$

that is, $F(f)$ contains the basin of attraction of z_0. Further, if f is a rational mapping, by Theorem 5.53 and Theorem 5.10, then $F(f)$ contains all attracting cycles and its basins of attraction.

Obviously, Theorem 5.55 holds for sets $F_{equ}(f)$. By Theorem 5.12 and (5.4), we see that Theorem 5.54 is true for $J_{equ}(f)$. From the proof of Theorem 5.23, we have

Theorem 5.56. *Let f be a rational mapping of degree at least two. Then the set $\text{Exc}(f)$ of exceptional points is contained in $F(f)$.*

Theorem 5.57. *The set $F_{equ}(f)$ contains all indifferent fixed points.*

Proof. Let z_0 be an indifferent fixed point of f. W. l. o. g., we may assume $z_0 = 0$. By (5.4), there is a $r \in \mathbb{R}^+$ such that

$$|f(z)| = |z|, \quad z \in \kappa[0; r].$$

Thus, by iterating, we obtain

$$|f^n(z)| = |z|, \quad n \in \mathbb{Z}^+, \quad z \in \kappa[0; r].$$

Hence $\{f^n(z)\}$ is equicontinuousis at $z = 0$, and so it is spherically equicontinuous at $z = 0$.   □

For attracting fixed points that are not superattracting, it is easy to prove the following linearization theorem which is a non-Archimedean analogue of Koenigs' theorem ([77]):

Theorem 5.58. *Suppose that f has an attracting fixed point at z_0, with multiplier λ satisfying $0 < |\lambda| < 1$. Then there is a holomorphic mapping $\zeta = \varphi(z)$ of a neighborhood of z_0 onto a neighborhood of 0 which is normalized so that $\varphi'(z_0) = 1$, and conjugates $f(z)$ to the linear function $g(\zeta) = \lambda\zeta$.*

Proof. Suppose $z_0 = 0$. We can write

$$f(z) = \lambda z + a_m z^m + \cdots \quad (a_m \neq 0, \ m \geq 2)$$

near 0, and define

$$\varphi_n(z) = \lambda^{-n} f^n(z) = z + \cdots .$$

Then φ_n satisfies

$$\varphi_n \circ f = \lambda^{-n} f^{n+1}(z) = \lambda\varphi_{n+1}.$$

Thus if $\varphi_n \to \varphi$, then $\varphi \circ f = \lambda\varphi$, so $\varphi \circ f \circ \varphi^{-1}(\zeta) = \lambda\zeta$, and φ is a conjugation. Obviously, we have $\varphi'(0) = 1$.

To show convergence note that for $r > 0$ small

$$|f(z) - \lambda z| = |a_m||z|^m \ (|z| \leq r).$$

Thus

$$|f(z)| \leq \max\{|\lambda||z|, |a_m||z|^m\} \leq \rho|z|,$$

where

$$\rho = \max\{|\lambda|, |a_m|r^{m-1}\}.$$

We choose $r > 0$ so small that ρ satisfies $\rho^m < |\lambda|$. Then when $|z| \leq r$,

$$|f^n(z)| \leq \rho^n|z|,$$

and we obtain

$$\begin{aligned}
|\varphi_{n+1}(z) - \varphi_n(z)| &= \left|\frac{f(f^n(z)) - \lambda f^n(z)}{\lambda^{n+1}}\right| = \frac{|a_m||f^n(z)|^m}{|\lambda|^{n+1}} \\
&\leq \frac{|a_m|\rho^{mn}}{|\lambda|^{n+1}}|z|^m = \frac{|a_m|}{|\lambda|}\left(\frac{\rho^m}{|\lambda|}\right)^n |z|^m \to 0 \ (n \to \infty).
\end{aligned}$$

Hence $\varphi_n(z)$ converges uniformly for $|z| \leq r$, and the conjugation exists.   □

We can interpret Theorem 5.58 in the context of repelling fixed points, for if z_0 is a repelling fixed point of f, then z_0 is a attracting fixed point of f^{-1} near z_0. Applying Theorem 5.58 to f^{-1}, we find that f is locally conjugate to the linear function $g(\zeta) = f'(z_0)\zeta$ in some neighborhood of z_0.

Problem 5.59. *Characterize the conditions such that f is linearizable near indifferent fixed points.*

5.7 Properties of the Julia set

Theorem 5.60. *Let f be a rational mapping with $\deg(f) \geq 2$, and suppose that E is a closed, completely invariant subset of $\bar{\kappa}$. Then either E has at most two elements and $E \subset \operatorname{Exc}(f) \subset F(f)$ or E is infinite and $J_{equ}(f) \subset E$.*

Proof. By Theorem 5.21, either E has at most two elements or it is infinite. If E is finite, Theorem 5.23 and Theorem 5.56 imply that $E \subset \operatorname{Exc}(f) \subset F(f)$. Suppose that E is infinite. Note that $E^c = \bar{\kappa} - E$ is also completely invariant. Thus f maps the open set E^c into E^c. By Corollary 5.45, the family $\mathcal{F} = \{f^n\}$ is spherically equicontinuous in E^c, and hence $E^c \subset F_{equ}(f)$. Therefore, it follows that $J_{equ}(f) \subset E$. $\square$

By Theorem 5.60, we have

Theorem 5.61. *Let f be a rational mapping with $\deg(f) \geq 2$. Then either $J(f)$ (resp., $J_{equ}(f)$) is empty or $J(f)$ (resp., $J_{equ}(f)$) is infinite.*

Actually $J_{equ}(f)$ may be empty. In fact, if $\operatorname{Exc}(f)$ contains two points, for instance, $f(z) = z^d$, by Corollary 5.45, the family $\{f^n\}$ is spherically equicontinuous in $\bar{\kappa} - \operatorname{Exc}(f)$ because $\bar{\kappa} - \operatorname{Exc}(f)$ is completely invariant, and thus $F_{equ}(f) = \bar{\kappa}$ by using Theorem 5.60. Here we do not know whether one can conclude that $J(f) \neq \emptyset$ or not. For the mapping $f(z) = z^d$, we have

$$J(f) \subset \kappa\langle 0; 1 \rangle,$$

but we can not confirm whether $J(f) = \kappa\langle 0; 1 \rangle$ or $J(f) = \emptyset$.

Theorem 5.62. *Let f be a rational mapping with $\deg(f) \geq 2$. Then either $J(f)$ (resp., $J_{equ}(f)$) $= \emptyset$ or $\bar{\kappa}$, or $J(f)$ (resp., $J_{equ}(f)$) has an empty interior.*

Proof. Here write $J = J(f)$ and $F = F(f)$. Note that we have a disjoint decomposition

$$\bar{\kappa} = J^o \cup \partial J \cup F.$$

Since F and J are completely invariant, by Corollary 5.4, J^o and ∂J are completely invariant. If F is not empty, then $\partial J \cup F$ is an infinite, closed, completely invariant set, and so it contains J by Theorem 5.60. Thus, it follows that $J \subset \partial J$, that is, $J = \emptyset$ or $J^o = \emptyset$.

Similarly, we can prove the assertion for the set $J_{equ}(f)$. The proof is completed. $\square$

Theorem 5.63. *Let f be a rational mapping with $\deg(f) \geq 2$. Then either the derived set of $J_{equ}(f)$ is empty or it is infinite and is equal to $J_{equ}(f)$.*

Proof. Let $J' = J'_{equ}(f)$ be the derived set of $J_{equ}(f)$, that is, the set of accumulation points of $J_{equ}(f)$. Then J' is closed. Suppose that $J' \neq \emptyset$. Since f is continuous, it is clear that $f(J') \subset J'$, and hence $J' \subset f^{-1}(J')$. Also, it is easy to see that $f^{-1}(J') \subset J'$ since f is an open mapping, and we deduce that J' is completely invariant. It follows from Theorem 5.60 that J' is infinite, and $J_{equ}(f) \subset J'$, and hence that $J_{equ}(f) = J'$. $\square$

Theorem 5.64. *Let f be a rational mapping with $\deg(f) \geq 2$. Suppose that $J_{equ}(f) \neq \emptyset$ and let D be any non-empty open set that meets $J_{equ}(f)$. Then*

$$\bar{\kappa} - \operatorname{Exc}(f) \subset O^+(D).$$

Proof. Write
$$S = \bar{\kappa} - O^+(D).$$

If S contains at least two distinct points, say z_1 and z_2, by Corollary 5.45, the family $\{f^n\}$ is spherically equicontinuous in D, and hence $D \subset J_{equ}(f)$. This is a contradiction. Thus S contains at most one point of $\bar{\kappa}$. Take $z \notin \text{Exc}(f)$. By Theorem 5.24, the backward orbit $O^-(z)$ of z is infinite, and hence
$$O^-(z) \cap O^+(D) \neq \emptyset$$

by the above argument. Thus, there exist some point w and some non-negative integers m and n such that $f^m(w) = z$ and $w \in f^n(D)$. It follows that $z \in f^{m+n}(D)$ which also proves the theorem. $\square$

Theorem 5.65. *Let f be a rational mapping with $\deg(f) \geq 2$. Suppose that $J_{equ}(f) \neq \emptyset$, and has no isolated points. Then $J_{equ}(f)$ is contained in the derived set of the set $\text{Per}(f)$ of periodic points of f. In particular, one has*
$$J_{equ}(f) \subset \overline{\text{Per}(f)}.$$

Proof. Let D be any open set that meets $J_{equ}(f)$. We will prove that D contains some element of $\text{Per}(f)$. We can choose a point $w_0 \in J_{equ}(f) \cap D$ such that w_0 is not a critical value of f^2. Thus, there are at least four distinct points in $f^{-2}(w)$. Take three of them, say w_1, w_2 and w_3, distinct from w, and construct neighborhoods D_i $(i = 0, 1, ..., 4)$ of w_i respectively, with disjoint closure in each other, such that $D_0 \subset D$ and that $f^2 : D_j \longrightarrow D_0$ is a homeomorphism for each $j = 1, 2, 3$.

Now let $\varphi_j : D_0 \longrightarrow D_j$ be the inverse of $f^2 : D_j \longrightarrow D_0$. If
$$f^n(z) \neq \varphi_j(z), \quad j = 1, 2, 3; \quad n \in \mathbb{Z}^+; \quad z \in D_0,$$

by Theorem 5.47 $\{f^n\}$ is spherically equicontinuous in D_0. This is impossible since $D_0 \cap J_{equ}(f) \neq \emptyset$. Hence, there exist some $z \in D_0$, some $j \in \{1, 2, 3\}$, and some $n \in \mathbb{Z}^+$, such that $f^n(z) = \varphi_j(z)$. This implies that
$$f^{2+n}(z) = f^2(\varphi_j(z)) = z$$

and, therefore, z is a periodic point in D, as required. $\square$

Theorem 5.66. *Let f be a rational mapping with $\deg(f) \geq 2$ and suppose that $J_{equ}(f) \neq \emptyset$.*
1) If $z \notin \text{Exc}(f)$, then $J_{equ}(f) \subset \overline{O^-(z)}$;
2) If $z \in J_{equ}(f)$, then $J_{equ}(f) = \overline{O^-(z)}$.

Proof. Take $z \notin \text{Exc}(f)$ and let D be any non-empty open set that meets $J_{equ}(f)$. Theorem 5.64 implies that $z \in f^n(D)$ for some n, and so $O^-(z) \cap D \neq \emptyset$. This proves (1). If $z \in J_{equ}(f)$, we obviously have
$$J_{equ}(f) \subset \overline{O^-(z)} \subset J_{equ}(f),$$

where the first relation follows from (1) and the second one comes from the complete invariance of the closed set $J_{equ}(f)$. Thus (2) is proved. $\square$

Theorem 5.67. *Let f and g be two rational mappings with $\deg(f) \geq 2$ and $\deg(g) \geq 2$. Suppose that*

$$f \circ g = g \circ f, \quad J_{equ}(f) \neq \emptyset, \quad J_{equ}(g) \neq \emptyset.$$

Then $J_{equ}(f) = J_{equ}(g)$.

Proof. We continue to follow Beardon's proof [5]. For any set $E \subset \bar{\kappa}$, define

$$\text{Diam}[E] = \sup_{z,w \in E} \chi(z, w).$$

By Theorem 5.46, f satisfies a Lipschitz condition

$$\chi(f(z), f(w)) \leq \lambda \chi(z, w)$$

on $\bar{\kappa}$. Take $w \in F_{equ}(g)$. Since $\{g^n\}$ is spherically equicontinuous at w, then for any $\varepsilon > 0$ there is a positive δ such that

$$\text{Diam}[g^n(\kappa(w; \delta))] < \frac{\varepsilon}{\lambda},$$

which implies

$$
\begin{aligned}
\text{Diam}[g^n \circ f(\kappa(w; \delta))] &= \text{Diam}[f \circ g^n(\kappa(w; \delta))] \\
&\leq \lambda \text{Diam}[g^n(\kappa(w; \delta))] < \varepsilon.
\end{aligned}
$$

It follows that $\{g^n\}$ is spherically equicontinuous at $f(w)$, and so, in particular, $f(w) \in F_{equ}(g)$. This proves that f and, hence, each f^n, maps $F_{equ}(g)$ into itself, so $\{f^n\}$ is spherically equicontinuous on $F_{equ}(g)$ by using Corollary 5.45. We conclude that $F_{equ}(g) \subset F_{equ}(f)$ and, by symmetry, $F_{equ}(g) = F_{equ}(f)$. $\qquad\square$

Conjecture 5.68. *Let f be a rational mapping with $\deg(f) \geq 2$. Then $J_{equ}(f)$ is the closure of the repelling periodic points of f.*

L. C. Hsia [56] also proved independently the results in Theorem 5.65 and Theorem 5.66.

5.8 Iteration of $z \mapsto z^d$

We study dynamical system of the following polynomial

$$f(z) = z^d, \quad d \geq 2$$

in an algebraically closed field κ of characteristic zero, complete for a non-trivial non-Archimedean absolute value $|\cdot|$. It is evident that the points 0 and ∞ are attractors with basins of attraction

$$\text{Att}(0) = \kappa(0; 1), \quad \text{Att}(\infty) = \kappa - \kappa[0; 1].$$

Thus main scenario is developed on the sphere $\kappa\langle 0; 1\rangle$. Obviously, we have

$$\text{Fix}(f) \cap \kappa\langle 0; 1\rangle = \Omega_{d-1}(\kappa).$$

Note that $|\cdot|$ is equivalent to a p-adic absolute value $|\cdot|_p$ for some prime p, that is, there exists a positive real number α satisfying

$$|z| = |z|_p^\alpha, \quad z \in \kappa.$$

Then

$$|f'(a)| = |d|_p^\alpha |a^{d-1}| = |d|_p^\alpha, \quad a \in \Omega_{d-1}(\kappa).$$

Therefore there are two essentially different cases: 1) d is not divisible by p; 2) d is divisible by p. Basic properties are surveyed in the following theorems (cf. [75]):

Theorem 5.69. *Assume* $(d,p) \neq 1$. *Then each point* a *of* $\Omega_{d-1}(\kappa)$ *is an attractor of* f *with* $\kappa(a;1) \subset \mathrm{Att}(a)$. *For any* $k \geq 2$, *each* k-*cycle* A *of* f *is also an attractor satisfying* $\kappa(b;1) \subset \mathrm{Att}(A)$ *for* $b \in A$.

Proof. Let $d = p^v d'$ with $v \geq 1$, $(d',p) = 1$. Then we have

$$|f'(a)| = p^{-\alpha v} < 1, \quad a \in \Omega_{d-1}(\kappa).$$

Thus each point a of $\Omega_{d-1}(\kappa)$ is an attractor of f. Note that when $z \in \kappa[a;r]$, we have

$$|f(z) - a| = \left| \sum_{j=1}^{d} \frac{1}{j!} \frac{d^j f}{dz^j}(a)(z-a)^j \right| \leq \lambda |z - a|,$$

where

$$\lambda = \max_{1 \leq j \leq d} \left| \frac{1}{j!} \frac{d^j f}{dz^j}(a) \right| r^{j-1} = \max_{1 \leq j \leq d} \left| \binom{d}{j} \right| r^{j-1}. \tag{5.29}$$

Then $\lambda < 1$ if $r < 1$, and hence

$$|f^n(z) - a| \leq \lambda^n |z - a|,$$

that is, $\kappa[a;r] \subset \mathrm{Att}(a)$ for any $r < 1$. Therefore $\kappa(a;1) \subset \mathrm{Att}(a)$.

To study k-cycles, we use the fact $(d^k,p) \neq 1$ if and only if $(d,p) \neq 1$. Thus each fixed point of f^k is an attractor, and the last claim in the theorem follows from the above argument. $\square$

Theorem 5.70. *Suppose* $(d,p) = 1$. *Then each point* a *of* $\Omega_{d-1}(\kappa)$ *is a center of a Siegel disc satisfying* $\mathrm{Sie}(a) = \kappa(a;1)$. *If* $d - 1 = p^l$ $(l \geq 1)$, *then*

$$\mathrm{Sie}(a) = \mathrm{Sie}(1) = \kappa(1;1), \quad a \in \Omega_{d-1}(\kappa).$$

If $(d-1,p) = 1$, *then*

$$\Omega_{d-1}(\kappa) - \{1\} \subset \kappa\langle 1;1 \rangle, \quad \mathrm{Sie}(a) \cap \mathrm{Sie}(b) = \emptyset \ (a,b \in \Omega_{d-1}(\kappa), a \neq b).$$

For any $k \geq 2$, *each element in any* k-*cycle of* f *is also a center of a Siegel disc of unit radius.*

Proof. Note that
$$|f'(a)| = |d|_p^\alpha = 1, \quad a \in \Omega_{d-1}(\kappa).$$

By Proposition 5.15, each $a \in \Omega_{d-1}(\kappa)$ is a center of a Siegel disc satisfying
$$\kappa[a; r] \subset \mathrm{Sie}(a),$$

where r is determined by
$$\max_{2 \leq j \leq d} \left| \frac{1}{j!} \frac{d^j f}{dz^j}(a) \right| r^{j-1} = \max_{2 \leq j \leq d} \left| \binom{d}{j} \right| r^{j-1} < 1,$$

which is satisfied when $r < 1$. Thus $\kappa(a; 1) \subset \mathrm{Sie}(a)$. Since we have
$$\mathrm{Att}(0) = \kappa(0; 1) \subset \kappa(a; r'), \quad \kappa(a; r') \cap \mathrm{Att}(\infty) \neq \emptyset$$

for any $r' > 1$, then it follows that $\kappa(a; 1) = \mathrm{Sie}(a)$.

Next we assume $d - 1 = p^l$ ($l \geq 1$). Take $a \in \Omega_{d-1}(\kappa)$. Then $a^{p^l} = a^{d-1} = 1$. Suppose $|a - 1| = 1$. We have
$$0 = |a^{p^l} - 1| = |a - 1| \cdot |(a-1)^{p^l-1} + \xi|,$$

where
$$\xi = \sum_{j=1}^{p^l-1} \binom{p^l}{j} (a-1)^{j-1}.$$

One claims that (cf. [75], Chapter VIII, Lemma 1.3)
$$\left| \binom{p^l}{j} \right|_p \leq \frac{1}{p}, \quad j = 1, ..., p^l - 1. \tag{5.30}$$

To prove (5.30), write $j = ip + q$ ($0 \leq q \leq p - 1$). If $q = 0$, one has
$$\left| \binom{p^l}{j} \right|_p = \left| \frac{p^l(p^l - p) \cdots (p^l - ip + p)}{p \cdots ip} \right|_p$$
$$= \left| \frac{p^l}{ip} \cdot \frac{p^l - p}{p} \cdots \frac{p^l - (i-1)p}{(i-1)p} \right|_p \leq \left| \frac{p^{l-1}}{i} \right|_p \leq \frac{1}{p}$$

as $i < p^{l-1}$. Now let $q \neq 0$. Then
$$\left| \binom{p^l}{j} \right|_p = \left| p^l \cdot \frac{p^l - p}{p} \cdots \frac{p^l - ip}{ip} \right|_p \leq \left| p^l \right|_p \leq \frac{1}{p}.$$

Hence the claim (5.30) is proved. Thus we have
$$|\xi| \leq \max_{1 \leq j \leq p^l - 1} \left| \binom{p^l}{j} \right| \leq \frac{1}{p^\alpha} < 1,$$

and so $|(a-1)^{p^l-1} + \xi| = 1$. This is a contradiction. Hence $|a - 1| < 1$, that is, $a \in \kappa(1; 1)$. Therefore
$$\mathrm{Sie}(a) = \kappa(a; 1) = \kappa(1; 1) = \mathrm{Sie}(1).$$

Finally we consider the case $(d-1,p) = 1$. According to Lemma 1.2 of Chapter VIII in [75], we prove the following fact: If $(n,p) = 1$, then

$$\Omega_{n,a}(\kappa) = \{y \mid y^n = a, \ y \neq a\} \subset \kappa\langle a; 1\rangle, \quad a \in \Omega_{n-1}(\kappa). \tag{5.31}$$

Take $y \in \Omega_{n,a}(\kappa)$. First we prove that $y \notin \kappa(a;1)$. Suppose that $y \in \kappa(a;1)$, i.e., $x = y - a$ satisfies $|x| < 1$. Since $a^n = a$, we have

$$0 = |y^n - a^n| = \left| \sum_{j=1}^{n} \binom{n}{j} x^j a^{n-j} \right| = |x| \cdot |na^{n-1} + x\eta|,$$

where

$$\eta = \sum_{j=2}^{n} \binom{n}{j} x^{j-2} a^{n-j}$$

satisfying

$$|\eta| \leq \max_{2 \leq j \leq d} \left| \binom{n}{j} \right| |x|^{j-2} \leq 1.$$

Then $|na^{n-1} + x\eta| = 1$ since $|na^{n-1}| = 1$. We obtain $x = 0$, however, this contradicts the condition $y \neq a$. It follows that $y \notin \kappa(a;1)$, and so $y \in \kappa\langle a;1\rangle$. This fact implies

$$\Omega_{d-1}(\kappa) - \{1\} = \Omega_{d-1,1}(\kappa) \subset \kappa\langle 1;1\rangle,$$

which further yields

$$\mathrm{Sie}(a) \cap \mathrm{Sie}(b) = \emptyset \ (a,b \in \Omega_{d-1}(\kappa), a \neq b).$$

Otherwise, we have $\kappa(a;1) = \kappa(b;1)$, that is, $|a - b| < 1$. Note that $c = b/a \in \Omega_{d-1}(\kappa) - \{1\}$. We obtain

$$|a - b| = |a| \cdot |c - 1| = 1.$$

This is a contradiction.

To study k-cycles, we use the fact $(d^k, p) = 1$ if and only if $(d, p) = 1$. Thus each fixed point of f^k is indifferent, and hence is a center of a Siegel disc of unit radius by the above argument. □

Corollary 5.71. *Suppose* $(d,p) = 1$, $d^k - 1 = p^l$ $(l \geq 1)$. *Then any* k-*cycle* $A = \{x_0, x_1, ..., x_{k-1}\}$ *of* f *is located in the ball* $\kappa(1;1)$, *which has the behaviour of a Siegel disc with*

$$\mathrm{Sie}(A) = \bigcup_{j=0}^{k-1} \kappa(x_j; 1) = \kappa(1;1).$$

Theorem 5.72. *Suppose* $(d,p) = 1$ *and write* $d-1 = mp^l$ *with* $(m,p) = 1$ *and* $l \geq 0$. *Then each point* ab *with* $a \in \Omega_m(\kappa), b \in \Omega_{p^l}(\kappa)$ *is a center of a Siegel disc satisfying*

$$\mathrm{Sie}(ab) = \kappa(ab; 1) = \kappa(a; 1).$$

If $a \in \Omega_m(\kappa) - \{1\}$, *then*

$$\mathrm{Sie}(ab) \subset \kappa\langle 1;1\rangle \cap \kappa\langle 0;1\rangle.$$

Further

$$\mathrm{Sie}(ab) \cap \mathrm{Sie}(a'b') = \emptyset \ (a,a' \in \Omega_m(\kappa), b,b' \in \Omega_{p^l}(\kappa), a \neq a').$$

Proof. Take $a \in \Omega_m(\kappa)$ and $b \in \Omega_{p^l}(\kappa)$. By Theorem 5.70, the point ab is a center of a Siegel disc satisfying $\mathrm{Sie}(ab) = \kappa(ab; 1)$. Next we show $c = ab \in \kappa(a; 1)$. Suppose, to the contrary, that $c \notin \kappa(a; 1)$. Then $|c - a| = 1$. Note that

$$0 = |c^{p^l} - a^{p^l}| = |c - a| \cdot |(c - a)^{p^l - 1} + \xi|,$$

where

$$\xi = \sum_{j=1}^{p^l - 1} \binom{p^l}{j} (c - a)^{j-1} a^{p^l - j}$$

with $|\xi| < 1$. This is a contradiction. Hence $ab \in \kappa(a; 1)$, and so we have $\mathrm{Sie}(ab) = \kappa(a; 1)$.

Secondly, we assume $a \neq 1$. By the fact (5.31), we have $a \in \Omega_{m,1}(\kappa) \subset \kappa\langle 1; 1 \rangle$ since $(m, p) = 1$. Therefore $\kappa(a; 1) \subset \kappa\langle 1; 1 \rangle$, that is, $\mathrm{Sie}(ab) \subset \kappa\langle 1; 1 \rangle$.

Finally, if $a \neq a'$, then $a/a' \in \kappa\langle 1; 1 \rangle$ by above argument. Thus we have

$$|a - a'| = |a'| \cdot \left| \frac{a}{a'} - 1 \right| = 1,$$

which means

$$\kappa(a; 1) \cap \kappa(a'; 1) = \emptyset \quad (a \neq a'),$$

that is,

$$\mathrm{Sie}(ab) \cap \mathrm{Sie}(a'b') = \emptyset \quad (a, a' \in \Omega_m(\kappa), b, b' \in \Omega_{p^l}(\kappa), a \neq a').$$

Hence the theorem is proved. $\square$

5.9 Iteration of $z \mapsto z^2 + c$

Let κ be an algebraically closed field of characteristic zero, complete for a non-trivial non-Archimedean absolute value $|\cdot|$. Then $|\cdot|$ is equivalent to a p-adic absolute value $|\cdot|_p$ for some prime p, that is, there exists a positive real number α satisfying

$$|z| = |z|_p^\alpha, \quad z \in \kappa.$$

In this section, we will study dynamical system of the following polynomial

$$f_c(z) = z^2 + c, \quad c \in \kappa.$$

It is evident that the point ∞ is an attractor. If $|z| > |c|^\vee = \max\{1, |c|\}$, then by induction, we can obtain

$$|f_c^n(z)| = |z|^{2^n} \to \infty \quad (n \to \infty).$$

Therefore

$$\kappa - \kappa[0; |c|^\vee] \subset \mathrm{Att}(\infty).$$

Theorem 5.73. *If* $p \neq 2$, *then* f_c *has only one attracting fixed point if and only if* $c \in \kappa(0; 1)$.

Proof. Note that f_c has two fixed points, say a and b, in κ, and these are solutions of

$$z^2 - z + c = 0, \tag{5.32}$$

they satisfy

$$a + b = 1, \quad ab = c.$$

This shows that

$$f_c'(a) + f_c'(b) = 2,$$

and this in turn shows that not both a and b can be attracting for if they are,

$$1 = |2| = |f_c'(a) + f_c'(b)| \leq \max\{|f_c'(a)|, |f_c'(b)|\} < 1,$$

which is impossible. It follows that f_c can have at most one attracting fixed point, and the condition that one of the fixed points, say a, is attracting is

$$|a| = |2a| = |f_c'(a)| < 1.$$

However, from (5.32), $c = a - a^2$, and so $|c| = |a - a^2| = |a| < 1$. Hence $c \in \kappa(0; 1)$. Conversely, if $c \in \kappa(0; 1)$, then

$$|f_c'(a) f_c'(b)| = |4ab| = |c| < 1.$$

Thus we have either $|f_c'(a)| < 1$ or $|f_c'(b)| < 1$, that is, one and only one of a and b is attracting. $\qquad\square$

If $p = 2$, we have

$$|f_c'(a) + f_c'(b)| = |2| = 2^{-\alpha} < 1.$$

Thus if a is attracting, then b also is attracting. Similarly, we can show the following fact:

Theorem 5.74. *If $p = 2$, then f_c has two attracting fixed points if and only if $c \in \kappa(0; |2|^{-2})$.*

Let us describe basins of attraction for f_c. Let a be an attracting fixed point of f_c. Take $r \in \mathbb{R}^+$ with

$$r = \max_{j \geq 2} \left| \frac{1}{j!} \frac{d^j f_c}{dz^j}(a) \right| r^{j-1} < |f_c'(a)| = |2a|.$$

According to the discussion in § 5.1, we have $\kappa[a; r] \subset \mathrm{Att}(a)$. It follows that

$$\kappa(a; |2a|) \subset \mathrm{Att}(a).$$

Theorem 5.75. *The quadratic polynomial f_c has an attracting 2-cycle if and only if $c \in \kappa(-1; |2|^{-2})$.*

Proof. Set $\mathrm{Fix}(f_c) = \{a, b\}$. Then a and b are also fixed by f_c^2. Thus $f_c(z) - z$ divides $f_c^2(z) - z$ and we can write

$$f_c^2(z) - z = (z^2 - z + c)(z^2 + z + 1 + c) = (z - a)(z - b)(z - x_0)(z - x_1).$$

By definition, $\{x_0, x_1\}$ is an attracting 2-cycle if and only if

$$f_c(x_0) = x_1, \ f_c(x_1) = x_0, \ x_0 \neq x_1, \ |(f_c^2)'(x_i)| < 1 \ (i = 0, 1).$$

Note that

$$(f_c^2)'(x_0) = f_c'(f_c(x_0))f_c'(x_0) = f_c'(x_1)f_c'(x_0) = 4x_0 x_1 = 4(1 + c).$$

Therefore the set of c we are seeking is just the disc $\kappa(-1; |2|^{-2})$. $\qquad\square$

To search for k-cycles, we can divide $f_c^k(z) - z$ by $f_c(z) - z$, but the resulting polynomial has degree $2^k - 2$ (and this is at least 6 when $k \geq 3$), so that this analysis does not seem to carry us any further. For the case $k = 3$, we can obtain a part of answer by the equation

$$\begin{aligned}
f_c^3(z) - z \ = \ & (z^2 - z + c)\{z^6 + z^5 + (3c + 1)z^4 + (2c + 1)z^3 \\
& + (3c^2 + 3c + 1)z^2 + (c + 1)^2 z + 1 + c(c + 1)^2\}.
\end{aligned}$$

For example, if f_c has two attracting 3-cycles, then c must satisfy

$$|1 + c(c + 1)^2| < |2|^{-6}.$$

Let $\mathcal{Y}_k$ be the subset of κ such that f_c has at least an attracting k-cycle when $c \in \mathcal{Y}_k$. Set

$$\mathcal{Y} = \bigcup_{k=1}^{\infty} \mathcal{Y}_k.$$

By the above argument, we find $\mathcal{Y}_1 \subset \kappa[0; 1]$ and $\mathcal{Y}_2 \subset \kappa[0; 1]$ when $(p, 2) = 1$. Here we suggest the following problem:

Conjecture 5.76. *If* $(p, 2) = 1$, *then* $\mathcal{Y} \subset \kappa[0; 1]$, *and further*

$$\mathcal{Y} \subset \mathcal{Y}^* := \{c \in \kappa \mid f_c^n(0) \text{ is bounded }\}.$$

Take $c \in \mathcal{Y}^*$. It is easy to prove $|c| \leq 1$, that is,

$$\mathcal{Y}^* \subset \kappa[0; 1].$$

Assume $(p, 2) = 1$ and write
$$\text{Fix}(f_c^n) = \{a_1, ..., a_{2^n}\}.$$

Then we have
$$|f_c'(a_1) \cdots f_c'(a_{2^n})| = |a_1 \cdots a_{2^n}| = |f_c^n(0)| \leq 1.$$

Thus if $c \notin \mathcal{Y}$, then all cycles of f_c are indifferent.

Theorem 5.77. *Assume* $(p, 2) = 1$. *If* $c \in \mathcal{Y}^* - \mathcal{Y}$, *then all cycles of* f_c *are indifferent.*

Chapter 6

Holomorphic curves

We will introduce the value distribution theory of holomorphic curves defined on a non-Archimedean algebraically closed field of characteristic zero.

6.1 Multilinear algebra

Let κ be a number field with a non-trivial absolute value $|\cdot|$. Let V be a vector space of finite dimension $n+1 > 0$ over κ. Write the projective space $\mathbb{P}(V) = V/\kappa_*$ and let $\mathbb{P} : V - \{0\} \longrightarrow \mathbb{P}(V)$ be the standard projection. If $S \subset V$, abbreviate

$$\mathbb{P}(S) = \mathbb{P}(S \cap V_*), \quad V_* = V - \{0\}.$$

The dual vector space V^* of V consists of all κ-linear functions $\alpha : V \longrightarrow \kappa$, and we shall call

$$\langle \xi, \alpha \rangle = \alpha(\xi)$$

the *inner product* of $\xi \in V$ and $\alpha \in V^*$. If $\alpha \neq 0$, the n-dimensional linear subspace

$$E[a] = E[\alpha] = \mathrm{Ker}(\alpha) = \alpha^{-1}(0)$$

depends on $a = \mathbb{P}(\alpha) \in \mathbb{P}(V^*)$ only, and $\ddot{E}[a] = \mathbb{P}(E[a])$ is a *hyperplane* in $\mathbb{P}(V)$. Thus $\mathbb{P}(V^*)$ bijectively parameterizes the hyperplances in $\mathbb{P}(V)$.

Take a base $e = (e_0, ..., e_n)$ for V. For $\xi = \xi_0 e_0 + \cdots + \xi_n e_n \in V$, define the norm

$$|\xi| = |\xi|_e = \begin{cases} \left(|\xi_0|^2 + \cdots + |\xi_n|^2 \right)^{\frac{1}{2}} & : \quad \text{if } |\cdot| \text{ is Archimedean} \\ \max_{0 \leq i \leq n} \{ |\xi_i| \} & : \quad \text{if } |\cdot| \text{ is non-Archimedean} . \end{cases}$$

Obviously, the norm depends on the base e, and will be called a *norm over the base e*. If $|\cdot|_{e'}$ is another norm over a base $e' = (e_0', ..., e_n')$, it is easy to prove

$$c|\xi|_e \leq |\xi|_{e'} \leq c'|\xi|_e$$

for all $\xi \in V$, i.e., norms over bases are equivalent.

177

Let $\epsilon = (\epsilon_0, ..., \epsilon_n)$ be the dual base of $e = (e_0, ..., e_n)$. Then the norm on V induces a norm on V^* defined by

$$|\alpha| = |\alpha|_e = \begin{cases} (|\alpha_0|^2 + \cdots + |\alpha_n|^2)^{\frac{1}{2}} & : \quad \text{if } |\cdot| \text{ is Archimedean} \\ \max_{0 \le i \le n}\{|\alpha_i|\} & : \quad \text{if } |\cdot| \text{ is non-Archimedean} \end{cases},$$

where $\alpha = \alpha_0 \epsilon_0 + \cdots + \alpha_n \epsilon_n$. Schwarz inequality

$$|\langle \xi, \alpha \rangle| \le |\xi| \cdot |\alpha|$$

holds for $\xi \in V$, $\alpha \in V^*$. The distance from $x = \mathbb{P}(\xi)$ to $\ddot{E}[a]$ with $a = \mathbb{P}(\alpha) \in \mathbb{P}(V^*)$ is defined by

$$0 \le |x, a| = \frac{|\langle \xi, \alpha \rangle|}{|\xi| \cdot |\alpha|} \le 1.$$

Take non-negative integers a and b with $a \le b$. Let J_a^b be the set of all increasing injective mappings $\lambda : \mathbb{Z}[0, a] \longrightarrow \mathbb{Z}[0, b]$, where

$$\mathbb{Z}[m, n] = \{i \in \mathbb{Z} \mid m \le i \le n\}.$$

Then $J_b^b = \{\iota\}$, where ι is the inclusion mapping. If $a < b$, there exists one and only one $\lambda^\perp \in J_{b-a-1}^b$ for each $\lambda \in J_a^b$ such that $\text{Im}\lambda \cap \text{Im}\lambda^\perp = \emptyset$. The mapping $\perp: J_a^b \longrightarrow J_{b-a-1}^b$ is bijective. A permutation $(\lambda, \lambda^\perp)$ of $\mathbb{Z}[0, b]$ is defined by

$$(\lambda, \lambda^\perp)(i) = \begin{cases} \lambda(i) & : \quad i \in \mathbb{Z}[0, a] \\ \lambda^\perp(i - a - 1) & : \quad i \in \mathbb{Z}[a + 1, b] \end{cases}.$$

The signature of the permutation is denoted by $\text{sign}(\lambda, \lambda^\perp)$.

Identify $V^{**} = V$ by $\langle \xi, \alpha \rangle = \langle \alpha, \xi \rangle$ and $\left(\bigwedge_{k+1} V \right)^* = \bigwedge_{k+1} V^*$ by

$$\langle \xi_0 \wedge \cdots \wedge \xi_k, \alpha_0 \wedge \cdots \wedge \alpha_k \rangle = \det((\langle \xi_i, \alpha_j \rangle)),$$

where $\bigwedge_{k+1} V$ is the *exterior product of* V *of order* $k+1$. The norm on V also induces norms on $\bigwedge_{k+1} V$ and $\bigwedge_{k+1} V^*$. Take $\xi \in \bigwedge_{k+1} V$, $\alpha \in \bigwedge_{k+1} V^*$ and write

$$\xi = \sum_{\lambda \in J_k^n} \xi_\lambda e_\lambda, \qquad \alpha = \sum_{\lambda \in J_k^n} \alpha_\lambda \epsilon_\lambda,$$

where

$$e_\lambda = e_{\lambda(0)} \wedge \cdots \wedge e_{\lambda(k)}.$$

Then we can define the norms

$$|\xi| = |\xi|_e = \begin{cases} \left(\sum_{\lambda \in J_k^n} |\xi_\lambda|^2 \right)^{\frac{1}{2}} & : \quad \text{if } |\cdot| \text{ is Archimedean} \\ \max_{\lambda \in J_k^n}\{|\xi_\lambda|\} & : \quad \text{if } |\cdot| \text{ is non-Archimedean} \end{cases}$$

and

$$|\alpha| = |\alpha|_e = \begin{cases} \left(\sum_{\lambda \in J_k^n} |\alpha_\lambda|^2\right)^{\frac{1}{2}} & : \quad \text{if } |\cdot| \text{ is Archimedean} \\ \max_{\lambda \in J_k^n}\{|\alpha_\lambda|\} & : \quad \text{if } |\cdot| \text{ is non-Archimedean} \ . \end{cases}$$

Generally, let $V_1, ..., V_m$ and W be normed vector spaces over κ. Let

$$\odot : V_1 \times \cdots \times V_m \longrightarrow W$$

be an m-linear mapping over κ. If $\xi = (\xi_1, ..., \xi_m) \in V_1 \times \cdots \times V_m$, we write

$$\odot(\xi) = \xi_1 \odot \cdots \odot \xi_m,$$

and say that ξ is *free for the operation* $\odot$ if $\odot(\xi) \neq 0$. Take $x_j \in \mathbb{P}(V_j)$ $(j = 1, ..., m)$. We will say that $x_1, ..., x_m$ are *free for* $\odot$ if there exist $\xi_j \in V_j$ such that $x_j = \mathbb{P}(\xi_j)$ and $\xi = (\xi_1, ..., \xi_m)$ is free for the operation $\odot$. For free $x_1, ..., x_m$, we can define

$$x_1 \odot \cdots \odot x_m = \mathbb{P}(\xi_1 \odot \cdots \odot \xi_m).$$

Also, the *gauge of* $x_1, ..., x_m$ *for* $\odot$ is defined to be

$$|x_1 \odot \cdots \odot x_m| = \frac{|\xi_1 \odot \cdots \odot \xi_m|}{|\xi_1| \cdots |\xi_m|}$$

which is well defined. If $x_1, ..., x_m$ are not free for $\odot$, we define $|x_1 \odot \cdots \odot x_m| = 0$. In the following, we will give several gauges.

Example 6.1. *Take* $V_1 = \cdots = V_{k+1} = V$ $(k \leq n)$, $W = \bigwedge_{k+1} V$ *the exterior product of* V *of order* $k+1$, *and let* $\odot = \wedge$ *be the exterior product. Let* $|\ |$ *be the norm of* V *over a base* e. *Thus, for* $x_j \in \mathbb{P}(V)$ $(j = 0, ..., k)$, *the gauge* $|x_0 \wedge \cdots \wedge x_k|$ *of* $x_0, ..., x_k$ *for* $\wedge$ *is well-defined with* $0 \leq |x_0 \wedge \cdots \wedge x_k| \leq 1$.

Take $k, l \in \mathbb{Z}[0, n]$ and take $\xi \in \bigwedge_{k+1} V$ and $\alpha \in \bigwedge_{l+1} V^*$. If $k \geq l$, the *interior product* $\xi \angle \alpha \in \bigwedge_{k-l} V$ is uniquely defined by

$$\langle \xi \angle \alpha, \beta \rangle = \langle \xi, \alpha \wedge \beta \rangle$$

for all $\beta \in \bigwedge_{k-l} V^*$. If $k = l$, then

$$\xi \angle \alpha = \langle \xi, \alpha \rangle \in \kappa = \bigwedge_0 V$$

by definition. If $k < l$, we define the *interior product* $\xi \angle \alpha \in \bigwedge_{l-k} V^*$ such that if $\eta \in \bigwedge_{l-k} V$,

$$\langle \eta, \xi \angle \alpha \rangle = \langle \xi \wedge \eta, \alpha \rangle.$$

Lemma 6.2. *Take* $k, l \in \mathbb{Z}[0, n]$ *with* $k \geq l$ *and take* $\xi \in \bigwedge_{k+1} V$ *and* $\alpha \in \bigwedge_{l+1} V^*$. *Then we have Schwarz's inequality*

$$|\xi \angle \alpha| \leq |\xi| \cdot |\alpha|.$$

Proof. Here we only consider non-Archimedean cases. The Archimedean cases can be proved similarly. Write

$$\xi = \sum_{\lambda \in J_k^n} \xi_\lambda e_\lambda, \quad \alpha = \sum_{\lambda \in J_l^n} \alpha_\lambda e_\lambda.$$

If $l = k$, noting that

$$\xi \angle \alpha = \langle \xi, \alpha \rangle = \sum_{\lambda \in J_k^n} \xi_\lambda \alpha_\lambda,$$

we have

$$|\xi \angle \alpha| \le \max_{\lambda \in J_k^n} |\xi_\lambda \alpha_\lambda| \le \left(\max_{\lambda \in J_k^n} |\xi_\lambda| \right) \cdot \left(\max_{\lambda \in J_k^n} |\alpha_\lambda| \right) = |\xi| \cdot |\alpha|$$

and so the inequality follows. If $l < k$, by the Laplace's theorem of determinant expansion

$$
\begin{aligned}
\langle (e_0 \wedge \cdots \wedge e_k) \angle \alpha, \beta \rangle &= \langle e_0 \wedge \cdots \wedge e_k, \alpha \wedge \beta \rangle \\
&= \sum_{\nu \in J_l^k} \mathrm{sign}(\nu, \nu^\perp) \langle e_\nu, \alpha \rangle \langle e_{\nu^\perp}, \beta \rangle \\
&= \left\langle \sum_{\nu \in J_l^k} \mathrm{sign}(\nu, \nu^\perp) \langle e_\nu, \alpha \rangle e_{\nu^\perp}, \beta \right\rangle
\end{aligned}
$$

holds for any $\beta \in \bigwedge_{k-l} V^*$, that is,

$$(e_0 \wedge \cdots \wedge e_k) \angle \alpha = \sum_{\nu \in J_l^k} \mathrm{sign}(\nu, \nu^\perp) \langle e_\nu, \alpha \rangle e_{\nu^\perp}. \tag{6.1}$$

Then

$$|(e_0 \wedge \cdots \wedge e_k) \angle \alpha| = \max_{\nu \in J_l^k} \{ |\langle e_\nu, \alpha \rangle| \} \le |\alpha|.$$

Thus, we have

$$|\xi \angle \alpha| = \left| \sum_{\lambda \in J_k^n} \xi_\lambda e_\lambda \angle \alpha \right| \le \max_{\lambda \in J_k^n} \{ |\xi_\lambda| |e_\lambda \angle \alpha| \} \le |\xi| \cdot |\alpha|.$$

$\square$

Now assume $\xi \ne 0$ and $\alpha \ne 0$ and set $x = \mathbb{P}(\xi) \in \mathbb{P}\left(\bigwedge_{k+1} V \right)$ and $a = \mathbb{P}(\alpha) \in \mathbb{P}\left(\bigwedge_{l+1} V^* \right)$. We can define the *gauge of x and a for $\angle$*

$$|x \angle a| = \frac{|\xi \angle \alpha|}{|\xi| \cdot |\alpha|} \le 1.$$

In particular, if $k = l = 0$, then $|x \angle a| = |x, a|$. The projective space $\mathbb{P}\left(\bigwedge_{n+1} V^* \right)$ consists of one and only one point denoted by ∞.

Lemma 6.3. For all $x \in \mathbb{P}(V)$, $|x \angle \infty| = 1$.

Proof. We only consider non-Archimedean cases. Take $\xi \in V - \{0\}$ with $x = \mathbb{P}(\xi)$. Put $\xi_j = \langle \xi, \epsilon_j \rangle$. For $j \in \mathbb{Z}[0, n]$, define

$$\hat{\epsilon}_j = (-1)^j \epsilon_0 \wedge \cdots \wedge \epsilon_{j-1} \wedge \epsilon_{j+1} \wedge \cdots \wedge \epsilon_n.$$

Therefore, we have

$$|\xi \angle (\epsilon_0 \wedge \cdots \wedge \epsilon_n)| = \left| \sum_{j=0}^{n} \langle \xi, \epsilon_j \rangle \hat{\epsilon}_j \right| = \max_{0 \leq j \leq n} \{|\xi_j|\} = |\xi|.$$

Since $\infty = \mathbb{P}(\epsilon_0 \wedge \cdots \wedge \epsilon_n)$, then

$$|x \angle \infty| = \frac{|\xi \angle (\epsilon_0 \wedge \cdots \wedge \epsilon_n)|}{|\xi| \cdot |\epsilon_0 \wedge \cdots \wedge \epsilon_n|} = 1.$$

$\square$

Lemma 6.4. *For $x \in \mathbb{P}(V)$, $a_j \in \mathbb{P}(V^*), j = 0, 1, ..., n$, then*

$$|a_0 \wedge \cdots \wedge a_n| \leq \varsigma \max_{0 \leq j \leq n} |x, a_j|,$$

where

$$\varsigma = \begin{cases} \sqrt{n+1} & : \quad if \, |\cdot| \text{ is Archimedean} \\ 1 & : \quad if \, |\cdot| \text{ is non-Archimedean} \end{cases}.$$

Proof. We only prove the lemma for non-Archimedean cases. If $|a_0 \wedge \cdots \wedge a_n| = 0$, the inequality is trivial. If $|a_0 \wedge \cdots \wedge a_n| > 0$, then $a_0 \wedge \cdots \wedge a_n = \infty$. Thus Lemma 6.3 implies $|x \angle (a_0 \wedge \cdots \wedge a_n)| = 1$. For each $j \in \mathbb{Z}[0, n]$, take $\alpha_j \in V^* - \{0\}$ with $\mathbb{P}(\alpha_j) = a_j$. Also take $\xi \in V - \{0\}$ with $\mathbb{P}(\xi) = x$. We have

$$\begin{aligned} |a_0 \wedge \cdots \wedge a_n| &= |a_0 \wedge \cdots \wedge a_n| \cdot |x \angle (a_0 \wedge \cdots \wedge a_n)| \\ &= \frac{|\alpha_0 \wedge \cdots \wedge \alpha_n|}{|\alpha_0| \cdots |\alpha_n|} \cdot \frac{|\xi \angle (\alpha_0 \wedge \cdots \wedge \alpha_n)|}{|\xi| |\alpha_0 \wedge \cdots \wedge \alpha_n|} \\ &= \frac{|\sum_{j=0}^{n} \langle \xi, \alpha_j \rangle \hat{\alpha}_j|}{|\xi| |\alpha_0| \cdots |\alpha_n|} \\ &\leq \max_{0 \leq j \leq n} \frac{|\langle \xi, \alpha_j \rangle| |\hat{\alpha}_j|}{|\xi| |\alpha_0| \cdots |\alpha_n|} \\ &= \max_{0 \leq j \leq n} |x, a_j| |a_0 \wedge \cdots \wedge a_{j-1} \wedge a_{j+1} \wedge \cdots \wedge a_n| \\ &\leq \max_{0 \leq j \leq n} |x, a_j|. \end{aligned}$$

This finishes the proof.

$\square$

Let $\mathcal{J}_m$ be the permutation group on $\mathbb{Z}[1, m]$ and let $\otimes_m V$ be the m-fold tensor product of V. For each $\lambda \in \mathcal{J}_m$, a linear isomorphism $\lambda : \otimes_m V \longrightarrow \otimes_m V$ is uniquely defined by

$$\lambda(\xi_1 \otimes \cdots \otimes \xi_m) = \xi_{\lambda^{-1}(1)} \otimes \cdots \otimes \xi_{\lambda^{-1}(m)}, \quad \xi_j \in V \ (j = 1, ..., m).$$

A vector $\xi \in \otimes_m V$ is said to be *symmetric* if $\lambda(\xi) = \xi$ for all $\lambda \in \mathcal{J}_m$. The set of all symmetric vectors in $\otimes_m V$ is a linear subspace of $\otimes_m V$, denoted by $\amalg_m V$, called the *m-fold symmetric tensor product of V*. Then

$$\dim \amalg_m V = \binom{n+m}{m}.$$

The linear mapping

$$S_m = \frac{1}{m!} \sum_{\lambda \in \mathcal{J}_m} \lambda : \otimes_m V \longrightarrow \otimes_m V$$

is called the *symmetrizer of $\otimes_m V$* with $\mathrm{Im} S_m = \amalg_m V$. If $\xi \in \amalg_m V$ and $\eta \in \amalg_l V$, the symmetric tensor product

$$\xi \amalg \eta = \eta \amalg \xi = S_{m+l}(\xi \otimes \eta)$$

is defined. For $\xi_j \in V$ $(j = 1, ..., m)$, let $\xi_1 \amalg \cdots \amalg \xi_m$ be the symmetric tensor product $S_m(\xi_1 \otimes \cdots \otimes \xi_m)$. For $\xi \in V$, let $\xi^{\amalg m}$ be the m-th symmetric tensor power, and define

$$x^{\amalg m} = \mathbb{P}(\xi^{\amalg m})$$

for $x = \mathbb{P}(\xi)$. We can identify $\amalg_m V^* = (\amalg_m V)^*$ by

$$\langle \xi_1 \amalg \cdots \amalg \xi_m, \alpha_1 \amalg \cdots \amalg \alpha_m \rangle = \frac{1}{m!} \sum_{\lambda \in \mathcal{J}_m} \langle \xi_1, \alpha_{\lambda(1)} \rangle \cdots \langle \xi_m, \alpha_{\lambda(m)} \rangle$$

for all $x_j \in V, \alpha_j \in V^*, j = 1, ..., m$.

Let $J_{n,m}$ be the set of all mappings $\lambda : \mathbb{Z}[0,n] \longrightarrow \mathbb{Z}[0,m]$ such that

$$|\lambda| = \lambda(0) + \cdots + \lambda(n) = m.$$

For $\lambda \in J_{n,m}, e = (e_0, ..., e_n) \in V^{n+1}$, define

$$\lambda! = \lambda(0)! \cdots \lambda(n)!, \quad e^{\amalg \lambda} = e_0^{\amalg \lambda(0)} \amalg \cdots \amalg e_n^{\amalg \lambda(n)} \in \amalg_m V.$$

If $e = (e_0, ..., e_n)$ is a base of V, then $\{e^{\amalg \lambda}\}_{\lambda \in J_{n,m}}$ is a base of $\amalg_m V$, and $\{\frac{m!}{\lambda!} \epsilon^{\amalg \lambda}\}_{\lambda \in J_{n,m}}$ is the dual base of $\amalg_m V^*$, where $\epsilon = (\epsilon_0, ..., \epsilon_n)$ is the dual of e. The norm on V induces norms on $\amalg_m V$ and $\amalg_m V^*$ as follows: For $\eta \in \amalg_m V, \beta \in \amalg_m V^*$ with

$$\eta = \sum_{\lambda \in J_{n,m}} \eta_\lambda e^{\amalg \lambda}, \quad \beta = \sum_{\lambda \in J_{n,m}} \beta_\lambda \frac{m!}{\lambda!} \epsilon^{\amalg \lambda},$$

define

$$|\eta| = |\eta|_e = \begin{cases} \left(\sum_{\lambda \in J_{n,m}} |\eta_\lambda|^2 \right)^{\frac{1}{2}} & : \quad \text{if } |\cdot| \text{ is Archimedean} \\ \max_{\lambda \in J_{n,m}} \{|\eta_\lambda|\} & : \quad \text{if } |\cdot| \text{ is non-Archimedean} \end{cases}$$

and

$$|\beta| = |\beta|_e = \begin{cases} \left(\sum_{\lambda \in J_{n,m}} |\beta_\lambda|^2 \right)^{\frac{1}{2}} & : \quad \text{if } |\cdot| \text{ is Archimedean} \\ \max_{\lambda \in J_{n,m}} \{|\beta_\lambda|\} & : \quad \text{if } |\cdot| \text{ is non-Archimedean} . \end{cases}$$

Note that

$$\xi^{\amalg m} = \sum_{\lambda \in J_{n,m}} \xi_0^{\lambda(0)} \cdots \xi_n^{\lambda(n)} e^{\amalg \lambda}, \quad \alpha^{\amalg m} = \sum_{\lambda \in J_{n,m}} \alpha_0^{\lambda(0)} \cdots \alpha_n^{\lambda(n)} \frac{m!}{\lambda!} \epsilon^{\amalg \lambda},$$

for

$$\xi = \xi_0 e_0 + \cdots + \xi_n e_n \in V, \quad \alpha = \alpha_0 \epsilon_0 + \cdots + \alpha_n \epsilon_n.$$

Then

$$|\xi^{\amalg m}| = |\xi|^m, \quad |\alpha^{\amalg m}| = |\alpha|^m.$$

Let $V_{[m]}$ be the vector space of all homogeneous polynomials of degree m on V. We obtain a linear isomorphism

$$\sim : \amalg_m V^* \longrightarrow V_{[m]}$$

defined by

$$\tilde{\alpha}(\xi) = \langle \xi^{\amalg m}, \alpha \rangle, \quad \xi \in V, \ \alpha \in \amalg_m V^*.$$

Thus if $\xi \neq 0$ and $\alpha \neq 0$, the distance $|x^{\amalg m}, a|$ is well defined for $x^{\amalg m} = \mathbb{P}(\xi^{\amalg m})$ and $a = \mathbb{P}(\alpha)$. If $\alpha \neq 0$, the n-dimensional subspace

$$E[a] = \mathrm{Ker}(\tilde{\alpha}) = \tilde{\alpha}^{-1}(0)$$

depends on a only, and $\ddot{E}[a] = \mathbb{P}(E[a])$ is a *hypersurface of degree m* in $\mathbb{P}(V)$. Thus $\mathbb{P}(\amalg_m V^*)$ bijectively parameterizes the hypersurfaces in $\mathbb{P}(V)$.

Take $\tilde{\alpha} \in V_{[m]}$ and $\xi \in V_*$. Then the property of $\tilde{\alpha}(\xi)$ being zero or not depends only on the equivalence class $x = \mathbb{P}(\xi)$ of ξ. Thus $\tilde{\alpha}$ gives a function from $\mathbb{P}(V)$ to $\{0, 1\}$ by

$$\tilde{\alpha}(x) = \begin{cases} 0 & : \quad \text{if } \tilde{\alpha}(\xi) = 0 \\ 1 & : \quad \text{if } \tilde{\alpha}(\xi) \neq 0 \end{cases}$$

Now we can talk about the zeros of a homogeneous polynomial. A subset $M \subset \mathbb{P}(V)$ is an *algebraic set* if there exists a set T of homogeneous polynomials on V such that

$$M = \{x \in \mathbb{P}(V) \mid \tilde{\alpha}(x) = 0 \text{ for all } \tilde{\alpha} \in T\}.$$

Obviously, the union of two algebraic sets is an algebraic set. The intersection of any family of algebraic sets is an algebraic set. The empty set and the whole space are algebraic sets. Thus we can define the *Zariski topology* on $\mathbb{P}(V)$ by taking the open sets to be the complements of algebraic sets. A *projective algebraic variety* (or simply *projective variety*) is an irreducible algebraic set in $\mathbb{P}(V)$ with the induced topology, where, by definition, a nonempty subset A of a topological space X is *irreducible* if it cannot be expressed as the union $A = A_1 \cup A_2$ of two proper subsets, each one of which is closed in X. A *projective subvariety* of a projective svariety is a subset given locally as the zeros of a finite collection of homogeneous polynomials on V.

Let $\mathscr{A} = \{a_0, a_1, ..., a_q\}$ be a family of points $a_j \in \mathbb{P}(V^*)$. Take $\alpha_j \in V^* - \{0\}$ with $\mathbb{P}(\alpha_j) = a_j$. For $\lambda \in J_l^q$, set $\mathscr{A}_\lambda = \{a_{\lambda(0)}, ..., a_{\lambda(l)}\}$, and let $E(\mathscr{A}_\lambda)$ be the linear subspace generated by $\{\alpha_{\lambda(0)}, ..., \alpha_{\lambda(l)}\}$ in V^* over κ. Define

$$J_l(\mathscr{A}) = \{\lambda \in J_l^q \mid \alpha_\lambda \neq 0\}.$$

Obviously, if $\lambda \in J_l(\mathscr{A})$, then $\dim E(\mathscr{A}_\lambda) = l + 1$, and $\alpha_{\lambda(0)}, ..., \alpha_{\lambda(l)}$ is a base of $E(\mathscr{A}_\lambda)$. For $1 \leq n \leq q$, then $\mathscr{A}$ is said to be in *general position* if $E(\mathscr{A}_\lambda) = V^*$ for any $\lambda \in J_n^q$. According to Ru-Stoll [112], we also use the following notation:

Definition 6.5. *Let $\mathscr{A} = \{a_0, a_1, ..., a_q\}$ be a finite family of points in $\mathbb{P}(V^*)$ with $q \geq n$ and $J_n(\mathscr{A}) \neq \emptyset$. A base $e = (e_0, ..., e_n)$ of V is said to be perfect if for every $\lambda \in J_n(\mathscr{A})$*

$$\langle e_0 \wedge \cdots \wedge e_j, \alpha_{\pi(0)} \wedge \cdots \wedge \alpha_{\pi(j)} \rangle \neq 0, \quad \pi \in \mathcal{J}_\lambda, \quad j = 0, 1, ..., n,$$

where $\alpha_j \in V^* - \{0\}$ with $\mathbb{P}(\alpha_j) = a_j$, $\mathcal{J}_\lambda$ the permutation group of $\{\lambda(0), ..., \lambda(n)\}$.

The existence of perfect bases was proved by Ru-Stoll [112] in the complex number field. It also works on κ as follows.

Lemma 6.6. *Let κ be an algebraically closed field of characteristic zero, complete for a non-trivial non-Archimedean absolute value $|\cdot|$. Given a finite family $\mathscr{A} = \{a_0, a_1, ..., a_q\}$ of points of $\mathbb{P}(V^*)$ with $q \geq n$ and $J_n(\mathscr{A}) \neq \emptyset$. Then there is a perfect base of V for $\mathscr{A}$.*

Proof. By Ostrowski's theorem (Theorem 1.6), the restriction of $|\cdot|$ on $\mathbb{Q}$ is equivalent to one p-adic absolute value $|\cdot|_p$ for some prime p. Hence, there exists a positive real number α such that for each $a \in \mathbb{Q}$, one has $|a| = |a|_p^\alpha$. For any $x \in V_*$, write

$$|x| = p^{-v}, \quad x' = p^{-\left[\frac{v}{\alpha}\right]}x,$$

where $[v/\alpha]$ is the maximal integral part of v/α. Then we have $p^{-\alpha} < |x'| \leq 1$, and x' will be referred as the normalization of x. Let M be the set of all bases $(v_0, ..., v_n)$ of V with $p^{-\alpha} \leq |v_j| \leq 1$ $(0 \leq j \leq n)$. Then M is a closed variety of $V^{n+1} = V \times \cdots \times V$ ($n+1$ times). For each $\lambda \in J_n(\mathscr{A})$ and $j = 0, ..., n$, an analytic set

$$K_\pi^j = \{(v_0, ..., v_n) \in M \mid \langle v_0 \wedge \cdots \wedge v_j, \alpha_{\pi(0)} \wedge \cdots \wedge \alpha_{\pi(j)} \rangle = 0\}$$

is defined for each $\pi \in \mathcal{J}_\lambda$, where $\alpha_j \in V^* - \{0\}$ with $\mathbb{P}(\alpha_j) = a_j$. Let $\xi_0, ..., \xi_n$ be the dual base of $\alpha_{\pi(0)}, ..., \alpha_{\pi(n)}$, then

$$\langle \xi_0 \wedge \cdots \wedge \xi_j, \alpha_{\pi(0)} \wedge \cdots \wedge \alpha_{\pi(j)} \rangle = 1, \quad j = 0, 1, ..., n.$$

Normalize the base $\xi_0, ..., \xi_n$ to a base $\xi_0', ..., \xi_n'$ of V, then $(\xi_0', ..., \xi_n') \in M$ with

$$\xi_j' = r_j \xi_j, \quad 0 < r_j \in \mathbb{Q}, \quad j = 0, 1, ..., n.$$

It follows, for $j = 0, 1, ..., n$,

$$\begin{aligned}
\langle \xi_0' \wedge \cdots \wedge \xi_j', \alpha_{\pi(0)} \wedge \cdots \wedge \alpha_{\pi(j)} \rangle &= r_0 \cdots r_j \langle \xi_0 \wedge \cdots \wedge \xi_j, \alpha_{\pi(0)} \wedge \cdots \wedge \alpha_{\pi(j)} \rangle \\
&= r_0 \cdots r_j \neq 0,
\end{aligned}$$

which means $(\xi'_0, ..., \xi'_n) \in M - K^j_\pi$. Thus, the analytic set K^j_π is nowhere dense in M and

$$K = \bigcup_{\lambda \in J_n(\mathscr{A})} \bigcup_{\pi \in \mathscr{J}_\lambda} \bigcup_{j=0}^{n} K^j_\pi$$

is of the first category while M is of the second category. Thus $K \neq M$, and so each element of $M - K$ is a perfect base of V for $\mathscr{A}$. $\qquad \square$

6.2 The first main theorem of holomorphic curves

Let κ be an algebraically closed field of characteristic zero, complete for a non-trivial non-Archimedean absolute value $|\cdot|$. Let V be a normed vector space of dimension $n+1 > 0$ over κ and let $|\ |$ denote a norm defined over a base $e = (e_0, ..., e_n)$ of V. By a *(non-Archimedean) holomorphic curve*

$$f : \kappa \longrightarrow \mathbb{P}(V),$$

we mean an equivalence class of $(n + 1)$-tuples of entire functions

$$(\tilde{f}_0, ..., \tilde{f}_n) : \kappa \longrightarrow \kappa^{n+1}$$

such that $\tilde{f}_0, ..., \tilde{f}_n$ have not common factors in the ring of entire functions on κ and such that not all of the $\tilde{f}_j$ are identically zero. Here

$$\tilde{f} = \tilde{f}_0 e_0 + \cdots + \tilde{f}_n e_n : \kappa \longrightarrow V$$

will be called a *reduced representation of f*. Define

$$\mu(r, \tilde{f}) = \max_{0 \leq k \leq n} \mu(r, \tilde{f}_k).$$

Note that

$$\|\quad \mu(|z|, \tilde{f}) = \max_{0 \leq k \leq n} |\tilde{f}_k(z)| = |\tilde{f}(z)|.$$

Then the *characteristic function*

$$T(r, f) = \log \mu(r, \tilde{f})$$

is well defined for all $r > 0$, up to $O(1)$.

Proposition 6.7 ([26]). *Let $f : \kappa \longrightarrow \mathbb{P}(V)$ be a holomorphic curve. Then*
1) f is constant if and only if $T(r, f) = o(\log r)$.
2) f is rational if and only if $T(r, f) = O(\log r)$.
3) f is non-constant if and only if there exist constants $c \in \mathbb{R}$ and $A \in \mathbb{R}^+$ such that $T(r, f) \geq \log r + c$ for $r > A$.

Proof. Here we follow the proof of Cherry-Ye [26]. Let $\tilde{f} = \tilde{f}_0 e_0 + \cdots + \tilde{f}_n e_n : \kappa \longrightarrow V$ be a reduced representation of f. To show 2), first suppose $T(r, f) = O(\log r)$, that is, there is a constant B such that $T(r, f) \leq B \log r$, for all r sufficiently large. By the definition and the Jensen formula, we have

$$B \log r \quad \geq \quad T(r, f) = \max_{0 \leq k \leq n} \log \mu(r, \tilde{f}_k)$$

$$= \quad \max_{0 \leq k \leq n} N\left(r, \frac{1}{\tilde{f}_k}\right) + O(1)$$

for all r sufficiently large. Hence

$$\lim_{r \to \infty} n\left(r, \frac{1}{\tilde{f}_k}\right) \leq B, \quad k = 0, ..., n,$$

that is, each f_k is a polynomial.

Now assume that f is rational, that is, each f_k is a polynomial. By Example 2.4, we obtain

$$T(r, f) = \max_{0 \leq k \leq n} \log \mu(r, \tilde{f}_k) = \max_{0 \leq k \leq n} \deg(\tilde{f}_k) \log r + O(1)$$

as was to be shown.

If f is constant, then clearly $T(r, f)$ also is constant. Thus 3) implies 1). To show 3), let f be non-constant. Then at least one of functions $\tilde{f}_k$, say $\tilde{f}_0$, has a zero. Therefore

$$T(r, f) \geq N\left(r, \frac{1}{\tilde{f}_0}\right) + O(1) \geq \log r + O(1)$$

for all r sufficiently large. $\square$

Definition 6.8. *A holomorphic curve $f : \kappa \longrightarrow \mathbb{P}(V)$ is said to be transcendental if*

$$\limsup_{r \to \infty} \frac{T(r, f)}{\log r} = \infty.$$

Generally, take a mapping

$$\tilde{h} = \tilde{h}_0 e_0 + \cdots + \tilde{h}_n e_n : \kappa \longrightarrow V,$$

where $(\tilde{h}_0, ..., \tilde{h}_n)$ are $(n + 1)$-tuples of meromorphic functions such that not all of the $\tilde{h}_j$ are identically zero. By Corollary 1.29, the greatest common divisor h of $\tilde{h}_0, ..., \tilde{h}_n$ exists such that $\tilde{f} = \frac{\tilde{h}}{h}$ is a reduced representation of a non-Archimedean holomorphic curve $f : \kappa \longrightarrow \mathbb{P}(V)$. We will call $\tilde{h}$ a *representation* of f. Also write $f = \mathbb{P} \circ \tilde{h}$. Note that

$$\mu(r, \tilde{h}) = \mu(r, h)\mu(r, \tilde{f}).$$

By the Jensen formula, we obtain

$$T(r, f) + N_f(r) = \log \mu(r, \tilde{h}) + O(1), \tag{6.2}$$

where

$$N_f(r) = N\left(r, \frac{1}{h}\right) - N(r, h).$$

We also write

$$N\left(r, \frac{1}{\tilde{h}}\right) = N\left(r, \frac{1}{h}\right), \quad N(r, \tilde{h}) = N(r, h).$$

Let $f : \kappa \longrightarrow \mathbb{P}(V)$ be a holomorphic curve and let $\tilde{f} = \tilde{f}_0 e_0 + \cdots + \tilde{f}_n e_n : \kappa \longrightarrow V$ be a reduced representation of f. Then

$$\tilde{f} \wedge \tilde{f}' \wedge \cdots \wedge \tilde{f}^{(n)} = \mathbf{W}\left[e, \tilde{f}\right] e_0 \wedge e_1 \wedge \cdots \wedge e_n,$$

where $\mathbf{W}\left[e, \tilde{f}\right]$ is the Wronski determinant of $\tilde{f}$ with respect to the base e, that is,

$$\mathbf{W}\left[e, \tilde{f}\right] = \mathbf{W}\left(\tilde{f}_0, ..., \tilde{f}_n\right) = \begin{vmatrix} \tilde{f}_0 & \tilde{f}_1 & \cdots & \tilde{f}_n \\ \tilde{f}_0' & \tilde{f}_1' & \cdots & \tilde{f}_n' \\ \cdots\cdots\cdots\cdots\cdots\cdots\cdots \\ \tilde{f}_0^{(n)} & \tilde{f}_1^{(n)} & \cdots & \tilde{f}_n^{(n)} \end{vmatrix},$$

and, hence,

$$\left|\mathbf{W}\left[e, \tilde{f}\right]\right| = \left|\tilde{f} \wedge \tilde{f}' \wedge \cdots \wedge \tilde{f}^{(n)}\right|.$$

Also, we obtain

$$\mathbf{W}\left[e, \tilde{f}\right] = \left\langle \tilde{f} \wedge \tilde{f}' \wedge \cdots \wedge \tilde{f}^{(n)}, \epsilon_0 \wedge \epsilon_1 \wedge \cdots \wedge \epsilon_n \right\rangle,$$

where $\epsilon = (\epsilon_0, ..., \epsilon_n)$ is the dual base of e. Define an entire function

$$\mathbf{K}\left[e, \tilde{f}\right] = \mathbf{K}\left(\tilde{f}_0, ..., \tilde{f}_n\right) = \tilde{f}_0 \cdots \tilde{f}_n,$$

and define a meromorphic function

$$\mathbf{S}\left[e, \tilde{f}\right] = \mathbf{S}\left(\tilde{f}_0, ..., \tilde{f}_n\right) = \frac{\mathbf{W}[e, \tilde{f}]}{\mathbf{K}[e, \tilde{f}]}.$$

By the lemma of logarithmic derivative, we have

$$\mu\left(r, \mathbf{S}\left[e, \tilde{f}\right]\right) \leq r^{-\frac{n(n+1)}{2}}, \quad m\left(r, \mathbf{S}\left[e, \tilde{f}\right]\right) = O(1). \tag{6.3}$$

Let $V_0, V_1, ..., V_s$ and W be normed vector spaces over κ. Let

$$\odot : V_0 \times \cdots \times V_s \longrightarrow W$$

be a $(s+1)$-linear mapping over κ. Take non-Archimedean holomorphic curves

$$f_j : \kappa \longrightarrow \mathbb{P}(V_j), \quad j = 0, 1, ..., s,$$

with reduced representation

$$\tilde{f}_j : \kappa \longrightarrow V_j, \quad j = 0, 1, ..., s.$$

Definition 6.9. *The $(s+1)$-fold $(f_0, ..., f_s)$ is said to be free for $\odot$ if $\tilde{f}_0 \odot \cdots \odot \tilde{f}_s \not\equiv 0$.*

Assume that $(f_0, ..., f_s)$ is free for $\odot$. Thus, we can define

$$\mu(r, f_0 \odot \cdots \odot f_s) = \frac{\mu(r, \tilde{f}_0 \odot \cdots \odot \tilde{f}_s)}{\mu(r, \tilde{f}_0) \cdots \mu(r, \tilde{f}_s)}.$$

Then

$$\| \quad \mu(|z|, f_0 \odot \cdots \odot f_s) = |f_0(z) \odot \cdots \odot f_s(z)|.$$

Assume that $(f_0, ..., f_s)$ is free for $\odot$ and define the *compensation function*

$$m_{f_0 \odot \cdots \odot f_s}(r) = -\log \mu(r, f_0 \odot \cdots \odot f_s).$$

Note that

$$f_0 \odot \cdots \odot f_s : \kappa \longrightarrow \mathbb{P}(W)$$

is a non-Archimedean holomorphic curve with a representation $\tilde{f}_0 \odot \cdots \odot \tilde{f}_s$. Thus by (6.2), we obtain the *first main theorem*

$$\sum_{j=0}^{s} T(r, f_j) = N_{f_0 \odot \cdots \odot f_s}(r) + m_{f_0 \odot \cdots \odot f_s}(r) + T(r, f_0 \odot \cdots \odot f_s) + O(1). \qquad (6.4)$$

If $\dim W = 1$, then $\mathbb{P}(W)$ is a point and $T(r, f_0 \odot \cdots \odot f_s) = $ constant.

Let $g : \kappa \longrightarrow \mathbb{P}(V^*)$ be a non-Archimedean holomorphic curve with a reduced representation

$$\tilde{g} = \tilde{g}_0 \epsilon_0 + \cdots + \tilde{g}_n \epsilon_n : \kappa \longrightarrow V^*,$$

where $\epsilon = (\epsilon_0, ..., \epsilon_n)$ is the dual base of e.

Definition 6.10. *The pair (f, g) is said to be free if it is free for $\angle$, i.e.,*

$$\tilde{f} \angle \tilde{g} = \langle \tilde{f}, \tilde{g} \rangle = \tilde{g}_0 \tilde{f}_0 + \cdots + \tilde{g}_n \tilde{f}_n \not\equiv 0.$$

Assume that the pair (f, g) is free. Then the first main theorem reads

$$T(r, f) + T(r, g) = N_f(r, g) + m_f(r, g) + O(1), \qquad (6.5)$$

where

$$N_f(r, g) = N_{f \angle g}(r) = N\left(r, \frac{1}{\langle \tilde{f}, \tilde{g} \rangle}\right), \quad m_f(r, g) = m_{f \angle g}(r).$$

The *defect of f for g* is defined by

$$\delta_f(g) = 1 - \limsup_{r \to \infty} \frac{N_f(r, g)}{T(r, f) + T(r, g)},$$

with $0 \leq \delta_f(g) \leq 1$. We say that *$g$ grows slower than f* if

$$\lim_{r \to \infty} \frac{T(r, g)}{T(r, f)} = 0.$$

If so, we have

$$\delta_f(g) = 1 - \limsup_{r \to \infty} \frac{N_f(r, g)}{T(r, f)}.$$

In particular, if $g = a$ is a constant, then (6.5) becomes (cf. [26])

$$T(r, f) = N_f(r, a) + m_f(r, a) + O(1), \tag{6.6}$$

and the defect of f for a is given by

$$\delta_f(a) = 1 - \limsup_{r \to \infty} \frac{N_f(r, a)}{T(r, f)}.$$

For this case, the pair (f, a) is free if and only if $f(\kappa) \not\subset \ddot{E}[a]$.

The holomorphic curve f induces a holomorphic curve

$$f^{\amalg m} : \kappa \longrightarrow \mathbb{P}(\amalg_m V)$$

with a reduced representation $\tilde{f}^{\amalg m} : \kappa \longrightarrow \amalg_m V$ defined by

$$\tilde{f}^{\amalg m}(z) = \tilde{f}(z)^{\amalg m}, \quad z \in \kappa.$$

Let $g : \kappa \longrightarrow \mathbb{P}(\amalg_m V^*)$ be a non-Archimedean holomorphic curve such that the pair $(f^{\amalg m}, g)$ is free. Since

$$\| \quad \mu\left(|z|, \tilde{f}^{\amalg m}\right) = \left|\tilde{f}(z)^{\amalg m}\right| = |\tilde{f}(z)|^m,$$

then the characteristic function of $f^{\amalg m}$ satisfies

$$T\left(r, f^{\amalg m}\right) = mT(r, f).$$

Applying (6.5) to $f^{\amalg m}$ and g, we have

$$mT(r, f) + T(r, g) = N_{f^{\amalg m}}(r, g) + m_{f^{\amalg m}}(r, g) + O(1). \tag{6.7}$$

Define the *defect of f for g* by

$$\delta_f(g) = 1 - \limsup_{r \to \infty} \frac{N_{f^{\amalg m}}(r, g)}{mT(r, f) + T(r, g)}$$

with $0 \le \delta_f(g) \le 1$. If g is a constant a, then (6.7) becomes

$$mT(r, f) = N_{f^{\amalg m}}(r, a) + m_{f^{\amalg m}}(r, a) + O(1), \tag{6.8}$$

and the defect of f for a is given by

$$\delta_f(a) = 1 - \limsup_{r \to \infty} \frac{N_{f^{\amalg m}}(r, a)}{mT(r, f)}.$$

6.3   The second main theorem of holomorphic curves

Let κ be an algebraically closed field of characteristic zero, complete for a non-trivial non-Archimedean absolute value $|\cdot|$. Let V be a normed vector space of dimension $n + 1 > 0$ over κ and let $|\;|$ be a norm defined over a base $e = (e_0, ..., e_n)$ of V.

Definition 6.11. *A holomorphic curve $f : \kappa \longrightarrow \mathbb{P}(V)$ is said to be linearly non-degenerate if $f(\kappa) \not\subset \ddot{E}[a]$ for all $a \in \mathbb{P}(V^*)$.*

Lemma 6.12. *A holomorphic curve $f : \kappa \longrightarrow \mathbb{P}(V)$ is linearly degenerate if and only if the Wronski determinant $\mathbf{W}\left[e, \tilde{f}\right]$ of a reduced representation $\tilde{f}$ of f with respect to the base e is identically zero.*

Proof. Here we follow the proof of Lemma 3.5 in Wu [136]. Suppose f is linearly degenerate, that is, $f(\kappa) \subset \ddot{E}[a]$ for some $a \in \mathbb{P}(V^*)$. Then $\tilde{f}(\kappa) \subset E[a]$. Take $\alpha \in V^* - \{0\}$ with $\mathbb{P}(\alpha) = a$, and write

$$\alpha = \alpha_0 \epsilon_0 + \cdots + \alpha_n \epsilon_n.$$

Then

$$\alpha_0 \tilde{f}_0 + \cdots + \alpha_n \tilde{f}_n \equiv 0,$$

and, hence,

$$\alpha_0 \tilde{f}_0^{(k)} + \cdots + \alpha_n \tilde{f}_n^{(k)} \equiv 0,$$

i.e., $\tilde{f}^{(k)}(\kappa) \subset E[a]$ for $k = 1, ..., n$. Since $\alpha \neq 0$, $\mathbf{W}\left[e, \tilde{f}\right] = \det\left(\tilde{f}_j^{(i)}\right)$ has to be identically zero in κ.

Conversely, suppose $\mathbf{W}\left[e, \tilde{f}\right] \equiv 0$. Thus there is an integer l, $0 \leq l < n$, such that

$$\mathbf{W}\left(\tilde{f}_0, ..., \tilde{f}_l\right) = \begin{vmatrix} \tilde{f}_0 & \tilde{f}_1 & \cdots & \tilde{f}_l \\ \tilde{f}_0' & \tilde{f}_1' & \cdots & \tilde{f}_l' \\ \cdots\cdots\cdots\cdots\cdots\cdots \\ \tilde{f}_0^{(l)} & \tilde{f}_1^{(l)} & \cdots & \tilde{f}_l^{(l)} \end{vmatrix} \not\equiv 0,$$

while

$$\mathbf{W}\left(\tilde{f}_0, ..., \tilde{f}_{l+1}\right) = \begin{vmatrix} \tilde{f}_0 & \tilde{f}_1 & \cdots & \tilde{f}_{l+1} \\ \tilde{f}_0' & \tilde{f}_1' & \cdots & \tilde{f}_{l+1}' \\ \cdots\cdots\cdots\cdots\cdots\cdots \\ \tilde{f}_0^{(l+1)} & \tilde{f}_1^{(l+1)} & \cdots & \tilde{f}_{l+1}^{(l+1)} \end{vmatrix} \equiv 0.$$

Expanding $\mathbf{W}\left(\tilde{f}_0, ..., \tilde{f}_{l+1}\right)$ by the last column, this gives a differential equation, homogeneous of order $(l + 1)$, which is satisfied by $\tilde{f}_{l+1}$. Obviously, the same differential equation is also satisfied by $\tilde{f}_0, ..., \tilde{f}_l$. By Corollary 4.27, $\tilde{f}_0, ..., \tilde{f}_{l+1}$ are linearly dependent. Consequently, $\tilde{f}(\kappa)$ has to lie in a proper subspace of V, and f is linearly degenerate. $\square$

Consider a linearly non-degenerate holomorphic curve

$$f : \kappa \longrightarrow \mathbb{P}(V),$$

with a reduced representation

$$\tilde{f} = \tilde{f}_0 e_0 + \cdots + \tilde{f}_n e_n : \kappa \longrightarrow V_*.$$

Then the *ramification term*

$$N_{\text{Ram}}(r, f) = N\left(r, \frac{1}{\mathbf{W}[e, \tilde{f}]}\right)$$

is well defined, where $\mathbf{W}\left[e, \tilde{f}\right]$ is the Wronski determinant of $\tilde{f}$ with respect to the base e. Ha-Mai [49] showed an analogue of the second main theorem due to Cartan by using heights. Here we follow the method due to Cartan, and use the characteristic functions to reformulate the result (see [26]).

Theorem 6.13. *Let* $f : \kappa \longrightarrow \mathbb{P}(V)$ *be a linearly non-degenerate holomorphic curve and let* $\mathscr{A} = \{a_0, a_1, ..., a_q\}$ *be a family of points* $a_j \in \mathbb{P}(V^*)$ *in general position. Then*

$$(q - n)T(r, f) \le \sum_{j=0}^{q} N_f(r, a_j) - N_{\text{Ram}}(r, f) - \frac{n(n+1)}{2} \log r + O(1).$$

Proof. Take $\tilde{a}_j \in V^* - \{0\}$ with $\mathbb{P}(\tilde{a}_j) = a_j$. Write

$$\tilde{a}_j = \tilde{a}_{j0}\epsilon_0 + \cdots + \tilde{a}_{jn}\epsilon_n, \quad j = 0, ..., q,$$

where $\epsilon = (\epsilon_0, ..., \epsilon_n)$ is the dual of e. For $i = 0, 1, ..., q$, set

$$F_i = \left\langle \tilde{f}, \tilde{a}_i \right\rangle = \tilde{a}_{i0}\tilde{f}_0 + \tilde{a}_{i1}\tilde{f}_1 + \cdots + \tilde{a}_{in}\tilde{f}_n.$$

Since f is linearly non-degenerate, then $F_i \not\equiv 0$. Because $\mathscr{A}$ is in general position, we have $\det\left(\tilde{a}_{\lambda(i)j}\right) \neq 0$ for any $\lambda \in J_n^q$, and hence

$$\tilde{f}_i = \tilde{a}^{\lambda(i)0} F_{\lambda(0)} + \tilde{a}^{\lambda(i)1} F_{\lambda(1)} + \cdots + \tilde{a}^{\lambda(i)n} F_{\lambda(n)}, \quad i = 0, 1, ..., n,$$

where $\left(\tilde{a}^{\lambda(i)j}\right)$ is the inverse matrix of $\left(\tilde{a}_{\lambda(i)j}\right)$. Thus for any $\lambda \in J_n^q$, we obtain

$$\left|\tilde{f}_i(z)\right| \le A \max_{0 \le j \le n} \left\{|F_{\lambda(j)}(z)|\right\}, \quad z \in \kappa, \ i = 0, 1, ..., n,$$

where

$$A = \max\left\{\left|\tilde{a}^{\lambda(i)j}\right| \mid \lambda \in J_n^q, \ 0 \le i, j \le n\right\}.$$

We abbreviate the Wronskian

$$\mathbf{W} = \mathbf{W}\left(\tilde{f}_0, \tilde{f}_1, ..., \tilde{f}_n\right) = \mathbf{W}\left[e, \tilde{f}\right],$$

$$\mathbf{W}_\lambda = \mathbf{W}\left(F_{\lambda(0)}, F_{\lambda(1)}, ..., F_{\lambda(n)}\right).$$

Then

$$\mathbf{W}_\lambda = c_\lambda \mathbf{W}, \quad c_\lambda = \det\left(\tilde{a}_{\lambda(i)j}\right).$$

Next we fix $z \in \kappa - \kappa[0; \rho_0]$ such that

$$\mathbf{W}(z) \neq 0, \quad f_i(z) \neq 0 \ (i = 0, ..., n), \quad F_j(z) \neq 0 \ (j = 0, 1, ..., q).$$

Thus we can take distinct indices $\alpha_0, ..., \alpha_n (= \beta_0), \beta_1, ..., \beta_{q-n}$ such that

$$0 < |F_{\alpha_0}(z)| \leq \cdots \leq |F_{\alpha_n}(z)| \leq |F_{\beta_1}(z)| \leq \cdots \leq |F_{\beta_{q-n}}(z)| < \infty.$$

Take $\lambda \in J_n^q$ with $\mathrm{Im}\lambda = \{\alpha_0, ..., \alpha_n\}$. Therefore, we have

$$|\tilde{f}_k(z)| \leq A \max_{0 \leq j \leq n} \{|F_{\lambda(j)}(z)|\} \leq A|F_{\beta_l}(z)|$$

for $k = 0, 1, ..., n; \ l = 0, 1, 2, ..., q - n$. Hence, we obtain

$$|\tilde{f}(z)| = \max_k \left|\tilde{f}_k(z)\right| \leq A|F_{\beta_l}(z)|, \quad l = 0, 1, ..., q - n.$$

By $\mathbf{W}_\lambda = c_\lambda \mathbf{W}$, we obtain

$$\log \frac{|F_0(z) \cdots F_q(z)|}{|\mathbf{W}(z)|} = \log |F_{\beta_1}(z) \cdots F_{\beta_{q-n}}(z)| - \log D_\lambda(z) + \log |c_\lambda|,$$

where

$$D_\lambda = \frac{|\mathbf{W}_\lambda|}{|F_{\lambda(0)} \cdots F_{\lambda(n)}|} = \left| \sum_{i \in \mathcal{J}^n} \mathrm{sign}(i) \frac{F_{\lambda(0)}^{(i(0))}}{F_{\lambda(0)}} \cdots \frac{F_{\lambda(n)}^{(i(n))}}{F_{\lambda(n)}} \right|,$$

in which $\mathcal{J}^n$ is the permutation group on $\mathbb{Z}[0, n]$ and $F_j^{(0)} = F_j$. Then

$$\log |F_{\beta_1}(z) \cdots F_{\beta_{q-n}}(z)| \leq \log \frac{|F_0(z) \cdots F_q(z)|}{|\mathbf{W}(z)|} + \log D_\lambda(z) - \log |c_\lambda|.$$

Therefore, we have

$$(q - n) \log |\tilde{f}(z)| \leq \log \frac{|F_0(z) \cdots F_q(z)|}{|\mathbf{W}(z)|} + \log D_\lambda(z) + (q - n) \log A - \log A', \qquad (6.9)$$

where A' is defined by

$$A' = \min_{\lambda \in J_n^q} |c_\lambda|.$$

Set $r = |z|$. By using Theorem 1.15, we obtain

$$D_\lambda(z) \leq \max_{i \in \mathcal{J}^n} \left\{ \frac{|F_{\lambda(0)}^{(i(0))}(z)|}{|F_{\lambda(0)}(z)|} \cdots \frac{|F_{\lambda(n)}^{(i(n))}(z)|}{|F_{\lambda(n)}(z)|} \right\} \leq r^{-\frac{n(n+1)}{2}},$$

and hence

$$\log D_\lambda(z) \leq -\frac{n(n + 1)}{2} \log r.$$

By using the Jensen formula again,

$$\log |\mathbf{W}(z)| = \log \mu(r, \mathbf{W}) = N\left(r, \frac{1}{\mathbf{W}}\right) + \log \mu(\rho_0, \mathbf{W}),$$

$$\log |F_i(z)| = \log \mu(r, F_i) = N_f(r, a_i) + \log \mu(\rho_0, F_i),$$

for $i = 0, 1, ..., q$, and noting that

$$\log |\tilde{f}(z)| = T(r, f) + O(1),$$

we obtain

$$(q - n)T(r, f) \leq \sum_{j=0}^{q} N_f(r, a_j) - N\left(r, \frac{1}{\mathbf{W}}\right) - \frac{n(n+1)}{2} \log r + O(1). \tag{6.10}$$

Note that the set of r for which (6.10) holds is dense in (ρ_0, ∞). Thus (6.10) also holds for all $\rho_0 < r < \infty$, by continuity of the functions involved in the inequality. □

Corollary 6.14. *Let* $f : \kappa \longrightarrow \mathbb{P}(V)$ *be a linearly non-degenerate holomorphic curve and let* $\mathscr{A} = \{a_0, a_1, ..., a_q\}$ *be a family of points* $a_j \in \mathbb{P}(V^*)$ *in general position. Then*

$$(q - n)T(r, f) \leq \sum_{j=0}^{q} N_{f,n}(r, a_j) - \frac{n(n+1)}{2} \log r + O(1),$$

where

$$N_{f,n}(r, a_j) = N_n\left(r, \frac{1}{F_j}\right).$$

For any $k \in \mathbb{Z}^+$, $a \in \mathbb{P}(V^*)$, we define

$$\delta_f(a, k) = 1 - \limsup_{r \to \infty} \frac{N_{f,k}(r, a)}{T(r, f)}$$

with $0 \leq \delta_f(a) \leq \delta_f(a, k) \leq 1$. We may think $\delta_f(a) = \delta_f(a, \infty)$. Now Corollary 6.14 means the following defect relation:

Corollary 6.15. *Let* $f : \kappa \longrightarrow \mathbb{P}(V)$ *be a linearly non-degenerate holomorphic curve and let* $\mathscr{A} = \{a_0, a_1, ..., a_q\}$ *be a family of points* $a_j \in \mathbb{P}(V^*)$ *in general position. Then*

$$\sum_{j=0}^{q} \delta_f(a_j) \leq \sum_{j=0}^{q} \delta_f(a_j, n) \leq n + 1.$$

Cherry-Ye [26] pointed out that the defect relation (Corollary 6.15) from the second main theorem (Theorem 6.13) is not sharp by proving that there can be at most n hyperplanes in general position that have non-zero defects.

Theorem 6.16. *Let* $f : \kappa \longrightarrow \mathbb{P}(V)$ *be a linearly non-degenerate holomorphic curve and let* $\mathscr{A} = \{a_0, a_1, ..., a_q\}$ *be a family of points* $a_j \in \mathbb{P}(V^*)$ *in general position such that*

$$\delta_f(a_j) > 0, \quad j = 0, 1, ..., q.$$

Then $q \leq n - 1$.

Proof. Assume, on the contrary, that $q \geq n$. By using the notations in the proof of Theorem 6.13, for any $\lambda \in J_n^q$, we obtain a linearly non-degenerate holomorphic curve $\mathbf{f}_\lambda : \kappa \longrightarrow \mathbb{P}(V)$ with a reduced representation

$$\tilde{\mathbf{f}}_\lambda = F_{\lambda(0)} e_0 + \cdots + F_{\lambda(n)} e_n : \kappa \longrightarrow V_*.$$

Obviously, we have

$$T(r, \mathbf{f}_\lambda) = T(r, f) + O(1).$$

Since $T(r, f) \to \infty$ as $r \to \infty$, we obtain

$$
\begin{aligned}
\delta_f(a_j) &= 1 - \limsup_{r \to \infty} \frac{N_f(r, a_j)}{T(r, f)} = 1 - \limsup_{r \to \infty} \frac{N\left(r, \frac{1}{F_{\lambda(j)}}\right)}{T(r, \mathbf{f}_\lambda)} \\
&= 1 - \limsup_{r \to \infty} \frac{\log \mu(r, F_{\lambda(j)})}{T(r, \mathbf{f}_\lambda)} > 0
\end{aligned}
$$

for $j = 0, ..., n$. This implies that for all sufficiently large r

$$\log \mu(r, F_{\lambda(j)}) < T(r, \mathbf{f}_\lambda), \quad j = 0, ..., n.$$

In particular,

$$|F_{\lambda(j)}(z)| < \max_{0 \leq j \leq n} |F_{\lambda(j)}(z)|, \quad j = 0, ..., n$$

hold for all sufficiently large $|z|$. Clearly this is absurd. $\square$

In fact, if $\mathscr{A} = \{a_0, a_1, ..., a_q\}$ is a family of points $a_j \in \mathbb{P}(V^*)$ in general position with $q \geq n$, we can obtain stronger result than Theorem 6.16. Note that

$$T(r, f) = T(r, \mathbf{f}_\lambda) + O(1) = \max_{0 \leq j \leq n} \log \mu(r, F_{\lambda(j)}) + O(1).$$

By using the Jensen formula, we obtain the formula:

$$T(r, f) = \max_{0 \leq j \leq n} N_f(r, a_{\lambda(j)}) + O(1), \quad \lambda \in J_n^q, \tag{6.11}$$

for any linearly non-degenerate holomorphic curve $f : \kappa \longrightarrow \mathbb{P}(V)$.

Corollary 6.17. *Let* $f : \kappa \longrightarrow \mathbb{P}(V)$ *be a linearly non-degenerate holomorphic curve and let* $\mathscr{A} = \{a_0, a_1, ..., a_q\}$ *be a family of points* $a_j \in \mathbb{P}(V^*)$ *in general position. Then*

$$\sum_{j=0}^{q} \delta_f(a_j) \leq n.$$

Finally, we formulate the Cartan's method in the proof of Theorem 6.13. Let $\mathscr{A} = \{a_0, a_1, ..., a_q\}$ $(q \geq n)$ be a family of points $a_j \in \mathbb{P}(V^*)$ in general position. Define the *gauge* $\Gamma(\mathscr{A})$ of $\mathscr{A}$ with respect to the norm $\rho = |\cdot|$ by

$$\Gamma(\mathscr{A}) = \Gamma(\mathscr{A}; \rho) = \inf_{\lambda \in J_n^q} \{|a_{\lambda(0)} \wedge \cdots \wedge a_{\lambda(n)}|\}$$

with $0 < \Gamma(\mathscr{A}) \leq 1$.

Lemma 6.18. For $x \in \mathbb{P}(V)$, $0 < b \in \mathbb{R}$, define

$$\mathscr{A}(x, b) = \mathscr{A}(x, b; \rho) = \{j \in \mathbb{Z}[0, q] \mid |x, a_j| < b\}.$$

If $0 < b \leq \Gamma(\mathscr{A})$, then $\#\mathscr{A}(x, b) \leq n$.

Proof. Assume that $\#\mathscr{A}(x, b) \geq n + 1$. Then $\lambda \in J_n^q$ exists such that $\operatorname{Im}\lambda \subseteq \mathscr{A}(x, b)$. Hence

$$|x, a_{\lambda(j)}| < b, \quad j = 0, ..., n.$$

Then Lemma 6.4 implies

$$\begin{aligned} 0 < \Gamma(\mathscr{A}) &\leq |a_{\lambda(0)} \wedge \cdots \wedge a_{\lambda(n)}| \\ &\leq \max_{0 \leq j \leq n} |x, a_{\lambda(j)}| < b \leq \Gamma(\mathscr{A}), \end{aligned}$$

which is impossible. □

Lemma 6.19. Take $x \in \mathbb{P}(V)$ such that $|x, a_j| > 0$ for $j = 0, ..., q$. Then

$$\prod_{j=0}^{q} \frac{1}{|x, a_j|} \leq \left(\frac{1}{\Gamma(\mathscr{A})}\right)^{q-n} \max_{\lambda \in J_n^q} \prod_{j=0}^{n} \frac{1}{|x, a_{\lambda(j)}|} \tag{6.12}$$

$$\leq \left(\frac{1}{\Gamma(\mathscr{A})}\right)^{q+1-n} \max_{\lambda \in J_{n-1}^q} \prod_{j=0}^{n-1} \frac{1}{|x, a_{\lambda(j)}|}. \tag{6.13}$$

Proof. Take $b = \Gamma(\mathscr{A})$. Lemma 6.18 implies $\#\mathscr{A}(x, b) \leq n$. Thus $\operatorname{Im}\lambda - \mathscr{A}(x, b) \neq \emptyset$ for any $\lambda \in J_n^q$, that is, for any $\lambda \in J_n^q$ there is some $j_0 \in \mathbb{Z}[0, n]$ such that $|x, a_{\lambda(j_0)}| \geq \Gamma(\mathscr{A})$. Further, $\sigma \in J_n^q$ exists such that $\mathscr{A}(x, b) \subset \operatorname{Im}\sigma$. Then

$$\begin{aligned} \prod_{j=0}^{q} \frac{1}{|x, a_j|} &\leq \left(\frac{1}{\Gamma(\mathscr{A})}\right)^{q-n} \prod_{j=0}^{n} \frac{1}{|x, a_{\sigma(j)}|} \\ &\leq \left(\frac{1}{\Gamma(\mathscr{A})}\right)^{q-n} \max_{\lambda \in J_n^q} \prod_{j=0}^{n} \frac{1}{|x, a_{\lambda(j)}|} \\ &\leq \left(\frac{1}{\Gamma(\mathscr{A})}\right)^{q+1-n} \max_{\lambda \in J_{n-1}^q} \prod_{j=0}^{n-1} \frac{1}{|x, a_{\lambda(j)}|}. \end{aligned}$$

Thus the lemma follows. □

Now we can prove Theorem 6.13 again. Since f is non-constant, then $\mu(r, \tilde{f}) \to \infty$ as $r \to \infty$. So we may assume $\mu(r, \tilde{f}) \geq 1$ for $r \geq r_0$. Lemma 6.19 implies

$$\prod_{j=0}^{q} \frac{1}{|f, a_j|} \leq \left(\frac{1}{\Gamma(\mathscr{A})}\right)^{q-n} \max_{\lambda \in J_n^q} \prod_{i=0}^{n} \frac{1}{|f, a_{\lambda(i)}|},$$

which yields, for $r \geq r_0$,

$$
\begin{aligned}
\Gamma(\mathscr{A})^{q-n} \prod_{j=0}^{q} \frac{1}{\mu(r, f \angle a_j)}
&\leq \max_{\lambda \in J_n^q} \prod_{i=0}^{n} \frac{1}{\mu(r, f \angle a_{\lambda(i)})} \\
&= c \max_{\lambda \in J_n^q} \prod_{i=0}^{n} \frac{\mu(r, \tilde{f})}{\mu(r, F_{\lambda(i)})} \\
&\leq c\mu(r, \tilde{f})^{n+1} \max_{\lambda \in J_n^q} \mu\left(r, \frac{\mathbf{W}_\lambda}{F_{\lambda(0)} \cdots F_{\lambda(n)}}\right) \mu\left(r, \frac{1}{\mathbf{W}_\lambda}\right) \\
&\leq c'\mu(r, \tilde{f})^{n+1} r^{-\frac{n(n+1)}{2}} \mu\left(r, \frac{1}{\mathbf{W}}\right),
\end{aligned}
$$

where c and c' are positive constants. Thus Theorem 6.13 follows again.

If the non-degeneracy of f in Theorem 6.13 is replaced by the condition $f(\kappa) \not\subset \ddot{E}[a_j]$ for $j = 0, ..., q$, then by Lemma 6.19, we have

$$
\prod_{j=0}^{q} \frac{1}{|f, a_j|} \leq \left(\frac{1}{\Gamma(\mathscr{A})}\right)^{q+1-n} \max_{\lambda \in J_{n-1}^q} \prod_{i=0}^{n-1} \frac{1}{|f, a_{\lambda(i)}|},
$$

which yields,

$$
\sum_{j=0}^{q} m_f(r, a_j) \leq \max_{\lambda \in J_{n-1}^q} \sum_{i=0}^{n-1} m_f(r, a_{\lambda(i)}) + O(1) \leq nT(r, f) + O(1).
$$

Hence by the first main theorem, we obtain the following result:

Theorem 6.20. *Let* $f : \kappa \longrightarrow \mathbb{P}(V)$ *be a holomorphic curve and let* $\mathscr{A} = \{a_0, a_1, ..., a_q\}$ *be a family of points* $a_j \in \mathbb{P}(V^*)$ *in general position. If* $f(\kappa) \not\subset \ddot{E}[a_j]$ *for* $j = 0, ..., q$, *then*

$$
(q + 1 - n)T(r, f) \leq \sum_{j=0}^{q} N_f(r, a_j) + O(1).
$$

Min Ru [111] also notices the fact. For related results, see Ha-Mai [49] and Cherry-Ye [26].

6.4 Nochka weight

The weights of Nochka [99],[100],[101] (also see Chen [21]) are proved originally for the complex number field. Nochka's original paper was quite sketchy; a complete proof can be found in Chen's thesis which, however, is quite lengthy. Ru and Wong [115] shortened the proof and extended the weights of Nochka to any field of characteristic zero. Here we will introduce the Ru and Wong's approaches. Following Chen, Ru, Stoll and Wong, and using the notation and terminology of § 6.1, we also need the concept of subgeneral position as follows:

Definition 6.21. *Let $\mathscr{A} = \{a_0, a_1, ..., a_q\}$ be a family of points $a_j \in \mathbb{P}(V^*)$. For $1 \leq n \leq u < q$, then $\mathscr{A}$ is said to be in u-subgeneral position if $E(\mathscr{A}_\lambda) = V^*$ for any $\lambda \in J_u^q$.*

For $u = n$, this concept agrees with the usual concept of hyperplanes in general position. Given a family $\mathscr{A} = \{a_0, a_1, ..., a_q\}$ in u-subgeneral position and an element $\lambda \in J_l^q$, denote by

$$1 \leq i(\lambda) = l + 1, \quad d_\lambda = \dim E(\mathscr{A}_\lambda) \leq n + 1.$$

If $l \leq u$, say $i(\lambda) = u + 1 - p$, then $n + 1 - p \leq d_\lambda$, and so one has

$$i(\lambda) + n - u \leq d_\lambda \leq i(\lambda). \tag{6.14}$$

Take $\lambda \in J_l^q$ and $\sigma \in J_k^q$. If

$$\{\lambda(0), ..., \lambda(l)\} \cup \{\sigma(0), ..., \sigma(k)\} = \{i_0, ..., i_p \mid 0 \leq i_0 < \cdots < i_p \leq q\},$$

we define the union $\lambda \cup \sigma \in J_p^q$ by

$$\lambda \cup \sigma(j) = i_j, \quad j = 0, ..., p.$$

Assume $\mathrm{Im}(\lambda) \cap \mathrm{Im}(\sigma) \neq \emptyset$ and write

$$\{\lambda(0), ..., \lambda(l)\} \cap \{\sigma(0), ..., \sigma(k)\} = \{j_0, ..., j_r \mid 0 \leq j_0 < \cdots < j_r \leq q\},$$

we define the intersection $\lambda \cap \sigma \in J_r^q$ by

$$\lambda \cap \sigma(i) = j_i, \quad i = 0, ..., r.$$

We also write $\lambda \cap \sigma = \emptyset$ if $\mathrm{Im}(\lambda) \cap \mathrm{Im}(\sigma) = \emptyset$, and denote $\lambda \subset \sigma$ if $\mathrm{Im}(\lambda) \subset \mathrm{Im}(\sigma)$. Similarly, we can define $\lambda - \sigma$. Set

$$J_{u)}^q = \bigcup_{l=0}^u J_l^q.$$

Proposition 6.22. *Given a family $\mathscr{A} = \{a_0, a_1, ..., a_q\}$ in u-subgeneral position. Then there exists a uniquely determined sequence of $J_{u-1)}^q \cup \{\emptyset\}$:*

$$\emptyset = \lambda_0 \subset \lambda_1 \subset \cdots \subset \lambda_s \tag{6.15}$$

and an associated sequence of sets

$$J_{u)}^q[\lambda_k] = \{\lambda \mid \lambda_k \subset \lambda \in J_{u)}^q, \; d_\lambda > d_{\lambda_k}\}, \quad k = 0, ..., s, \tag{6.16}$$

where $d_{\lambda_0} = 0$, such that the index of independence of $J_{u)}^q[\lambda_k]$

$$\iota(\lambda_k) = \min_{\lambda \in J_{u)}^q[\lambda_k]} \frac{d_\lambda - d_{\lambda_k}}{i(\lambda) - i(\lambda_k)} \tag{6.17}$$

satisfies

$$0 < \iota(\lambda_k) = \frac{d_{\lambda_{k+1}} - d_{\lambda_k}}{i(\lambda_{k+1}) - i(\lambda_k)} < \frac{n + 1 - d_{\lambda_k}}{2u - n + 1 - i(\lambda_k)}, \quad k = 0, ..., s - 1, \tag{6.18}$$

$$\iota(\lambda_s) \geq \frac{n+1-d_{\lambda_s}}{2u-n+1-i(\lambda_s)} > 0, \tag{6.19}$$

and for $\lambda \in J_{u)}^q[\lambda_k]$ the equality

$$\frac{d_\lambda - d_{\lambda_k}}{i(\lambda) - i(\lambda_k)} = \iota(\lambda_k), \quad k = 0, ..., s, \tag{6.20}$$

implies $\lambda \subset \lambda_{k+1}$.

Proof. Inductively, if $\mathscr{A}_{\lambda_k}$ is constructed, there are two alternatives, either

$$\iota(\lambda_k) \geq \frac{n+1-d_{\lambda_k}}{2u-n+1-i(\lambda_k)} \text{ or } \iota(\lambda_k) < \frac{n+1-d_{\lambda_k}}{2u-n+1-i(\lambda_k)}.$$

In the first case the sequence terminates at $s = k$. In the second case, let

$$J_{u)\lambda_k}^q = \left\{ \lambda \in J_{u)}^q[\lambda_k] \mid \frac{d_\lambda - d_{\lambda_k}}{i(\lambda) - i(\lambda_k)} = \iota(\lambda_k) \right\}.$$

One claims that $\lambda, \sigma \in J_{u)\lambda_k}^q$ implies $\lambda \cup \sigma \in J_{u)\lambda_k}^q$. Note that $\lambda, \sigma \in J_{u)\lambda_k}^q$ mean $d_\sigma - d_{\lambda_k} = c(d_\lambda - d_{\lambda_k})$ with $c = (i(\sigma) - i(\lambda_k))/(i(\lambda) - i(\lambda_k))$. The basic inequality in linear algebra

$$d_{\lambda \cup \sigma} \leq d_\lambda + d_\sigma - d_{\lambda \cap \sigma}$$

together with the elementary equation

$$i(\lambda \cup \sigma) = i(\lambda) + i(\sigma) - i(\lambda \cap \sigma)$$

yield

$$\frac{d_{\lambda \cup \sigma} - d_{\lambda_k}}{i(\lambda \cup \sigma) - i(\lambda_k)} \leq \frac{d_\lambda + d_\sigma - d_{\lambda \cap \sigma} - d_{\lambda_k}}{i(\lambda) + i(\sigma) - i(\lambda \cap \sigma) - i(\lambda_k)}$$
$$= \frac{(c+1)(d_\lambda - d_{\lambda_k}) - (d_{\lambda \cap \sigma} - d_{\lambda_k})}{(c+1)(i(\lambda) - i(\lambda_k)) - (i(\lambda \cap \sigma) - i(\lambda_k))}.$$

Now we distinguish two cases, either $d_{\lambda \cap \sigma} = d_{\lambda_k}$, or $d_{\lambda \cap \sigma} > d_{\lambda_k}$. The former means $i(\lambda \cap \sigma) = i(\lambda_k)$. In fact, this is obvious if $k = 0$. For the case $k \geq 1$, then $\lambda \cap \sigma \in J_{u)}^q[\lambda_{k-1}]$, and so we have

$$\frac{d_{\lambda_k} - d_{\lambda_{k-1}}}{i(\lambda_k) - i(\lambda_{k-1})} = \iota(\lambda_{k-1}) \leq \frac{d_{\lambda \cap \sigma} - d_{\lambda_{k-1}}}{i(\lambda \cap \sigma) - i(\lambda_{k-1})}$$
$$= \frac{d_{\lambda_k} - d_{\lambda_{k-1}}}{i(\lambda \cap \sigma) - i(\lambda_{k-1})},$$

which gives $i(\lambda \cap \sigma) \leq i(\lambda_k)$, and hence $i(\lambda \cap \sigma) = i(\lambda_k)$ since $\lambda_k \subset \lambda \cap \sigma$. Thus

$$\frac{d_{\lambda \cup \sigma} - d_{\lambda_k}}{i(\lambda \cup \sigma) - i(\lambda_k)} \leq \frac{d_\lambda - d_{\lambda_k}}{i(\lambda) - i(\lambda_k)},$$

and we are done by the minimality of $\iota(\lambda_k)$.

If $d_{\lambda \cap \sigma} > d_{\lambda_k}$, then $\lambda \cap \sigma \in J^q_{u)}[\lambda_k]$, and so we have

$$d_{\lambda \cap \sigma} - d_{\lambda_k} \geq \iota(\lambda_k)(i(\lambda \cap \sigma) - i(\lambda_k)) = \frac{d_\lambda - d_{\lambda_k}}{i(\lambda) - i(\lambda_k)}(i(\lambda \cap \sigma) - i(\lambda_k)).$$

Hence

$$\frac{d_{\lambda \cup \sigma} - d_{\lambda_k}}{i(\lambda \cup \sigma) - i(\lambda_k)} \leq \frac{(c+1)(d_\lambda - d_{\lambda_k}) - \frac{d_\lambda - d_{\lambda_k}}{i(\lambda) - i(\lambda_k)}(i(\lambda \cap \sigma) - i(\lambda_k))}{(c+1)(i(\lambda) - i(\lambda_k)) - (i(\lambda \cap \sigma) - i(\lambda_k))}$$

$$= \frac{d_\lambda - d_{\lambda_k}}{i(\lambda) - i(\lambda_k)} = \iota(\lambda_k),$$

which implies $\lambda \cup \sigma \in J^q_{u)\lambda_k}$ if $i(\lambda \cup \sigma) \leq u + 1$.

Now we prove $i(\lambda \cup \sigma) < u + 1$. Noting that

$$\frac{d_\lambda - d_{\lambda_k} + d_\sigma - d_{\lambda_k}}{i(\lambda) - i(\lambda_k) + i(\sigma) - i(\lambda_k)} = \frac{(c+1)(d_\lambda - d_{\lambda_k})}{(c+1)(i(\lambda) - i(\lambda_k))} = \iota(\lambda_k),$$

and by (6.14), we get

$$\frac{d_\lambda + d_\sigma - 2d_{\lambda_k}}{d_\lambda + d_\sigma + 2(u-n) - 2i(\lambda_k)} \leq \iota(\lambda_k) < \frac{n + 1 - d_{\lambda_k}}{2u - n + 1 - i(\lambda_k)},$$

which, by a direct computation, implies that

$$\begin{aligned}
2(u-n)(d_\lambda + d_\sigma - d_{\lambda_k}) &< 2(u-n)(n+1) \\
&\quad + (d_\lambda + d_\sigma - 2(n+1))(i(\lambda_k) - d_{\lambda_k}) \\
&\leq 2(u-n)(n+1),
\end{aligned}$$

where we have used the fact that $d_\lambda, d_\sigma \leq n + 1$. In particular, one obtains

$$d_{\lambda \cup \sigma} \leq d_\lambda + d_\sigma - d_{\lambda \cap \sigma} \leq d_\lambda + d_\sigma - d_{\lambda_k} < n + 1,$$

and so $i(\lambda \cup \sigma) < u + 1$ by the subgeneral position assumption.

Thus we can define

$$\lambda_{k+1} = \bigcup_{\lambda \in J^q_{u)\lambda_k}} \lambda.$$

Then $\lambda_{k+1} \in J^q_{u)\lambda_k} \cap J^q_{u-1)}$, and so

$$\frac{d_{\lambda_{k+1}} - d_{\lambda_k}}{i(\lambda_{k+1}) - i(\lambda_k)} = \iota(\lambda_k) < \frac{n + 1 - d_{\lambda_k}}{2u - n + 1 - i(\lambda_k)},$$

which obviously implies the inequality:

$$d_{\lambda_k} < d_{\lambda_{k+1}} < n + 1.$$

Since $\mathscr{A}$ is a finite set, this process cannot be continued indefinitely, and hence there is an integer s such that (6.19) is verified so that the process terminates. $\square$

Lemma 6.23. *Let V be a vector space of dimension $n+1$ over κ and let $\mathscr{A} = \{a_0, a_1, ..., a_q\}$ be a family of points $a_j \in \mathbb{P}(V^*)$ in u-subgeneral position with $1 \leq n \leq u < q$. Then there exists a function $\omega : \mathscr{A} \longrightarrow \mathbb{R}(0, 1]$ called a Nochka weight and a real number $\theta \geq 1$ called Nochka constant satisfying the following properties:*

1) $0 < \omega(a_j)\theta \leq 1, \quad j = 0, 1, ..., q;$

2) $q - 2u + n = \theta(\sum_{j=0}^{q} \omega(a_j) - n - 1);$

3) $1 \leq \frac{u+1}{n+1} \leq \theta \leq \frac{2u-n+1}{n+1};$

4) $\sum_{j=0}^{k} \omega(a_{\sigma(j)}) \leq d_\sigma$ if $\sigma \in J_k^q$ with $0 \leq k \leq u;$

5) Let $r_0, ..., r_q$ be a sequence of real numbers with $r_j \geq 1$ for all j. Then for any $\sigma \in J_k^q$ with $0 \leq k \leq u$, setting $d_\sigma = l + 1$, then there exists $\lambda \in J_l(\mathscr{A})$ with $\text{Im}\lambda \subset \text{Im}\sigma$ such that $E(\mathscr{A}_\lambda) = E(\mathscr{A}_\sigma)$, and such that

$$\prod_{j=0}^{k} r_{\sigma(j)}^{\omega(a_{\sigma(j)})} \leq \prod_{j=0}^{l} r_{\lambda(j)}.$$

Proof. Let $\{\mathscr{A}_{\lambda_k} \mid 0 \leq k \leq s\}$ be the sequence of subsets of $\mathscr{A}$ obtained in Proposition 6.22. Denote by

$$\theta = \frac{2u - n + 1 - i(\lambda_s)}{n + 1 - d_{\lambda_s}} \geq 1,$$

and define the Nochka weight by

$$\omega(a_j) = \begin{cases} \frac{1}{\theta} & : \quad a_j \notin \mathscr{A}_{\lambda_s} \\ \frac{d_{\lambda_{k+1}} - d_{\lambda_k}}{i(\lambda_{k+1}) - i(\lambda_k)} & : \quad a_j \in \mathscr{A}_{\lambda_{k+1}} - \mathscr{A}_{\lambda_k} \ (0 \leq k < s) \end{cases}.$$

Since $\lambda_{k+2} \in J_{u)}^q[\lambda_{k+1}] \subset J_{u)}^q[\lambda_k]$, we have

$$\frac{d_{\lambda_{k+2}} - d_{\lambda_k}}{i(\lambda_{k+2}) - i(\lambda_k)} \geq \iota(\lambda_k) = \frac{d_{\lambda_{k+1}} - d_{\lambda_k}}{i(\lambda_{k+1}) - i(\lambda_k)}$$

so that

$$\begin{aligned} (d_{\lambda_{k+1}} - d_{\lambda_k})(i(\lambda_{k+2}) - i(\lambda_{k+1})) &= (d_{\lambda_{k+1}} - d_{\lambda_k})(i(\lambda_{k+2}) - i(\lambda_k)) \\ &\quad - (d_{\lambda_{k+1}} - d_{\lambda_k})(i(\lambda_{k+1}) - i(\lambda_k)) \\ &\leq (d_{\lambda_{k+2}} - d_{\lambda_k})(i(\lambda_{k+1}) - i(\lambda_k)) \\ &\quad - (d_{\lambda_{k+1}} - d_{\lambda_k})(i(\lambda_{k+1}) - i(\lambda_k)) \\ &= (d_{\lambda_{k+2}} - d_{\lambda_{k+1}})(i(\lambda_{k+1}) - i(\lambda_k)), \end{aligned}$$

that is,

$$\iota(\lambda_k) \leq \iota(\lambda_{k+1}),$$

and so inductively one has $\iota(\lambda_k) \leq \iota(\lambda_{s-1})$. By (6.18) of Proposition 6.22,

$$(d_{\lambda_s} - d_{\lambda_{s-1}})(2u - n + 1 - i(\lambda_{s-1})) \leq (n + 1 - d_{\lambda_{s-1}})(i(\lambda_s) - i(\lambda_{s-1})),$$

and since

$$\begin{aligned} (d_{\lambda_s} - d_{\lambda_{s-1}})(2u - n + 1 - i(\lambda_s)) &= (d_{\lambda_s} - d_{\lambda_{s-1}})(2u - n + 1 - i(\lambda_{s-1})) \\ &\quad - (d_{\lambda_s} - d_{\lambda_{s-1}})(i(\lambda_s) - i(\lambda_{s-1})), \end{aligned}$$

we get

$$(d_{\lambda_s} - d_{\lambda_{s-1}})(2u - n + 1 - i(\lambda_s)) \le (n + 1 - d_{\lambda_s})(i(\lambda_s) - i(\lambda_{s-1}));$$

that is, $\iota(\lambda_{s-1}) \le \frac{1}{\theta}$. Combining this inequality with the previous one, we get (1), i.e., $\iota(\lambda_k) \le \frac{1}{\theta}$.

By the definition of the weights, we have

$$\sum_{j=0}^{q} \omega(a_j) = \sum_{k=0}^{s-1} \sum_{a_j \in \mathscr{A}_{\lambda_{k+1}} - \mathscr{A}_{\lambda_k}} \omega(a_j) + \sum_{a_j \in \mathscr{A} - \mathscr{A}_{\lambda_s}} \omega(a_j)$$

$$= \sum_{k=0}^{s-1} (d_{\lambda_{k+1}} - d_{\lambda_k}) + \frac{q + 1 - i(\lambda_s)}{\theta}$$

$$= d_{\lambda_s} + \frac{q + 1 - i(\lambda_s)}{\theta}.$$

Thus, by the definition of θ, one obtains

$$\theta\left(\sum_{j=0}^{q} \omega(a_j) - n - 1\right) = \theta(d_{\lambda_s} - n - 1) + q + 1 - i(\lambda_s) = q - 2u + n,$$

and (2) follows.

For (3), (1) and (2) imply that

$$(n + 1)\theta = \left(\sum_{j=0}^{q} \omega(a_j)\theta - q - 1\right) + 2u - n + 1 \le 2u - n + 1.$$

On the other hand

$$(n + 1)\theta = \theta d_{\lambda_s} + 2u - n + 1 - i(\lambda_s)$$
$$\ge d_{\lambda_s} + 2u - n + 1 - i(\lambda_s)$$
$$\ge 2u - n + 1 - (u - n) = u + 1.$$

Now we prove (4). First of all, we consider the case $i(\lambda_s \cup \sigma) \ge u + 1$. Then (1) and (6.14) imply

$$\theta\sum_{j=0}^{k} \omega\left(a_{\sigma(j)}\right) \le k + 1 = i(\sigma) \le u - n + d_\sigma.$$

Since $\mathscr{A}$ is in u-subgeneral position, then

$$n + 1 = d_{\lambda_s \cup \sigma} \le d_{\lambda_s} + d_\sigma,$$

that is, $d_\sigma \ge n + 1 - d_{\lambda_s} > 0$. Thus

$$\theta\sum_{j=0}^{k} \omega\left(a_{\sigma(j)}\right) \le d_\sigma + u - n = \left(1 + \frac{u - n}{d_\sigma}\right)d_\sigma$$

$$\le \left(1 + \frac{u - n}{n + 1 - d_{\lambda_s}}\right)d_\sigma = \frac{u + 1 - d_{\lambda_s}}{n + 1 - d_{\lambda_s}}d_\sigma$$

$$\le \frac{u + 1 + u - n - i(\lambda_s)}{n + 1 - d_{\lambda_s}}d_\sigma = \theta d_\sigma,$$

and hence (4) is proved.

Next we will assume $i(\lambda_s \cup \sigma) \le u$. For this case, we first prove the following relation:

$$\frac{i(\sigma - \lambda_s)}{\theta} = \sum_{j \in \mathrm{Im}(\sigma - \lambda_s)} \omega(a_j) \le d_\sigma - d_{\lambda_s \cap \sigma}. \qquad (6.21)$$

This is trivial if $\sigma \subset \lambda_s$. Suppose $\sigma \not\subset \lambda_s$. The equality is a direct consequence of the definition of $\omega(a_j)$. Thus, we need only to prove the inequality in detail. To do this, we first claim that

$$d_{\lambda_s \cup \sigma} > d_{\lambda_s}.$$

Assume, on the contrary, that $d_{\lambda_s \cup \sigma} = d_{\lambda_s}$. Then $\lambda_s \ne \emptyset$ which implies $s > 0$. Hence

$$d_{\lambda_s \cup \sigma} - d_{\lambda_{s-1}} = d_{\lambda_s} - d_{\lambda_{s-1}} > 0,$$

noting that $\lambda_{s-1} \subset \lambda_s \cup \sigma$, $i(\lambda_s \cup \sigma) < u+1$, and, hence,

$$\frac{d_{\lambda_s \cup \sigma} - d_{\lambda_{s-1}}}{i(\lambda_s \cup \sigma) - i(\lambda_{s-1})} \ge \iota(\lambda_{s-1}).$$

If the equality holds, then Proposition 6.22 means $\lambda_s \cup \sigma \subset \lambda_s$, i.e., $\sigma \subset \lambda_s$, which is impossible. Thus

$$\iota(\lambda_{s-1}) < \frac{d_{\lambda_s \cup \sigma} - d_{\lambda_{s-1}}}{i(\lambda_s \cup \sigma) - i(\lambda_{s-1})} \le \frac{d_{\lambda_s} - d_{\lambda_{s-1}}}{i(\lambda_s) - i(\lambda_{s-1})} = \iota(\lambda_{s-1}),$$

which is a contradiction, and hence our claim is proved. It follows that

$$\begin{aligned}
\frac{1}{\theta} &= \frac{n+1-d_{\lambda_s}}{2u-n+1-i(\lambda_s)} \le \iota(\lambda_s) \\
&\le \frac{d_{\lambda_s \cup \sigma} - d_{\lambda_s}}{i(\lambda_s \cup \sigma) - i(\lambda_s)} \le \frac{d_\sigma - d_{\lambda_s \cap \sigma}}{i(\sigma - \lambda_s)},
\end{aligned}$$

or

$$\frac{i(\sigma - \lambda_s)}{\theta} \le d_\sigma - d_{\lambda_s \cap \sigma}.$$

This is the inequality in (6.21).

We continue to prove (4). If $\lambda_s \cap \sigma = \emptyset$, the estimate follows from (6.21). Assume $\lambda_s \cap \sigma \ne \emptyset$, and define

$$\sigma_l = \sigma \cap \lambda_l, \qquad l = 0, ..., s,$$

with

$$\emptyset = \sigma_0 \subset \sigma_1 \subset \cdots \subset \sigma_s \subset \sigma.$$

Applying the inequality (6.21), we have

$$\begin{aligned}
\sum_{j=0}^{k} \omega\left(a_{\sigma(j)}\right) &= \sum_{l=0}^{s-1} \sum_{j \in \mathrm{Im}(\sigma_{l+1}-\sigma_l)} \omega(a_j) + \sum_{j \in \mathrm{Im}(\sigma - \sigma_s)} \omega(a_j) \\
&\le \sum_{l=0}^{s-1} (i(\sigma_{l+1}) - i(\sigma_l))\, \iota(\lambda_l) + d_\sigma - d_{\sigma_s}. \qquad (6.22)
\end{aligned}$$

We claim that: if $i(\sigma_{l+1}) > i(\sigma_l)$ $(0 \le l < s)$, then

$$d_{\lambda_l \cup \sigma_{l+1}} > d_{\lambda_l}.$$

If $l = 0$, then $i(\sigma_1) > 0$ implies

$$d_{\lambda_0 \cup \sigma_1} = d_{\sigma_1} > 0 = d_{\lambda_0}.$$

Suppose $l > 0$, and assume, on the contrary, that $d_{\lambda_l \cup \sigma_{l+1}} = d_{\lambda_l}$. Then

$$d_{\lambda_l \cup \sigma_{l+1}} - d_{\lambda_{l-1}} = d_{\lambda_l} - d_{\lambda_{l-1}} > 0.$$

We also have

$$\lambda_{l-1} \subset \lambda_l \cup \sigma_{l+1}, \quad i(\lambda_l \cup \sigma_{l+1}) \le i(\lambda_{l+1}) < u + 1.$$

Therefore

$$\iota(\lambda_{l-1}) \le \frac{d_{\lambda_l \cup \sigma_{l+1}} - d_{\lambda_{l-1}}}{i(\lambda_l \cup \sigma_{l+1}) - i(\lambda_{l-1})} \le \frac{d_{\lambda_l} - d_{\lambda_{l-1}}}{i(\lambda_l) - i(\lambda_{l-1})} = \iota(\lambda_{l-1}),$$

that is,

$$\iota(\lambda_{l-1}) = \frac{d_{\lambda_l \cup \sigma_{l+1}} - d_{\lambda_{l-1}}}{i(\lambda_l \cup \sigma_{l+1}) - i(\lambda_{l-1})}.$$

From Proposition 6.22, it follows that $\lambda_l \cup \sigma_{l+1} \subset \lambda_l$, i.e., $\sigma_{l+1} \subset \lambda_l$, and hence

$$\sigma_{l+1} = \sigma \cap \lambda_{l+1} \cap \lambda_l = \sigma_l.$$

Thus, we would obtain $i(\sigma_{l+1}) = i(\sigma_l)$, which is excluded by assumptions. The claim is proved. Further, we claim that for $l = 0, ..., s - 1$,

$$(i(\sigma_{l+1}) - i(\sigma_l)) \iota(\lambda_l) \le d_{\sigma_{l+1}} - d_{\sigma_l}.$$

If $i(\sigma_{l+1}) = i(\sigma_l)$, the estimate is trivial. Hence, we may assume $i(\sigma_{l+1}) > i(\sigma_l)$. The claim above implies $d_{\lambda_l \cup \sigma_{l+1}} > d_{\lambda_l}$. Noting that

$$\lambda_l \subset \lambda_l \cup \sigma_{l+1}, \quad i(\lambda_l \cup \sigma_{l+1}) \le i(\lambda_s \cup \sigma) < u + 1,$$

we have

$$\iota(\lambda_l) \le \frac{d_{\lambda_l \cup \sigma_{l+1}} - d_{\lambda_l}}{i(\lambda_l \cup \sigma_{l+1}) - i(\lambda_l)} \le \frac{d_{\sigma_{l+1}} - d_{\lambda_l \cap \sigma_{l+1}}}{i(\sigma_{l+1}) - i(\lambda_l \cap \sigma_{l+1})} = \frac{d_{\sigma_{l+1}} - d_{\sigma_l}}{i(\sigma_{l+1}) - i(\sigma_l)},$$

and then the claim is true. Finally, the claim and (6.22) yield

$$\sum_{j=0}^{k} \omega(a_{\sigma(j)}) \le \sum_{l=0}^{s-1} (d_{\sigma_{l+1}} - d_{\sigma_l}) + d_\sigma - d_{\sigma_s} = d_\sigma,$$

and (4) follows.

To prove (5), w. l. o. g., we may assume that $r_0 \geq r_1 \geq \cdots \geq r_q \geq 1$. Set $I_0 = \{\sigma(i) \mid a_{\sigma(i)} = a_{\sigma(0)}, 0 \leq i \leq k\}$, if $I_0 \neq \mathrm{Im}(\sigma)$ define $\lambda \in J_1^q$ by

$$\lambda(0) = \sigma(0), \quad \lambda(1) = \min\{p \mid p \in \mathrm{Im}(\sigma) - I_0\},$$

and set $I_1 = \{\sigma(i) \mid a_{\sigma(i)} \in E(\mathscr{A}_\lambda), 0 \leq i \leq k\}$. Inductively, if I_{j-1} is defined and if $I_{j-1} \neq \mathrm{Im}(\sigma)$, define $\lambda \in J_j^q$ by

$$\lambda(i) = \min\{p \mid p \in \mathrm{Im}(\sigma) - I_{i-1}\}, \quad i = 0, ..., j,$$

where $I_{-1} = \emptyset$, and write $I_j = \{\sigma(i) \mid a_{\sigma(i)} \in E(\mathscr{A}_\lambda), 0 \leq i \leq k\}$. This process stops at I_l with $l + 1 = d_\sigma$ such that $\lambda \in J_l(\mathscr{A})$ with $\mathrm{Im}\lambda \subset \mathrm{Im}\sigma$ satisfies $E(\mathscr{A}_\lambda) = E(\mathscr{A}_\sigma)$. Then, by construction,

$$\mathrm{Im}(\sigma) = \bigcup_{j=0}^{l}(I_j - I_{j-1})$$

is a disjoint union and

$$\prod_{j=0}^{k} r_{\sigma(j)}^{\omega(a_{\sigma(j)})} = \prod_{j=0}^{l} \prod_{i \in I_j - I_{j-1}} r_i^{\omega(a_i)} \leq \prod_{j=0}^{l} r_{\lambda(j)}^{\alpha_j},$$

because $r_0 \geq r_1 \geq \cdots \geq r_q \geq 1$, and by the definition,

$$r_{\lambda(j)} = \max_{i \in I_j - I_{j-1}} r_i, \quad \alpha_j = \sum_{i \in I_j - I_{j-1}} \omega(a_i).$$

For any $0 \leq p \leq l$, part (4) implies that

$$\sum_{j=0}^{p} \alpha_j = \sum_{j=0}^{p} \sum_{i \in I_j - I_{j-1}} \omega(a_i) = \sum_{i \in I_p} \omega(a_i) \leq p + 1.$$

Now by induction on l, we show that

$$\prod_{j=0}^{l} r_{\lambda(j)}^{\alpha_j} \leq \prod_{j=0}^{l} r_{\lambda(j)}.$$

For $l = 0$, $\alpha_0 \leq 1$ and one has $r_{\lambda(0)}^{\alpha_0} \leq r_{\lambda(0)}$ since $r_{\lambda(0)} \geq 1$. If the inequality is true for $l = p$, then

$$\alpha_{p+1} \leq p + 2 - \sum_{j=0}^{p} \alpha_j,$$

and hence

$$\prod_{j=0}^{p+1} r_{\lambda(j)}^{\alpha_j} \leq r_{\lambda(p+1)}^{p+2} \prod_{j=0}^{p} \left(\frac{r_{\lambda(j)}}{r_{\lambda(p+1)}} \right)^{\alpha_j}$$

$$\leq r_{\lambda(p+1)}^{p+2} \prod_{j=0}^{p} \frac{r_{\lambda(j)}}{r_{\lambda(p+1)}} \leq \prod_{j=0}^{p+1} r_{\lambda(j)}.$$

This completes the proof of the lemma. $\square$

For more detail, see Chen [21], Ru-Wong [115] or Fujimoto [36].

6.5 Degenerate holomorphic curves

Let κ be an algebraically closed field of characteristic zero, complete for a non-trivial non-Archimedean absolute value $|\cdot|$. Let V be a normed vector space of dimension $n+1 > 0$ over κ and let $|\cdot|$ be a norm defined over a base $e = (e_0, ..., e_n)$ of V. Let u be an integer with $n \leq u < q$ and let $\mathscr{A} = \{a_0, a_1, ..., a_q\}$ be a family of points $a_j \in \mathbb{P}(V^*)$ in u-subgeneral position. Define the *gauge* $\Gamma(\mathscr{A})$ of $\mathscr{A}$ with respect to the norm $\rho = |\cdot|$ by

$$\Gamma(\mathscr{A}) = \Gamma(\mathscr{A}; \rho) = \inf_{\lambda \in J_n(\mathscr{A})} \{|a_{\lambda(0)} \wedge \cdots \wedge a_{\lambda(n)}|\}$$

with $0 < \Gamma(\mathscr{A}) \leq 1$.

Lemma 6.24. *For* $x \in \mathbb{P}(V)$, $0 < b \in \mathbb{R}$, *define*

$$\mathscr{A}(x, b) = \mathscr{A}(x, b; \rho) = \{j \in \mathbb{Z}[0, q] \mid |x, a_j| < b\}.$$

If $0 < b \leq \Gamma(\mathscr{A})$, *then* $\#\mathscr{A}(x, b) \leq u$.

Proof. Assume that $\#\mathscr{A}(x, b) \geq u+1$. Then $\lambda \in J_n(\mathscr{A})$ exists such that $\mathrm{Im}\lambda \subseteq \mathscr{A}(x, b)$. Hence

$$|x, a_{\lambda(j)}| < b, \quad j = 0, ..., n.$$

Then Lemma 6.4 implies

$$0 < \Gamma(\mathscr{A}) \leq |a_{\lambda(0)} \wedge \cdots \wedge a_{\lambda(n)}| \\ \leq \max_{0 \leq j \leq n} |x, a_{\lambda(j)}| < b \leq \Gamma(\mathscr{A}),$$

which is impossible. $\qquad \square$

Lemma 6.25. *Take* $x \in \mathbb{P}(V)$ *such that* $|x, a_j| > 0$ *for* $j = 0, ..., q$. *Then*

$$\prod_{j=0}^{q} \left(\frac{1}{|x, a_j|}\right)^{\omega(a_j)} \leq \left(\frac{1}{\Gamma(\mathscr{A})}\right)^{q-u} \max_{\lambda \in J_n(\mathscr{A})} \prod_{j=0}^{n} \frac{1}{|x, a_{\lambda(j)}|},$$

where $\omega : \mathscr{A} \longrightarrow \mathbb{R}(0, 1]$ *is the Nochka weight.*

Proof. Take $b = \Gamma(\mathscr{A})$. Lemma 6.24 implies $\#\mathscr{A}(x, b) \leq u$. Thus $\sigma \in J_u^q$ exists such that $\mathscr{A}(x, b) \subset \mathrm{Im}\sigma$. Note that $E(\mathscr{A}_\sigma) = V^*$. By Lemma 6.23, there exists $\lambda \in J_n(\mathscr{A})$ with $\mathrm{Im}\lambda \subset \mathrm{Im}\sigma$ such that $E(\mathscr{A}_\lambda) = E(\mathscr{A}_\sigma)$, and such that

$$\prod_{j=0}^{u} \left(\frac{1}{|x, a_{\sigma(j)}|}\right)^{\omega(a_{\sigma(j)})} \leq \prod_{j=0}^{n} \frac{1}{|x, a_{\lambda(j)}|} \\ \leq \max_{\lambda \in J_n(\mathscr{A})} \prod_{j=0}^{n} \frac{1}{|x, a_{\lambda(j)}|}.$$

Set $C = \mathbb{Z}[0, q] - \mathrm{Im}\sigma$. Thus $|x, a_j| \geq b$ for $j \in C$. Hence

$$\prod_{j \in C} \left(\frac{1}{|x, a_j|} \right)^{\omega(a_j)} \leq \prod_{j \in C} \frac{1}{|x, a_j|}$$

$$\leq \left(\frac{1}{b} \right)^{\#C} = \left(\frac{1}{\Gamma(\mathscr{A})} \right)^{q-u}$$

Thus the lemma follows. $\square$

Theorem 6.26. *Let $\mathscr{A} = \{a_j\}_{j=0}^q$ be a finite family of points $a_j \in \mathbb{P}(V^*)$ in u-subgeneral position with $u \leq 2u - n < q$. Let $f : \kappa \longrightarrow \mathbb{P}(V)$ be a non-Archimedean holomorphic curve which is linearly non-degenerate. Then*

$$(q - 2u + n)T(r, f) \leq \sum_{j=0}^q N_f(r, a_j) - \theta N_{\mathrm{Ram}}(r, f) - \frac{n(n+1)}{2} \theta \log r + O(1),$$

where $\theta \geq 1$ is the Nochka constant.

Proof. We will adept the notations that were used in the proof of Theorem 6.13, and w. l. o. g., assume $|\tilde{a}_j| = 1$ for $j = 0, ..., q$. Lemma A.23 implies

$$\prod_{j=0}^q \left(\frac{1}{|f, a_j|} \right)^{\omega(a_j)} \leq \left(\frac{1}{\Gamma(\mathscr{A})} \right)^{q-u} \max_{\lambda \in J_n(\mathscr{A})} \prod_{j=0}^n \frac{1}{|f, a_{\lambda(j)}|},$$

which yields

$$\prod_{j=0}^q \left(\frac{1}{\mu(r, f \angle a_j)} \right)^{\omega(a_j)} \leq \left(\frac{1}{\Gamma(\mathscr{A})} \right)^{q-u} \max_{\lambda \in J_n(\mathscr{A})} \prod_{j=0}^n \frac{1}{\mu(r, f \angle a_{\lambda(j)})}$$

$$= \left(\frac{1}{\Gamma(\mathscr{A})} \right)^{q-u} \max_{\lambda \in J_n(\mathscr{A})} \prod_{j=0}^n \frac{\mu(r, \tilde{f})}{\mu(r, F_{\lambda(j)})}$$

$$= \left(\frac{1}{\Gamma(\mathscr{A})} \right)^{q-u} \frac{\mu(r, \tilde{f})^{n+1}}{\mu(r, \mathbf{W})} \max_{\lambda \in J_n(\mathscr{A})} \frac{1}{|c_\lambda|} \mu\left(r, \frac{\mathbf{W}_\lambda}{F_{\lambda(0)} \cdots F_{\lambda(n)}} \right)$$

$$\leq \left(\frac{1}{\Gamma(\mathscr{A})} \right)^{q-u} \frac{\mu(r, \tilde{f})^{n+1}}{\mu(r, \mathbf{W})} r^{-\frac{n(n+1)}{2}} \max_{\lambda \in J_n(\mathscr{A})} \frac{1}{|c_\lambda|}.$$

We obtain

$$\sum_{j=0}^q \omega(a_j) m_f(r, a_j) \leq (n+1)T(r, f) - N\left(r, \frac{1}{\mathbf{W}} \right) - \frac{n(n+1)}{2} \log r + O(1). \qquad (6.23)$$

By the properties of the Nochka weights, we have

$$\sum_{j=0}^q m_f(r, a_j) = \sum_{j=0}^q \theta \omega(a_j) m_f(r, a_j) + \sum_{j=0}^q (1 - \theta \omega(a_j)) m_f(r, a_j)$$

$$\leq \ \theta(n+1)T(r,f) - \theta N\left(r,\frac{1}{\mathbf{W}}\right) - \frac{n(n+1)}{2}\theta\log r$$

$$+ \left(1 + q - \theta\sum_{j=0}^{q}\omega(a_j)\right)T(r,f) + O(1)$$

$$\leq \ \left(1 + q - \theta\left(\sum_{j=0}^{q}\omega(a_j) - n - 1\right)\right)T(r,f)$$

$$-\theta N\left(r,\frac{1}{\mathbf{W}}\right) - \frac{n(n+1)}{2}\theta\log r + O(1)$$

$$= \ (2u - n + 1)T(r,f) - \theta N\left(r,\frac{1}{\mathbf{W}}\right)$$

$$-\frac{n(n+1)}{2}\theta\log r + O(1),$$

and, hence, the theorem follows from this and the first main theorem. $\qquad\square$

Corollary 6.27. *Assumptions as in Theorem 6.26. Then*

$$(q - 2u + n)T(r,f) \leq \sum_{j=0}^{q}N_{f,n}(r,a_j) - \frac{n(u+1)}{2}\log r + O(1).$$

Corollary 6.28. *Assumptions as in Theorem 6.26. Then*

$$\sum_{j=0}^{q}\delta_f(a_j,n) \leq 2u - n + 1.$$

Now we eliminate the restriction of non-degeneracy on f. Take a reduced representation $\tilde{f} : \kappa \longrightarrow V_*$ of a non-constant holomorphic curve $f : \kappa \longrightarrow \mathbb{P}(V)$ and define a linear subspace of V^*

$$E[f] = \{\alpha \in V^* \mid \langle \tilde{f}, \alpha \rangle \equiv 0\},$$

and write

$$\ell_f = \dim E[f], \quad k = n - \ell_f.$$

Then V^* is decomposed into a direct sum

$$V^* = W^* \oplus E[f],$$

where W^* is a $k+1$ dimensional subspace of V^*. Then f is said to be k-*flat*. In order to simplify our notation, we define $\ell_f = 0$ if f is linearly non-degenerate, that is, $E[f] = \emptyset$, and say that f is n-*flat*.

Lemma 6.29. *The number k is nonnegative, i.e., $0 \leq \ell_f \leq n$.*

Proof. If $k < 0$, that is, $\ell_f = n + 1$, there is $\{\alpha_0, ..., \alpha_n\} \subset E[f]$ such that

$$\alpha_0 \wedge \cdots \wedge \alpha_n \neq 0; \quad \langle \tilde{f}, \alpha_j \rangle \equiv 0 \ (0 \leq j \leq n).$$

By Cramer's rule, $\tilde{f} \equiv 0$, which is impossible.   □

From now on, we assume that $\mathscr{A} = \{a_j\}_{j=0}^q$ is in general position and assume that f is non-constant and k-flat with $0 \leq k \leq n < q$ such that each pair (f, a_j) is free for $j = 0, ..., q$. We take a base $\epsilon = (\epsilon_0, ..., \epsilon_n)$ of V^* such that $\epsilon_0, ..., \epsilon_k$ and $\epsilon_{k+1}, ..., \epsilon_n$ is a base of W^* and $E[f]$, respectively. Let $e = (e_0, ..., e_n)$ be the dual base of ϵ. Let W be the vector space spanned by $e_0, ..., e_k$ over κ. Thus the reduced representation $\tilde{f} : \kappa \longrightarrow V_*$ is given by

$$\tilde{f} = \sum_{j=0}^k \tilde{f}_j e_j = \sum_{j=0}^k \langle \tilde{f}, \epsilon_j \rangle e_j$$

such that $\langle \tilde{f}, \epsilon_0 \rangle, ..., \langle \tilde{f}, \epsilon_k \rangle$ are holomorphic and linearly independent over κ. Hence a linearly non-degenerate holomorphic curve $\hat{f} : \kappa \longrightarrow \mathbb{P}(W)$ is defined with a reduced representation

$$\tilde{\hat{f}} = \tilde{f} = \sum_{j=0}^k \langle \tilde{f}, \epsilon_j \rangle e_j : \kappa \longrightarrow W_*.$$

Therefore, we obtain

$$T(r, \hat{f}) = \log \mu(r, \tilde{f}) + O(1) = T(r, f) + O(1). \tag{6.24}$$

If $k = 0$, then $T(r, \hat{f})$ is constant. The inequality (6.24) will be impossible. Thus, we must have $k \geq 1$.

Take $\tilde{a}_j \in V^* - \{0\}$ with $\mathbb{P}(\tilde{a}_j) = a_j$ and write

$$\tilde{a}_j = \sum_{i=0}^n \langle e_i, \tilde{a}_j \rangle \epsilon_i, \quad j = 0, ..., q.$$

Define

$$\tilde{\hat{a}}_j = \sum_{i=0}^k \langle e_i, \tilde{a}_j \rangle \epsilon_i \in W^* - \{0\}, \quad \hat{a}_j = \mathbb{P}\left(\tilde{\hat{a}}_j \right) \in \mathbb{P}(W^*), \quad j = 0, ..., q.$$

Lemma 6.30. *The family $\hat{\mathscr{A}} = \{\hat{a}_j\}_{j=0}^q$ is in n-subgeneral position.*

Proof. Take $\sigma \in J_n^q$. Then $\tilde{a}_\sigma \neq 0$ since $\mathscr{A}$ is in general position, and hence

$$\det(\langle e_i, \tilde{a}_{\sigma(j)} \rangle) \neq 0 \quad (0 \leq i, j \leq n).$$

Therefore, there is a $\lambda \in J_k^q$ with $\mathrm{Im}\lambda \subseteq \mathrm{Im}\sigma$ such that

$$\det(\langle e_s, \tilde{a}_{\lambda(t)} \rangle) \neq 0 \quad (0 \leq s, t \leq k).$$

We have

$$\tilde{\hat{a}}_\lambda = \det(\langle e_s, \tilde{g}_{\lambda(t)} \rangle) \epsilon_0 \wedge \cdots \wedge \epsilon_k \neq 0.$$

Hence $\lambda \in J_k(\hat{\mathscr{A}})$. Thus $\hat{\mathscr{A}}$ is in n-subgeneral position.   □

Theorem 6.31. *Let $\mathscr{A} = \{a_j\}_{j=0}^{q}$ be a finite family of points $a_j \in \mathbb{P}(V^*)$ in general position. Take an integer k with $1 \le k \le n \le 2n - k < q$. Let $f : \kappa \longrightarrow \mathbb{P}(V)$ be a non-constant holomorphic curve that is k-flat such that each pair (f, a_j) is free for $j = 0, ..., q$. Then*

$$(q - 2n + k)T(r, f) \le \sum_{j=0}^{q} N_{f,k}(r, a_j) - \frac{k(n+1)}{2} \log r + O(1).$$

Proof. Note that

$$\langle \tilde{f}, \tilde{a}_j \rangle = \sum_{i=0}^{n} \langle \tilde{f}, \epsilon_i \rangle \langle e_i, \tilde{a}_j \rangle = \sum_{i=0}^{k} \langle \tilde{f}, \epsilon_i \rangle \langle e_i, \tilde{a}_j \rangle = \langle \hat{f}, \tilde{\tilde{a}}_j \rangle.$$

We obtain

$$N_f(r, a_j) = N_{\hat{f}}(r, \hat{a}_j).$$

By applying Theorem 6.26 to $\hat{f}$, then

$$(q - 2n + k)T(r, \hat{f}) \le \sum_{j=0}^{q} N_{\hat{f},k}(r, \hat{a}_j) - \frac{k(n+1)}{2} \log r + O(1),$$

and the theorem follows from the facts above. $\qquad\square$

Corollary 6.32. *With the assumptions as in Theorem 6.31,*

$$\sum_{j=0}^{q} \delta_f(a_j, k) \le 2n - k + 1.$$

6.6 Uniqueness of holomorphic curves

Let κ be an algebraically closed field of characteristic zero, complete for a non-trivial non-Archimedean absolute value $|\cdot|$. Let V be a normed vector space of dimension $n+1 > 0$ over κ and let $|\cdot|$ be a norm defined over a base $e = (e_0, ..., e_n)$ of V. We consider a holomorphic curve $f : \kappa \longrightarrow \mathbb{P}(V)$ with a reduced representation

$$\tilde{f} = \tilde{f}_0 e_0 + \cdots + \tilde{f}_n e_n : \kappa \longrightarrow V_*.$$

Take $a \in \mathbb{P}(V^*)$. Then there exists a $\tilde{a} \in V^* - \{0\}$ with $\mathbb{P}(\tilde{a}) = a$. Write

$$\tilde{a} = \tilde{a}_0 \epsilon_0 + \cdots + \tilde{a}_n \epsilon_n,$$

where $\epsilon = (\epsilon_0, ..., \epsilon_n)$ is the dual of e, and set

$$F = \left\langle \tilde{f}, \tilde{a} \right\rangle = \tilde{a}_0 \tilde{f}_0 + \tilde{a}_1 \tilde{f}_1 + \cdots + \tilde{a}_n \tilde{f}_n.$$

We define the *pull-back divisor* of $\ddot{E}[a]$ under f by

$$E_f(a) = f^* \ddot{E}[a] = E_F(0) = \{(\mu_F^0(z), z) \mid z \in \kappa\},$$

and denote the *preimage* of $\ddot{E}[a]$ under f by

$$\overline{E}_f(a) = f^{-1}(\ddot{E}[a]) = \{z \in \kappa \mid \mu_F^0(z) > 0\}.$$

Theorem 6.33. *Let $\mathscr{A} = \{a_0, a_1, ..., a_q\}$ be a family of $q+1$ points $a_j \in \mathbb{P}(V^*)$ in general position, and consider two holomorphic curves f, $g : \kappa \longrightarrow \mathbb{P}(V)$ satisfying the following conditions:*

$$f(\kappa) \not\subset \ddot{E}[a_j], \quad g(\kappa) \not\subset \ddot{E}[a_j], \quad j = 0, ..., q. \tag{6.25}$$

Suppose that

$$E_f(a_j) = E_g(a_j), \quad j = 0, 1, ..., q. \tag{6.26}$$

Then the following results hold:
*   1) if $q = n$, there is a projective linear transformation L of $\mathbb{P}(V)$ such that $L \circ g = f$;*
*   2) if $q = n+1$, and either f or g is linearly non-degenerate, then $f = g$.*

Proof. Take a reduced representation

$$\tilde{f} = \tilde{f}_0 e_0 + \cdots + \tilde{f}_n e_n : \kappa \longrightarrow V_*$$

and

$$\tilde{g} = \tilde{g}_0 e_0 + \cdots + \tilde{g}_n e_n : \kappa \longrightarrow V_*,$$

of f and g, respectively, and take $\tilde{a}_j \in V^* - \{0\}$ with $\mathbb{P}(\tilde{a}_j) = a_j$. Write

$$\tilde{a}_j = \tilde{a}_{j0}\epsilon_0 + \cdots + \tilde{a}_{jn}\epsilon_n, \quad j = 0, ..., q.$$

For $j = 0, 1, ..., q$, set

$$F_j = \left\langle \tilde{f}, \tilde{a}_j \right\rangle = \tilde{a}_{j0}\tilde{f}_0 + \tilde{a}_{j1}\tilde{f}_1 + \cdots + \tilde{a}_{jn}\tilde{f}_n$$

and

$$G_j = \langle \tilde{g}, \tilde{a}_j \rangle = \tilde{a}_{j0}\tilde{g}_0 + \tilde{a}_{j1}\tilde{g}_1 + \cdots + \tilde{a}_{jn}\tilde{g}_n.$$

The condition (6.26) leads to

$$E_{F_j}(0) = E_{G_j}(0), \quad j = 0, 1, ..., q. \tag{6.27}$$

Thus, there are constants $c_j \in \kappa_*$ such that

$$F_j = c_j G_j, \quad j = 0, 1, ..., q. \tag{6.28}$$

For $\lambda \in J_n^q$ $(q \geq n)$, define

$$A_\lambda = \begin{pmatrix} \tilde{a}_{\lambda(0)0} & \tilde{a}_{\lambda(0)1} & \cdots & \tilde{a}_{\lambda(0)n} \\ \tilde{a}_{\lambda(1)0} & \tilde{a}_{\lambda(1)1} & \cdots & \tilde{a}_{\lambda(1)n} \\ \cdots\cdots\cdots\cdots\cdots\cdots\cdots \\ \tilde{a}_{\lambda(n)0} & \tilde{a}_{\lambda(n)1} & \cdots & \tilde{a}_{\lambda(n)n} \end{pmatrix}, \quad C_\lambda = \begin{pmatrix} c_{\lambda(0)} & \cdots & 0 \\ \vdots & \ddots & \vdots \\ 0 & \cdots & c_{\lambda(n)} \end{pmatrix},$$

and write

$$\hat{f} = \begin{pmatrix} \tilde{f}_0 \\ \vdots \\ \tilde{f}_n \end{pmatrix}, \quad \hat{g} = \begin{pmatrix} \tilde{g}_0 \\ \vdots \\ \tilde{g}_n \end{pmatrix}.$$

Then (6.28) gives

$$A_\lambda \hat{f} = C_\lambda A_\lambda \hat{g}, \quad \lambda \in J_n^q,$$

that is,

$$\hat{f} = A_\lambda^{-1} C_\lambda A_\lambda \hat{g}, \quad \lambda \in J_n^q, \tag{6.29}$$

and hence part (1) of the theorem follows.

To prove (2), w. l. o. g., we assume that g is linearly non-degenerate. For any λ, $\sigma \in J_n^q$, then (6.29) implies

$$\hat{f} = A_\lambda^{-1} C_\lambda A_\lambda \hat{g} = A_\sigma^{-1} C_\sigma A_\sigma \hat{g},$$

which yields

$$A_\lambda^{-1} C_\lambda A_\lambda = A_\sigma^{-1} C_\sigma A_\sigma,$$

since g is linearly non-degenerate. Hence we have

$$\det(C_\lambda) = \det(C_\sigma), \quad \lambda, \ \sigma \in J_n^q. \tag{6.30}$$

If $q = n + 1$, then (6.30) yields easily

$$c_0 = c_1 = \cdots = c_q.$$

Hence $\hat{f} = c_0 \hat{g}$ from (6.29) and then $f = g$. $\qquad\square$

Theorem 6.34. *Let $\mathscr{A} = \{a_0, a_1, ..., a_q\}$ be a family of $q + 1$ points $a_j \in \mathbb{P}(V^*)$ in general position, and consider two k-flat non-constant holomorphic curves $f, \ g : \kappa \longrightarrow \mathbb{P}(V)$ satisfying the conditions:*

$$f(\kappa) \not\subset \ddot{E}[a_j], \ g(\kappa) \not\subset \ddot{E}[a_j], \ j = 0, ..., q, \tag{6.31}$$

and

$$f^{-1}\left(\ddot{E}[a_i] \cap \ddot{E}[a_j]\right) = \emptyset \ (0 \le i < j \le q). \tag{6.32}$$

Suppose that

$$\overline{E}_f(a_j) = \overline{E}_g(a_j), \quad j = 0, 1, ..., q, \tag{6.33}$$

and

$$f(z) = g(z), \quad z \in \bigcup_{j=0}^q \overline{E}_f(a_j). \tag{6.34}$$

If $q = 2n + k$, then $f = g$.

Proof. Assume, on the contrary, that $f \not\equiv g$. Then we have

$$(q - 2n + k)T(r, f) \ \le \ \sum_{j=0}^q N_{f,k}(r, a_j) - \frac{k(n+1)}{2} \log r + O(1)$$

$$\le \ k \sum_{j=0}^q N_{f,1}(r, a_j) - \frac{k(n+1)}{2} \log r + O(1),$$

which also holds if f is replaced by g, and hence

$$(q - 2n + k)T(r) \leq 2k \sum_{j=0}^{q} N_{f,1}(r, a_j) - k(n+1)\log r + O(1),$$

where $T(r) = T(r, f) + T(r, g)$. Now from the conditions (6.32) and (6.34), we have

$$\sum_{j=0}^{q} N_{f,1}(r, a_j) \leq N_{f \wedge g}(r).$$

By applying (6.4) to the operator $\odot = \wedge$, we obtain

$$T(r) = N_{f \wedge g}(r) + m_{f \wedge g}(r) + T(r, f \wedge g) + O(1).$$

Thus, we have

$$(q - 2n + k)T(r) \leq 2kT(r) - k(n+1)\log r + O(1),$$

that is

$$(q - 2n - k)T(r) + k(n+1)\log r \leq O(1).$$

This is impossible for $q = 2n + k$. $\qquad\qquad\qquad\qquad\qquad\qquad\qquad\qquad\qquad\square$

For the complex variables case, the results (1) and (2) in Theorem 6.33 are true for $q = 3n$ and $q = 3n + 1$, respectively, and Theorem 6.34 holds if both $q = 3n + 1$ and f, g are linearly non-degenerate. These theorems are due to H. Fujimoto [35] and L. M. Smiley [124], respectively.

6.7   Second main theorem for hypersurfaces

Let κ be an algebraically closed field of characteristic zero, complete for a non-trivial non-Archimedean absolute value $|\cdot|$. In this section, we will need the Hilbert's Nullstellensatz (cf. [132]):

Theorem 6.35. *Take polynomials* $P_1, ..., P_r$ *and* P *in* $\kappa[\xi_0, \cdots, \xi_n]$. *If* P *vanishes at all the common zeros of* $P_1, ..., P_r$, *then there exist polynomials* $Q_1, ..., Q_r$ *in* $\kappa[\xi_0, \cdots, \xi_n]$ *such that*

$$P^m = Q_1 P_1 + \cdots + Q_r P_r$$

holds for some natural number m.

Let V be a normed vector space of dimension $n+1 > 0$ over κ and let $|\cdot|$ be a norm defined over a base $e = (e_0, ..., e_n)$ of V. Take a sequence $\{m_0, m_1, ..., m_q\}$ of positive integers. Let $\mathscr{A} = \{a_0, a_1, ..., a_q\}$ be a family of points $a_j \in \mathbb{P}\left(\amalg_{m_j} V^*\right)$. Take $\alpha_j \in \amalg_{m_j} V^* - \{0\}$ with $\mathbb{P}(\alpha_j) = a_j$, and define

$$\tilde{\alpha}_j(\xi) = \left\langle \xi^{\amalg m_j}, \alpha_j \right\rangle, \quad \xi \in V, \ j = 0, 1, ..., q.$$

Definition 6.36. *The family* $\mathscr{A} = \{a_0, a_1, ..., a_q\}$ $(q \geq n)$ *is said to be admissible if, for every* $\lambda \in J_n^q$, *the system*

$$\tilde{\alpha}_{\lambda(i)}(\xi) = 0, \quad i = 0, 1, ..., n \tag{6.35}$$

has only the trivial solution $\xi = 0$ *in* V.

Assume that the family $\mathscr{A} = \{a_0, a_1, ..., a_q\}$ $(q \geq n)$ is admissible. Write $\xi = \xi_0 e_0 + \cdots \xi_n e_n$. By Theorem 6.35, for all $k, 0 \leq k \leq n$, the identity

$$\xi_k^m = \sum_{i=0}^{n} b_{ik}^{\lambda}(\xi) \tilde{\alpha}_{\lambda(i)}(\xi) \ (\lambda \in J_n^q) \tag{6.36}$$

is satisfied for some natural number m with

$$m \geq \max_{0 \leq j \leq q} m_j,$$

where $b_{ik}^{\lambda} \in \kappa[\xi_0, \cdots, \xi_n]$ are homogeneous polynomials of degree $m - m_{\lambda(i)}$ whose coefficients are integral-valued polynomials at the coefficients of $\tilde{\alpha}_{\lambda(i)}$ $(i = 0, ..., n)$. Write

$$b_{ik}^{\lambda}(\xi) = \sum_{\sigma \in J_{n,m-m_{\lambda(i)}}} b_{\sigma ik}^{\lambda} \xi_0^{\sigma(0)} \cdots \xi_n^{\sigma(n)}, \quad b_{\sigma ik}^{\lambda} \in \kappa. \tag{6.37}$$

From (6.36) and (6.37), we have

$$|\xi_k|^m \leq \left(\max_{i,\sigma} \left| b_{\sigma ik}^{\lambda} \right| \cdot |\alpha_{\lambda(i)}| \right) \max_{0 \leq i \leq n} \left\{ \frac{|\tilde{\alpha}_{\lambda(i)}(\xi)|}{|\xi|^{m_{\lambda(i)}} |\alpha_{\lambda(i)}|} \right\} |\xi|^m. \tag{6.38}$$

Note that

$$\max_{0 \leq k \leq n} |\xi_k|^m = |\xi|^m, \quad |\tilde{\alpha}_j(\xi)| \leq |\xi|^{m_j} |\alpha_j|. \tag{6.39}$$

By maximizing the inequalities (6.38) over $k, 0 \leq k \leq n$, and using (6.39), we obtain

$$1 \leq \max_{k,i,\sigma} \left| b_{\sigma ik}^{\lambda} \right| \cdot |\alpha_{\lambda(i)}|. \tag{6.40}$$

Define the *gauge*

$$\Gamma(\mathscr{A}) = \min_{\lambda \in J_n^q} \min_{k,i,\sigma} \left\{ \frac{1}{\left| b_{\sigma ik}^{\lambda} \right| \cdot |\alpha_{\lambda(i)}|} \right\}, \tag{6.41}$$

with $0 < \Gamma(\mathscr{A}) \leq 1$. From (6.38), (6.39) and (6.41), we obtain

$$\Gamma(\mathscr{A}) \leq \max_{0 \leq i \leq n} \left\{ \frac{|\tilde{\alpha}_{\lambda(i)}(\xi)|}{|\xi|^{m_{\lambda(i)}} |\alpha_{\lambda(i)}|} \right\},$$

that is,

$$\Gamma(\mathscr{A}) \leq \max_{0 \leq i \leq n} |x^{\mathrm{II}m_{\lambda(i)}}, a_{\lambda(i)}|, \quad \lambda \in J_n^q, \quad x \in \mathbb{P}(V). \tag{6.42}$$

For $x \in \mathbb{P}(V), 0 < r \in \mathbb{R}$, define

$$\mathscr{A}(x,r) = \left\{ j \mid |x^{\mathrm{II}m_j}, a_j| < r, \quad 0 \leq j \leq q \right\}. \tag{6.43}$$

Proposition 6.37. *If $x \in \mathbb{P}(V)$ and $0 < r \leq \Gamma(\mathscr{A})$, then*

$$\#\mathscr{A}(x,r) \leq n.$$

Proof. Assume that $\#\mathscr{A}(x,r) \geq n+1$. Then $\lambda \in J_n^q$ exists such that

$$\{\lambda(0), ..., \lambda(n)\} \subseteq \mathscr{A}(x,r).$$

Hence

$$|x^{\amalg m_{\lambda(i)}}, a_{\lambda(i)}| < r \leq \Gamma(\mathscr{A}), \quad i = 0, \cdots, n.$$

which is impossible according to (6.42). $\square$

Proposition 6.38. *Take $x \in \mathbb{P}(V)$ such that $|x^{\amalg m_j}, a_j| > 0$ for $j = 0, \cdots, q$. Then*

$$\prod_{j=0}^{q} \frac{1}{|x^{\amalg m_j}, a_j|} \leq \left(\frac{1}{\Gamma(\mathscr{A})}\right)^{q-n} \max_{\lambda \in J_n^q} \left\{ \prod_{i=0}^{n} \frac{1}{|x^{\amalg m_{\lambda(i)}}, a_{\lambda(i)}|} \right\} \tag{6.44}$$

$$\leq \left(\frac{1}{\Gamma(\mathscr{A})}\right)^{q+1-n} \max_{\lambda \in J_{n-1}^q} \left\{ \prod_{i=0}^{n-1} \frac{1}{|x^{\amalg m_{\lambda(i)}}, a_{\lambda(i)}|} \right\}. \tag{6.45}$$

Proof. Take $r = \Gamma(\mathscr{A})$. Proposition 6.37 implies $\#\mathscr{A}(x,r) \leq n$. Thus $\sigma \in J_n^q$ exists such that $\mathscr{A}(x,r) \subseteq \{\sigma(0), ..., \sigma(n)\}$. Note that $\mathrm{Im}\lambda - \mathscr{A}(x,r) \neq \emptyset$ for any $\lambda \in J_n^q$. Then we have

$$\prod_{j=0}^{q} \frac{1}{|x^{\amalg m_j}, a_j|} \leq r^{n-q} \prod_{i=0}^{n} \frac{1}{|x^{\amalg m_{\sigma(i)}}, a_{\sigma(i)}|}$$

$$\leq \left(\frac{1}{\Gamma(\mathscr{A})}\right)^{q-n} \max_{\lambda \in J_n^q} \left\{ \prod_{i=0}^{n} \frac{1}{|x^{\amalg m_{\lambda(i)}}, a_{\lambda(i)}|} \right\}$$

$$\leq \left(\frac{1}{\Gamma(\mathscr{A})}\right)^{q+1-n} \max_{\lambda \in J_{n-1}^q} \left\{ \prod_{i=0}^{n-1} \frac{1}{|x^{\amalg m_{\lambda(i)}}, a_{\lambda(i)}|} \right\}.$$

$\square$

Definition 6.39. *A holomorphic curve $f : \kappa \longrightarrow \mathbb{P}(V)$ is said to be algebraically non-degenerate of degree m if $f(\kappa) \not\subset \ddot{E}[a]$ for any $a \in \mathbb{P}(\amalg_m V^*)$, and is called algebraically non-degenerate if for each positive integer m, it is algebraically non-degenerate of degree m.*

Theorem 6.40. *Let $\mathscr{A} = \{a_j\}_{j=0}^{q}$ be a finite admissible family of points $a_j \in \mathbb{P}\left(\amalg_{m_j} V^*\right)$ with $n < q$. Let $f : \kappa \longrightarrow \mathbb{P}(V)$ be a holomorphic curve which is algebraically non-degenerate. Set*

$$d = \max_{\lambda \in J_n^q} \left\{ \frac{1}{n+1} \sum_{i=0}^{n} m_{\lambda(i)} \right\}.$$

Then

$$\left(\sum_{j=0}^{q} m_j - d(n+1) \right) T(r,f) \leq \sum_{j=0}^{q} N_{f^{\amalg m_j}}(r, a_j) - \frac{n(n+1)}{2} \log r + O(1).$$

Proof. W. l. o. g., assume $|\alpha_j| = 1$ for $j = 0, ..., q$. Write

$$F_j = \tilde{\alpha}_j \circ \tilde{f} = \left\langle \tilde{f}^{\text{II}m_j}, \alpha_j \right\rangle, \quad j = 0, 1, ..., q,$$

where $\tilde{f} : \kappa \longrightarrow V_*$ is a reduced representation of f. Since f is algebraically non-degenerate, then for each $\lambda \in J_n^q$, $F_{\lambda(0)}, ..., F_{\lambda(n)}$ are linearly independent, and hence

$$\mathbf{W}_\lambda = \mathbf{W}(F_{\lambda(0)}, ..., F_{\lambda(n)}) \not\equiv 0,$$

by using Lemma 6.12. Since f is non-constant, then $\mu(r, \tilde{f}) \to \infty$ as $r \to \infty$. So we may assume $\mu(r, \tilde{f}) \geq 1$ for $r \geq r_0$. Proposition 6.38 implies

$$\prod_{j=0}^{q} \frac{1}{|f^{\text{II}m_j}, a_j|} \leq \left(\frac{1}{\Gamma(\mathscr{A})}\right)^{q-n} \max_{\lambda \in J_n^q} \prod_{i=0}^{n} \frac{1}{|f^{\text{II}m_{\lambda(i)}}, a_{\lambda(i)}|},$$

which yields, for $r \geq r_0$,

$$
\begin{aligned}
\Gamma(\mathscr{A})^{q-n} \prod_{j=0}^{q} \frac{1}{\mu(r, f^{\text{II}m_j} \angle a_j)} &\leq \max_{\lambda \in J_n^q} \prod_{i=0}^{n} \frac{1}{\mu(r, f^{\text{II}m_{\lambda(i)}} \angle a_{\lambda(i)})} \\
&= \max_{\lambda \in J_n^q} \prod_{i=0}^{n} \frac{\mu(r, \tilde{f}^{\text{II}m_{\lambda(i)}})}{\mu(r, F_{\lambda(i)})} \\
&\leq \mu(r, \tilde{f})^{(n+1)d} \max_{\lambda \in J_n^q} \mu\left(r, \frac{\mathbf{W}_\lambda}{F_{\lambda(0)} \cdots F_{\lambda(n)}}\right) \mu\left(r, \frac{1}{\mathbf{W}_\lambda}\right) \\
&\leq \mu(r, \tilde{f})^{(n+1)d} r^{-\frac{n(n+1)}{2}} \max_{\lambda \in J_n^q} \mu\left(r, \frac{1}{\mathbf{W}_\lambda}\right).
\end{aligned}
$$

It follows that

$$
\begin{aligned}
\sum_{j=0}^{q} m_{f^{\text{II}m_j}}(r, a_j) &\leq d(n+1)T(r, f) - \min_{\lambda \in J_n^q} N\left(r, \frac{1}{\mathbf{W}_\lambda}\right) \\
&\quad - \frac{n(n+1)}{2} \log r + O(1).
\end{aligned}
\tag{6.46}
$$

Then theorem 6.40 follows from (6.46) and the first main theorem. $\square$

Corollary 6.41. *With the assumptions as in Theorem 6.40 ,*

$$\sum_{j=0}^{q} m_j \delta_f(a_j) \leq d(n+1).$$

Particularly, if $m_0 = m_1 = \cdots = m_q$, then

$$\sum_{j=0}^{q} \delta_f(a_j) \leq n+1. \tag{6.47}$$

Corollary 6.42. *Let* $\mathscr{A} = \{a_j\}_{j=0}^q$ *be a finite admissible family of points* $a_j \in \mathbb{P}(\amalg_m V^*)$ *with* $n < q$. *Let* $f : \kappa \longrightarrow \mathbb{P}(V)$ *be a holomorphic curve which is algebraically non-degenerate of degree* m. *If the integer* m *is of the property in (6.36), then*

$$m(q - n)T(r, f) \leq \sum_{j=0}^q N_{f^{\amalg m}}(r, a_j) - N\left(r, \frac{1}{\mathbf{W}}\right) - \frac{n(n+1)}{2}\log r + O(1),$$

where $\mathbf{W} = \mathbf{W}(\tilde{f}_0^m, ..., \tilde{f}_n^m)$.

Proof. For this case, b_{ik}^λ in (6.36) are constants. Thus

$$\mathbf{W} = \det\left(b_{ik}^\lambda\right)\mathbf{W}_\lambda, \quad \lambda \in J_n^q,$$

and then Corollary 6.42 follows from (6.46). $\square$

For the integer m in Corollary 6.42 and from (6.36), we can obtain

$$mT(r, f) = \max_{0 \leq i \leq n} \log \mu\left(r, F_{\lambda(i)}\right) + O(1), \quad \lambda \in J_n^q.$$

By using the Jensen formula, we derive the following:

$$mT(r, f) = \max_{0 \leq i \leq n} N_{f^{\amalg m}}(r, a_{\lambda(i)}) + O(1), \quad \lambda \in J_n^q, \tag{6.48}$$

which implies the following fact:

Corollary 6.43. *Let a holomorphic curve* $f : \kappa \longrightarrow \mathbb{P}(V)$ *be a algebraically non-degenerate of degree* m *and let* $\mathscr{A} = \{a_0, a_1, ..., a_q\}$ *be an admissible family of points* $a_j \in \mathbb{P}(\amalg_m V^*)$ *such that (6.36) and*

$$\delta_f(a_j) > 0, \quad j = 0, 1, ..., q$$

hold. Then $q \leq n - 1$.

. In the proof of Theorem 6.40, by using (6.45) in Proposition 6.38, we have

$$\prod_{j=0}^q \frac{1}{|f^{\amalg m_j}, a_j|} \leq \left(\frac{1}{\Gamma(\mathscr{A})}\right)^{q+1-n} \max_{\lambda \in J_{n-1}^q} \prod_{i=0}^{n-1} \frac{1}{|f^{\amalg m_{\lambda(i)}}, a_{\lambda(i)}|}.$$

Hence

$$\sum_{j=0}^q m_{f^{\amalg m_j}}(r, a_j) \leq \max_{\lambda \in J_{n-1}^q} \sum_{i=0}^{n-1} m_{f^{\amalg m_{\lambda(i)}}}(r, a_{\lambda(i)}) + O(1)$$

$$\leq \left(\max_{\lambda \in J_{n-1}^q} \sum_{i=0}^{n-1} m_{\lambda(i)}\right) T(r, f) + O(1).$$

Further, by the first main theorem, we can obtain the following result:

Theorem 6.44. Let $\mathscr{A} = \{a_j\}_{j=0}^q$ be a finite admissible family of points $a_j \in \mathbb{P}\left(\amalg_{m_j} V^*\right)$ with $n < q$. Let $f : \kappa \longrightarrow \mathbb{P}(V)$ be a holomorphic curve satisfying $f(\kappa) \not\subset \ddot{E}[a_j]$ for $j = 0, ..., q$. Set

$$d = \max_{\lambda \in J_{n-1}^q} \left\{ \frac{1}{n} \sum_{i=0}^{n-1} m_{\lambda(i)} \right\}.$$

Then

$$\left(\sum_{j=0}^q m_j - dn \right) T(r, f) \le \sum_{j=0}^q N_{f^{\amalg m_j}}(r, a_j) + O(1).$$

Min Ru [111] also obtains the result independently. Since we do not use derivatives in the proof of above theorem, the moving target version follows trivially (see Theorem A.27). For the complex variables case, in 1979 Shiffman [122] proved a defect relation with defect bound $2n$ for a class of holomorphic curve $f : \mathbb{C} \longrightarrow \mathbb{P}(\mathbb{C}^{n+1})$ and for hypersurface targets of degree m. He conjectured that this defect relation remains valid for general non-constant holomorphic curves. Special cases of this conjecture, under much stronger restrictions on f were proved in [8], and [2]. Finally, in 1991 Eremenko-Sodin [31] succeeded with an elegant proof by virtue of potential theory, without any nondegeneracy conditions on f. Also in [122], Shiffman conjectured that the defect relation with defect bound $n + 1$ holds for algebraically non-degenerate holomorphic curve $f : \mathbb{C} \longrightarrow \mathbb{P}(\mathbb{C}^{n+1})$ and for hypersurface targets of degree m. Some notes on this conjecture were referred by Hu-Yang [70].

S. Mori [94], [95] proved that for a given integer d and for a transcendental holomorphic mapping g of $\mathbb{C}^m$ into $\mathbb{P}(\mathbb{C}^{n+1})$, all deficient hypersurfaces of degree $\le d$ in $\mathbb{P}(\mathbb{C}^{n+1})$ can be eliminated by a small deformation of the mapping. The main theorems in [94] and [95] are true for non-Archimedean case.

Theorem 6.45 ([60]). Let $g : \kappa \longrightarrow \mathbb{P}(V)$ be a transcendental holomorphic curve. Assume that there exists some $a \in \mathbb{P}(V^*)$ such that

$$\lim_{r \to \infty} \frac{N_g(r, a)}{T(r, g)} = 1. \tag{6.49}$$

Take non-negative integers $d, k_0, k_1, ..., k_n$ satisfying

$$k_0 = 0, \ dk_{i-1} < k_i \ (i = 1, 2, ..., n).$$

Then there exists a matrix $L = (l_{ij})$ of the form $l_{ij}(z) = c_{ij} z^{k_i} + d_{ij}$ $(c_{ij}, d_{ij} \in \kappa, \ 0 \le i, j \le n)$ such that $\det(L) \ne 0$ and $f = L \circ g : \kappa \longrightarrow \mathbb{P}(V)$ is a linearly non-degenerate holomorphic curve without Nevanlinna deficient hypersurfaces of degree $\le d$. Further, we have

$$T(r, f) = T(r, g) + O(\log r). \tag{6.50}$$

6.8 Holomorphic curves into projective varieties

Let κ be an algebraically closed field of characteristic zero, complete for a non-trivial non-Archimedean absolute value $|\cdot|$. Let $M \subset \mathbb{P}(\kappa^N)$ be a projective variety of dimension m. A mapping $f : \kappa \longrightarrow M$ is said to be *holomorphic* if $f : \kappa \longrightarrow \mathbb{P}(\kappa^N)$ is a holomorphic curve with $f(\kappa) \subset M$.

A function $f : U \longrightarrow \kappa$ on an open subset U of M is *regular at a point* x if there is an open neighborhood U_x with $x \in U_x \subset U$, and homogeneous polynomials g, h of the same degree on V, such that h is nowhere zero on U_x, and $f = g/h$ on U_x. We say that f is *regular on* U if it is regular at every point of U. We denote by $\mathcal{O}(U)$ the ring of all regular functions on U. If x is a point of U, we define the *local ring of* x *on* U, $\mathcal{O}(x) = \mathcal{O}_M(x)$ to be the ring of germs of regular functions on U near x. In other words, an element of $\mathcal{O}(x)$ is a pair (U_x, f) where U_x is an open subset of U containing x, and f is a regular function on U_x, and where we identify two such pairs (U_x, f) and (W_x, g) if $f = g$ on $U_x \cap W_x$. Note that $\mathcal{O}(x)$ is indeed a local ring: its maximal ideal $\mathbf{m}$ is the set of germs of regular functions which vanish at x. For if $f(x) \neq 0$, then $1/f$ is regular in some neighborhood of x. The residue field $\mathcal{O}(x)/\mathbf{m}$ is isomorphic to κ. Also we define the *function field* $\kappa(U)$ of U as follows: an element of $\kappa(U)$ is an equivalence class of pairs (W, f) where W is a nonempty open subset of U, f is a regular function on W, and where we identify two pairs (W, f) and (W', g) if $f = g$ on $W \cap W'$. The elements of $\kappa(U)$ are called *rational functions* on U.

Any hypersurface $S \subset M$ is a projective subvariety of dimension $m - 1$, i.e., for any point $x \in S$, S can be given in a neighborhood of x as the zeros of a single homogeneous polynomial f on V. Moreover, any homogeneous polynomial g on V vanishing near p is divisible by f in a neighborhood of x. The polynomial f is called a *local defining function* for S near x, and is unique, up to multiplication by a function nonzero at x. By using the unique factorization of defining functions for S, thus S can be expressed uniquely as the union of irreducible hypersurfaces

$$S = S_1 \cup \cdots \cup S_l.$$

A *divisor* D on M is a locally finite formal linear combination

$$D = \sum_i n_i S_i$$

of irreducible hypersurfaces of M, where "locally finite" means that for any $x \in M$, there exists a neighborhood of x meeting only finitely many S_i's appearing in D. There is an open cover $\{U_\alpha\}$ of M such that in each U_α, every S_i appearing in D has a local defining function $g_{i\alpha}$. We can then set

$$f_\alpha = \prod_i g_{i\alpha}^{n_i}$$

called *local defining functions* for D. A divisor $D = \sum_i n_i S_i$ is called *effective* if $n_i \geq 0$ for all i; we write $D \geq 0$ for D effective. Any hypersurface S will be identified with the divisor $\sum_i S_i$, where the S_i's are the irreducible components of S.

Let $S \subset M$ be an irreducible hypersurface, $x \in S$ any point, and f a local defining function for S near x. For any regular function g at x, we define the *order* $\mathrm{ord}_{S,x}(g)$ *of* g *along* S *at* x to be the largest integer u such that in the local ring $\mathcal{O}_M(x)$,

$$g = f^u \cdot h.$$

If x is a generic point of S, then $\mathcal{O}_M(x)$ is a discrete valuation ring with quotient field $\kappa(M)$. Thus for g a rational function on M, $\mathrm{ord}_{S,x}(g)$ is independent of a generic point x. So we can define the *order* $\mathrm{ord}_S(g)$ *of* g *along* S to be the order of g along S at a generic point x. The *divisor* (g) of the rational function g on M is defined by

$$(g) = \sum_S \mathrm{ord}_S(g) S.$$

Usually we say that g has a *zero of order* u *along* S if $\mathrm{ord}_S(g) = u > 0$, and that g has a *pole of order* u *along* S if $\mathrm{ord}_S(g) = -u < 0$.

For any line bundle $\pi : L \longrightarrow M$ on the variety M, we can find an open covering $\{U_\alpha\}$ of M and trivializations

$$\varphi_\alpha : L_{U_\alpha} \longrightarrow U_\alpha \times \kappa$$

of $L_{U_\alpha} = \pi^{-1}(U_\alpha)$. The transition functions $g_{\alpha\beta} : U_\alpha \cap U_\beta \longrightarrow \kappa_*$ for L relative to the trivializations $\{\varphi_\alpha\}$ are defined by

$$g_{\alpha\beta}(x) = \left(\varphi_\alpha \circ \varphi_\beta^{-1} \right) |_{L_x} \in \kappa_*.$$

A *section* s of L over $U \subset M$ is given by a collection of rational functions $s_\alpha \in \kappa(U \cap U_\alpha)$ satisfying $s_\alpha = g_{\alpha\beta} s_\beta$ in $U \cap U_\alpha \cap U_\beta$. The *order* of s along any irreducible hypersurface $S \subset M$ can be defined by

$$\mathrm{ord}_S(s) = \mathrm{ord}_S(s_\alpha)$$

for any α such that $S \cap U_\alpha \neq \emptyset$, and the *divisor* (s) *of the section* s is given by

$$(s) = \sum_S \mathrm{ord}_S(s) S.$$

A *metric* ρ on the line bundle L is a set of functions $\rho_\alpha : U_\alpha \longrightarrow \mathbb{R}^+$ satisfying

$$\rho_\alpha = |g_{\alpha\beta}| \rho_\beta$$

on $U_\alpha \cap U_\beta$. Then we have

$$|s(x)|_\rho = \frac{|s_\alpha(x)|}{\rho_\alpha(x)}, \quad x \in U_\alpha.$$

Now let D be a divisor on M with local defining functions f_α over some open covering $\{U_\alpha\}$ of M. Then the functions

$$g_{\alpha\beta} = \frac{f_\alpha}{f_\beta}$$

are nonzero in $U_\alpha \cap U_\beta$. The line bundle given by the transition functions $\{g_{\alpha\beta}\}$ is called the *associated line bundle* of D, and written as $[D]$. The functions f_α clearly give a section $\hat{s}$ of $[D]$ with $(\hat{s}) = D$. Let $\mathscr{L}(D)$ denote the space of rational functions f on M such that $D + (f) \geq 0$ and write

$$\Gamma(M, [D]) = \{ f\hat{s} \mid f \in \mathscr{L}(D) \}.$$

Let L be a line bundle on the variety M. It is easy to see that $L = [(s)]$ for any section s of L. Thus $\Gamma(M, L)$ is well defined. Let V be the dual vector space of $\Gamma(M, L)$, and so $V^* = \Gamma(M, L)$. Let $\eta : M \times V^* \longrightarrow L$ be the *evaluation mapping* defined by $\eta(x, s) = s(x)$ for all $(x, s) \in M \times V^*$. It is said to be *ample* if $\eta(\{x\} \times V^*) = L_x$ for all $x \in M$. Assume that η is ample. Write $\dim V = n + 1 \geq 0$. The kernel E of η is a vector bundle of fiber dimension n on M. An exact sequence

$$0 \longrightarrow E \longrightarrow M \times V^* \longrightarrow L \longrightarrow 0 \qquad (6.51)$$

is defined. Thus if $x \in M$, one and only one $\varphi(x) \in \mathbb{P}(V)$ exists such that $E_x = E[\varphi(x)]$. The mapping $\varphi : M \longrightarrow \mathbb{P}(V)$ is called the *dual classification mapping*. We can describe the mapping φ more explicitly as follows. Choose a basis $s_0, ..., s_n$ for V^*. If $s_i = \{s_{i,\alpha}\}$ for a collection of trivialization φ_α of L over an open set $U_\alpha \subset M$, it is clear that the point $[s_{0,\alpha}(x), ..., s_{n,\alpha}(x)] \in \mathbb{P}(\kappa^{n+1})$ is independent of the collection of trivialization φ_α chosen; we denote this point by $[s_0(x), ..., s_n(x)]$. In terms of the identifications $\mathbb{P}(V^*) \cong \mathbb{P}(\kappa^{n+1})$ corresponding to the choice of basis $s_0, ..., s_n$, then the mapping φ is given by

$$\varphi(x) = [s_0(x), ..., s_n(x)].$$

We call L a *very ample line bundle* if η is ample and if $\varphi : M \longrightarrow \mathbb{P}(V)$ is an embedding. If some k-th tensor power L^k of L is very ample, then L is said to be *ample*. Also a divisor D is said to be *very ample* (resp., *ample*) if $[D]$ is a very ample (resp., *ample*) line bundle. If a divisor D is very ample and if $f_0, ..., f_n$ is a basis for $\mathscr{L}(D)$, then the dual classification mapping φ of $[D]$ is given by

$$\varphi(x) = [(f_0 \hat{s})(x), ..., (f_n \hat{s})(x)],$$

where $\hat{s}$ is a section of $[D]$ with $(\hat{s}) = D$. We write

$$\varphi(x) = [f_0(x), ..., f_n(x)],$$

according to the identifications $\mathbb{P}(V^*) \cong \mathbb{P}(\mathscr{L}(D)) \cong \mathbb{P}(\kappa^{n+1})$.

Now we assume that the line bundle L is very ample. Recall that the absolute value $\rho = |\cdot|$ on κ induces a norm of V^*, which further induces a metric on $M \times V^*$, and restricts to E. Thus a quotient metric ρ on L is defined by (6.51). For any $s \in V^* - \{0\}$, we have

$$|\varphi(x), a| = \frac{|s(x)|_\rho}{|s|}, \quad x \in M, \ a = \mathbb{P}(s). \qquad (6.52)$$

Write $D = (s)$. For a holomorphic curve $f : \kappa \longrightarrow M$ satisfying $f(\kappa) \not\subset D$, the functions

$$N_f(r, D) = N_{\varphi \circ f}(r, a), \quad m_f(r, D) = m_{\varphi \circ f}(r, a)$$

are well defined. Also the *characteristic function* of f for L

$$T_f(r, L) = T(r, \varphi \circ f)$$

is well defined, up to $O(1)$. These functions are related by the first main theorem

$$T_f(r, L) = N_f(r, D) + m_f(r, D) + O(1). \qquad (6.53)$$

Proposition 6.46. *If L and L' are very ample line bundles, then*

$$T_f(r, L \otimes L') = T_f(r, L) + T_f(r, L').$$

Proof. Let $\varphi : M \longrightarrow \mathbb{P}(V)$ and $\varphi' : M \longrightarrow \mathbb{P}(V')$ be the dual classification mappings of L and L', respectively, given by projective coordinates

$$\tilde{\varphi} = \sum_{i=0}^{n} \tilde{\varphi}_i e_i, \quad \tilde{\varphi}' = \sum_{i=0}^{n'} \tilde{\varphi}'_i e'_i,$$

where $e_0, ..., e_n$ and $e'_0, ..., e'_{n'}$ are a base of V and V', respectively. Then the dual classification mapping of $L \otimes L'$, denoted by $\varphi \otimes \varphi' : M \longrightarrow \mathbb{P}(V \otimes V')$, is given by projective coordinate

$$\tilde{\varphi} \otimes \tilde{\varphi}' = \sum_{i,j} \tilde{\varphi}_i \tilde{\varphi}'_j e_i \otimes e'_j.$$

Hence

$$
\begin{aligned}
T_f(r, L \otimes L') &= T(r, (\varphi \otimes \varphi') \circ f) = \log \mu(r, \tilde{\varphi} \circ f \otimes \tilde{\varphi}' \circ f) \\
&= \log \left(\max_{i,j} \mu(r, \tilde{\varphi}_i \circ f \tilde{\varphi}'_j \circ f) \right) \\
&= \log \left(\max_i \mu(r, \tilde{\varphi}_i \circ f) \max_j \mu(r, \tilde{\varphi}'_j \circ f) \right) \\
&= \log \left(\mu(r, \tilde{\varphi} \circ f) \mu(r, \tilde{\varphi}' \circ f) \right) \\
&= T_f(r, L) + T_f(r, L').
\end{aligned}
$$

$\square$

For a very ample divisor D, we also write

$$T_D(r) = T_{f,D}(r) = T_f(r, [D]).$$

If D' is another very ample divisor, then the formula $[D + D'] = [D] \otimes [D']$ and Proposition 6.46 imply

$$T_{D+D'}(r) = T_D(r) + T_{D'}(r). \qquad (6.54)$$

Given any divisor D, we can write $D = E - E'$, where E and E' are very ample, and define

$$N_f(r, D) = N_f(r, E) - N_f(r, E'),$$

$$m_f(r, D) = m_f(r, E) - m_f(r, E'),$$

and finally set

$$T_D(r) = T_E(r) - T_{E'}(r).$$

It follows that (6.54) holds for arbitrary divisors D and D'. If two divisors D and D' on M are linearly equivalent, that is, $D = D' + (h)$ for some rational function h on M, then

$$T_{D'}(r) = T_D(r) + O(1)$$

since $[(h)]$ is trivial.

Lemma 6.47. *If D is an effective divisor, then*

$$T_D(r) \geq 0.$$

Proof. There exist very ample divisors E and E' such that $D = E - E'$. Let $\{f_0, ..., f_n\}$ be a basis for $\mathscr{L}(E')$. Then we can extend this choice of functions to a basis $\{f_0, ..., f_{n+l}\}$ for $\mathscr{L}(E)$ because D is effective. Let $\varphi' = [f_0, ..., f_n]$ and $\varphi = [f_0, ..., f_{n+l}]$ be the dual classification mapping of E' and $[E]$, respectively. It then follows from the definition of $T_D(r)$ that

$$T_D(r) = T_E(r) - T_{E'}(r) \geq 0.$$

$\square$

Corollary 6.48. *If D is ample, then for any other divisor E,*

$$T_E(r) = O(T_D(r)) + O(1).$$

Definition 6.49. *Let D be a divisor on a nonsingular variety M. The dimension of D is the integer $d = \dim D$ such that*

$$cn^d < \dim \mathscr{L}(nD) < c'n^d$$

for n sufficiently divisible, where c and c' are some positive constants. If $\mathscr{L}(nD)$ is always empty, then let $d = -\infty$. The divisor D is said to be almost ample if $\dim D = \dim M$.

Some basic properties of the dimension are surveyed in [133]. For example, it satisfies $\dim D \leq \dim M$, $\dim D = \dim M$ if D is ample, and $\dim(D + E) \geq \dim D$ if E is an effective divisor. Kodaira's theorem shows that D is almost ample if and only if nD can be written as a sum of an ample divisor and an effective divisor, for some sufficiently large integer n. We say that a divisor D has *normal crossings* if locally it is of the form $z_1 \cdots z_i = 0$ for some choice of local coordinates $z_1, ..., z_m$. If D has normal crossings, and if after expressing $D = \sum D_j$ as a sum of irreducible components, all D_j are non-singular, then we say that D has *simple normal crossings*. Finally, we suggest the following problem:

Conjecture 6.50. *Let M be a nonsingular projective variety of dimension m over κ. Let K be the canonical divisor of M, and let D be a simple normal crossings divisor on M. Let E be an almost ample divisor. Let $f : \kappa \longrightarrow M$ be an holomorphic curves. Then there exists a proper algebraic subset Z_D having the following property: when $f(\kappa) \not\subset Z_D$,*

$$m_f(r, D) + T_K(r) \leq o(T_E(r)).$$

In value distribution theory of complex variables, this is corresponding to the conjecture due to P. Griffiths [40], S. Lang [83] and J. Noguchi [102].

Example 6.51. *Each hyperplane E of $\mathbb{P}(\kappa^{n+1})$ is very ample with $\dim \Gamma(\mathbb{P}(\kappa^{n+1}), [E]) = n + 1$, and hence the dual classification mapping φ is the identity. Thus we have $T_E(r) = T(r, f)$, and hence*

$$T_K(r) = -(n + 1)T(r, f)$$

since $K = -(n+1)E$. Let $D = \sum_i \ddot{E}[a_i]$ be a sum of hyperplanes in general position. Then the conjecture reduces to

$$\sum_i m_f(r, a_i) < (n+1)T(r, f) + o(T(r, f))$$

which follows from Theorem 6.13.

Chapter 7

Diophantine approximations

In this chapter, by using the methods developed in above chapters we introduce the generalizations of Schmidt's subspace theorems in Diophantine approximations given by Vojta, Ru, Wong, and so on.

7.1 Schmidt's subspace theorems

We assume that κ is a number field (i.e., a subfield of $\mathbb{C}$) with an absolute value ρ. The completion of κ relative to the topology induced by ρ is a field which is denoted by κ_ρ. If ρ is Archimedean, then ρ is equivalent to the infinite place $|\cdot|_\infty$, and hence $\kappa_\rho = \mathbb{R}$ or $\mathbb{C}$. If ρ is non-Archimedean, then ρ is equivalent to a p-adic absolute value $|\cdot|_p$, and so κ_ρ is a finite field extension of $\mathbb{Q}_p$. We will denote the normalization of ρ by $\|\cdot\|_\rho$, which is defined for $z \in \kappa$ by

$$\|z\|_\rho = \begin{cases} |z|_\infty & : \quad \text{if } \kappa_\rho = \mathbb{R} \\ |z|_\infty^2 & : \quad \text{if } \kappa_\rho = \mathbb{C} \\ |z|_p^{[\kappa_\rho : \mathbb{Q}_p]} & : \quad \text{if } \rho \sim |\cdot|_p \text{ is non-Archimedean .} \end{cases} \tag{7.1}$$

Under these conventions there exists a set M_κ of inequivalent places of κ satisfying the *product formula*,

$$\prod_{\rho \in M_\kappa} \|z\|_\rho = 1, \quad z \in \kappa_*. \tag{7.2}$$

Let V be a $(n+1)$-dimensional vector space over κ. Take a base $e = (e_0, ..., e_n)$ for V. For $\xi = \xi_0 e_0 + \cdots + \xi_n e_n \in V$, the norm $\|\xi\|_\rho = \|\xi\|_{\rho,e}$ over the base e is well-defined, and it induces a norm on V^*. Thus the distance from $x = \mathbb{P}(\xi)$ to $\ddot{E}[a]$ with $a = \mathbb{P}(\alpha) \in \mathbb{P}(V^*)$ is well-defined by

$$0 \leq \|x, a\|_\rho = \frac{\|\langle \xi, \alpha \rangle\|_\rho}{\|\xi\|_\rho \cdot \|\alpha\|_\rho} \leq 1.$$

225

Take $x \in \mathbb{P}(V)$. Then there exists $\xi \in V_*$ such that $x = \mathbb{P}(\xi)$. The *relative (multiplicative) height* of x is defined by

$$H(x) = \prod_{\rho \in M_\kappa} \|\xi\|_\rho,$$

and the *absolute (logarithmic) height* of x is defined as

$$h(x) = \frac{1}{[\kappa : \mathbb{Q}]} \sum_{\rho \in M_\kappa} \log \|\xi\|_\rho.$$

By the product formula, this does not depend on the choice of ξ. Let S be a finite subset of M_κ containing the set S_∞ of all Archimedean places of κ. Denote by $\mathcal{O}_{\kappa,S}$ the ring of S-integers of κ, i.e.,

$$\mathcal{O}_{\kappa,S} = \{z \in \kappa \mid \|z\|_\rho \leq 1, \ \rho \notin S\}.$$

A point $\xi = \xi_0 e_0 + \cdots + \xi_n e_n \in V$ is said to be a S-integral point if $\xi_i \in \mathcal{O}_{\kappa,S}$ for all $0 \leq i \leq n$. Denote by $\mathcal{O}_{V,S}$ the set of S-integral points of V, that is,

$$\mathcal{O}_{V,S} = \{\xi \in V \mid \|\xi\|_\rho \leq 1, \ \rho \notin S\}.$$

Lemma 7.1. *For $x \in \mathbb{P}(V)$, we can choose $\xi \in \mathcal{O}_{V,S}$ such that $x = \mathbb{P}(\xi)$, and*

$$\max\left\{\max_{\rho \in S} \|\xi\|_\rho, \prod_{\rho \in S} \|\xi\|_\rho\right\} \leq cH(x) \leq c\left\{\max_{\rho \in S} \|\xi\|_\rho\right\}^{\#S},$$

where c is a constant depending only on S but is independent of x.

Proof. Take $\xi \in V_*$ with $x = \mathbb{P}(\xi)$. Write $\xi = \xi_0 e_0 + \cdots + \xi_n e_n$ and let $\mathbf{a}$ be the fractional ideal generated by $\{\xi_0, ..., \xi_n\}$. Then (cf. Lang [82], p.54)

$$H(x) = \mathbf{N}\mathbf{a}^{-1} \prod_{\rho \in S} \|\xi\|_\rho,$$

where $\mathbf{N}$ denotes the absolute norm. By Theorem 6.3 of Lang [82], there exists a constant $c' > 0$ and an integral ideal $\mathbf{b}$ linearly equivalent to $\mathbf{a}$ so that $\mathbf{N}\mathbf{b} \leq c'$. In other words, the ξ_i can be chosen to be integers such that

$$\prod_{\rho \in S} \|\xi\|_\rho \leq c'H(x).$$

Now we assume $\xi \in \mathcal{O}_{V,S}$ with $x = \mathbb{P}(\xi)$. Then $\|\xi\|_\rho \leq 1$ for all $\rho \notin S$, and so

$$H(x) \leq \prod_{\rho \in S} \|\xi\|_\rho \leq \left\{\max_{\rho \in S} \|\xi\|_\rho\right\}^{\#S}.$$

On the other hand, by Vojta ([133], Lemma 2.2.2) or Lang ([82], Theorem 6.5 and Theorem 6.6), the coordinate ξ of x can be chosen so that

$$\max_{\rho \in S} \|\xi\|_\rho \leq cH(x),$$

where $c > c'$ is a constant depending only on S but is independent of x. Therefore, Lemma 7.1 follows. $\qquad\square$

The following version of the subspace theorem is due to Schlickewei [117] (see also Schmidt [118] and [119]).

Theorem 7.2. *For $\rho \in S$, $i = 0, ..., n$, take $\alpha_{\rho,i} \in V^*$ such that the $n + 1$ linear forms $\langle \xi, \alpha_{\rho,0} \rangle, ..., \langle \xi, \alpha_{\rho,n} \rangle$ are linearly independent for each ρ. Then for any $\varepsilon > 0$ there exists a finite set $b_1, ..., b_s$ of $\mathbb{P}(V^*)$ such that the inequality*

$$\prod_{\rho \in S} \prod_{i=0}^{n} \frac{1}{\|\langle \xi, \alpha_{\rho,i} \rangle\|_\rho} \leq \left\{ \max_{\rho \in S} \|\xi\|_\rho \right\}^\varepsilon$$

holds for all S-integral points $\xi \in \mathcal{O}_{V,S} - \bigcup_i E[b_i]$.

The *Weil function* of x with respect to $a \in \mathbb{P}(V^*)$ with $x \notin \ddot{E}[a]$ at a place $\rho \in M_\kappa$ is just $-\log \|x, a\|_\rho$. The *proximity function* $m_{x,S}(a)$ is defined by

$$m_x(a) = m_{x,S}(a) = \frac{1}{[\kappa : \mathbb{Q}]} \sum_{\rho \in S} \log \frac{1}{\|x, a\|_\rho}.$$

Set $S^c = M_\kappa - S$. Then

$$m_{x,S^c}(a) = \frac{1}{[\kappa : \mathbb{Q}]} \sum_{\rho \in M_\kappa - S} \log \frac{1}{\|x, a\|_\rho}$$

serves as the *valence function*. By the product formula, one obtains the first main theorem:

$$m_{x,S}(a) + m_{x,S^c}(a) = h(x) + h(a),$$

and therefore,

$$m_x(a) \leq h(x) + h(a).$$

The following Schmidt's subspace theorem is well-known (see Vojta [133]):

Theorem 7.3. *Take $\varepsilon > 0$. Assume that for each $\rho \in S$, the family $\mathscr{A}_\rho = \{a_{\rho,0}, ..., a_{\rho,q}\} \subset \mathbb{P}(V^*)$ $(q \geq n)$ is in general position. Then*

$$\frac{1}{[\kappa : \mathbb{Q}]} \sum_{\rho \in S} \sum_{j=0}^{q} \log \frac{1}{\|x, a_{\rho,j}\|_\rho} < (n + 1 + \varepsilon)h(x) + O(1)$$

holds for all $x \in \mathbb{P}(V)$ outside a finite union of proper linear subspaces of $\mathbb{P}(V)$. In particular, if $\mathscr{A} = \{a_0, ..., a_q\} \subset \mathbb{P}(V^*)$ is in general position, then

$$\sum_{j=0}^{q} m_x(a_j) < (n + 1 + \varepsilon)h(x) + O(1)$$

holds for all $x \in \mathbb{P}(V)$ outside a finite union of proper linear subspaces of $\mathbb{P}(V)$.

Proof. Define the *gauge* Γ of $\{\mathscr{A}_\rho\}_{\rho \in S}$ by

$$\Gamma = \inf_{\rho \in S} = \inf_{\rho \in S} \Gamma(\mathscr{A}_\rho) = \inf_{\lambda \in J_n^q} \{\|a_{\rho, \lambda(0)} \wedge \cdots \wedge a_{\rho, \lambda(n)}\|_\rho\}.$$

Then $0 < \Gamma \leq 1$ since the family $\mathscr{A}_\rho$ is in general position for each $\rho \in S$. For $x \in \mathbb{P}(V)$, $0 < r \in \mathbb{R}$, define

$$\mathscr{A}_\rho(x, r) = \{j \in \mathbb{Z}[0, q] \mid \|x, a_{\rho, j}\|_\rho < r\}.$$

Set

$$\varsigma_\rho = \begin{cases} \sqrt{n+1} & : \quad \text{if } \kappa_\rho = \mathbb{R} \\ n+1 & : \quad \text{if } \kappa_\rho = \mathbb{C} \\ 1 & : \quad \text{if } \rho \text{ is non-Archimedean} \end{cases}.$$

We claim that if $0 < r \leq \Gamma/\varsigma_\rho$, then $\#\mathscr{A}_\rho(x, r) \leq n$ for each $\rho \in S$. In fact, if $\#\mathscr{A}_\rho(x, r) \geq n+1$, then $\lambda_\rho \in J_n^q$ exists such that $\text{Im}\lambda_\rho \subseteq \mathscr{A}_\rho(x, r)$. Hence

$$\|x, a_{\rho, \lambda_\rho(j)}\|_\rho < r, \quad j = 0, ..., n.$$

Then Lemma 6.4 implies

$$0 < \Gamma \leq \|a_{\rho, \lambda_\rho(0)} \wedge \cdots \wedge a_{\rho, \lambda_\rho(n)}\|_\rho$$
$$\leq \varsigma_\rho \max_{0 \leq j \leq n} \|x, a_{\rho, \lambda_\rho(j)}\|_\rho < \varsigma_\rho r \leq \Gamma,$$

which is impossible, and so the claim is proved. Take $r = \Gamma/(n+1)$. Then we have $\#\mathscr{A}_\rho(x, r) \leq n$ for each $\rho \in S$. Thus $\sigma_\rho \in J_n^q$ exists such that $\mathscr{A}_\rho(x, r) \subset \text{Im}\sigma_\rho$. Hence by Theorem 7.2, there exists a finite set $\{b_1, ..., b_s\}$ of $\mathbb{P}(V^*)$ such that the inequality

$$\prod_{\rho \in S} \prod_{j=0}^q \frac{1}{\|x, a_{\rho, j}\|_\rho} \leq \prod_{\rho \in S} \left\{ \left(\frac{n+1}{\Gamma}\right)^{q-n} \prod_{j=0}^n \frac{1}{\|x, a_{\rho, \sigma_\rho(j)}\|_\rho} \right\}$$

$$\leq \left(\frac{n+1}{\Gamma}\right)^{(q-n)\#S} \prod_{\rho \in S} \left\{ \prod_{j=0}^n \frac{\|\xi\|_\rho \|\alpha_{\rho, \sigma_\rho(j)}\|_\rho}{\|\langle\xi, \alpha_{\rho, \sigma_\rho(j)}\rangle\|_\rho} \right\}$$

$$= c_1 \left(\prod_{\rho \in S} \|\xi\|_\rho\right)^{n+1} \left(\prod_{\rho \in S} \prod_{j=0}^n \frac{1}{\|\langle\xi, \alpha_{\rho, \sigma_\rho(j)}\rangle\|_\rho}\right)$$

$$\leq c_1 \left(\prod_{\rho \in S} \|\xi\|_\rho\right)^{n+1} \left(\max_{\rho \in S} \|\xi\|_\rho\right)^\varepsilon,$$

holds for all points $x = \mathbb{P}(\xi) \in \mathbb{P}(V) - \bigcup_i \ddot{E}[b_i]$, where c_1 is constant, and $\alpha_{\rho, j} \in V^* - \{0\}$ with $a_{\rho, j} = \mathbb{P}(\alpha_{\rho, j})$. By Lemma 7.1, there exists a constant c_2 such that

$$\prod_{\rho \in S} \prod_{j=0}^q \frac{1}{\|x, a_{\rho, j}\|_\rho} \leq c_2 H(x)^{n+1+\varepsilon},$$

and hence Theorem 7.3 follows. $\square$

Roth's theorem follows from Theorem 7.3 when $n = 1$.

7.2 Vojta's conjecture

Let κ be an algebraic number field. Let M be a projective variety defined over κ and let D be a very ample divisor on M. Denote the dual classification mapping by $\varphi : M \longrightarrow \mathbb{P}(V)$, where $V = \Gamma(M, [D])^*$. Then the *relative (multiplicative) height* of x for D is defined by

$$H_D(x) = H(\varphi(x)),$$

and the *absolute (logarithmic) height* of x for D is defined as

$$h_D(x) = h(\varphi(x)).$$

Take $a \in \mathbb{P}(V^*)$ with $x \notin \ddot{E}[a]$. The *proximity function* $m_{x,S}(D)$ is defined by

$$m_x(D) = m_{x,S}(D) = m_{\varphi(x),S}(a) = \frac{1}{[\kappa : \mathbb{Q}]} \sum_{\rho \in S} \log \frac{1}{\|\varphi(x), a\|_\rho}.$$

Set $S^c = M_\kappa - S$. Then

$$m_{x,S^c}(D) = m_{\varphi(x),S^c}(a) = \frac{1}{[\kappa : \mathbb{Q}]} \sum_{\rho \in M_\kappa - S} \log \frac{1}{\|\varphi(x), a\|_\rho}$$

serves as the *valence function*. By the product formula, one obtains the first main theorem:

$$m_{x,S}(D) + m_{x,S^c}(D) = h_D(x) + h(a).$$

Given any divisor D, one can express $D = E - E'$ where E and E' are very ample, and define

$$m_x(D) = m_x(E) - m_x(E'),$$

and finally set

$$h_D(x) = h_E(x) - h_{E'}(x).$$

P. Vojta [133] translated the conjecture due to P. Griffiths [40], S. Lang [83] and J. Noguchi [102] in value distribution theory of complex variables into the following form:

Conjecture 7.4. *Let M be a nonsingular projective variety over κ. Let K be the canonical divisor of M, and let D be a normal crossings divisor on M. If $\varepsilon > 0$ and E is an almost ample divisor, then there exists a proper Zariski closed subset Z such that for all $x \in M(\kappa) - Z$,*

$$m_x(D) + h_K(x) \le \varepsilon h_E(x) + O(1),$$

where $M(\kappa)$ is the set of κ-valued points of M.

Example 7.5. *Each hyperplane E of $\mathbb{P}(\kappa^{n+1})$ is very ample with $\dim \Gamma(\mathbb{P}(\kappa^{n+1}), [E]) = n + 1$, and hence the dual classification mapping φ is the identity. Thus $h_E(x) = h(x)$, and hence*

$$h_K(x) = -(n+1)h(x)$$

since $K = -(n+1)E$. Let $D = \sum_i \ddot{E}[a_i]$ be a sum of hyperplanes in general position. Then the conjecture reduces to

$$\sum_i m_x(a_i) < (n + 1 + \varepsilon)h(x)$$

which follows from Schmidt's subspace theorem (Theorem 7.3).

Influenced by Mason's theorem (Corollary 2.25), and considerations of Szpiro and Frey, Masser and Oesterlé formulated the "abc" conjecture for integers as follows:

Conjecture 7.6. *Let a be a non-zero integer. Define the radical of a to be*

$$\bar{n}\left(\frac{1}{a}\right) = \prod_{p|a} p$$

i.e. the product of the distinct primes dividing a. Given $\varepsilon > 0$, there exists a number $C(\varepsilon)$ having the following property. For any nonzero relatively prime integers a, b, c such that $a + b = c$,

$$\max\{|a|, |b|, |c|\} \le C(\varepsilon)\bar{n}\left(\frac{1}{abc}\right)^{1+\varepsilon}$$

This conjecture is a consequence of Conjecture 7.4 (see Vojta [133]).

Let a be a non-zero integer. Then a can be expressed as

$$|a| = p_1^{i_1} \cdots p_n^{i_n},$$

for distinct primes $p_1, ..., p_n$. Define

$$n_k\left(\frac{1}{a}\right) = \prod_{\nu=1}^{n} p_\nu^{\min\{i_\nu, k\}}.$$

Then an analogue of Corollary 2.29 for integers would be the following problem:

Conjecture 7.7. *Let $a_j (j = 0, \cdots, k)$ be nonzero integers such that a_i, a_j $(i \neq j)$ are relatively prime, and*

$$a_1 + \cdots + a_k = a_0. \tag{7.3}$$

Then for $\varepsilon > 0$, there exists a number $C(\varepsilon)$ such that

$$\max_{0 \le j \le k}\{|a_j|\} \le C(\varepsilon)\left\{n_{k-1}\left(\frac{1}{a_1 \cdots a_k}\right) \cdot \bar{n}\left(\frac{1}{a_0}\right)^{\iota_k}\right\}^{1+\varepsilon},$$

where

$$\iota_k = \begin{cases} 1 & : \quad k = 2 \\ \frac{k(k-1)}{2} - 1 & : \quad k \ge 3 \,. \end{cases}$$

7.3 General subspace theorems

Theorem 7.3 was generalized by M. Ru and P. M. Wong [115]. In this section, we introduce their results.

Theorem 7.8. *Take $\varepsilon > 0$. Assume that for each $\rho \in S$, the family $\mathscr{A}_\rho = \{a_{\rho,0}, ..., a_{\rho,q}\} \subset \mathbb{P}(V^*)$ is in u-subgeneral position with $q > u \ge n \ge 1$. Let ω_ρ be the associated Nochka weight of $\mathscr{A}_\rho$. Then*

$$\frac{1}{[\kappa : \mathbb{Q}]} \sum_{\rho \in S} \sum_{j=0}^{q} \omega_\rho(a_{\rho,j}) \log \frac{1}{\|x, a_{\rho,j}\|_\rho} < (n + 1 + \varepsilon)h(x) + O(1)$$

holds for all $x \in \mathbb{P}(V)$ outside a finite union of proper linear subspaces of $\mathbb{P}(V)$. In particular, if $\mathscr{A} = \{a_0, ..., a_q\} \subset \mathbb{P}(V^*)$ is in u-subgeneral position, then

$$\sum_{j=0}^{q} \omega(a_j) m_x(a_j) < (n + 1 + \varepsilon) h(x) + O(1)$$

holds for all $x \in \mathbb{P}(V)$ outside a finite union of proper linear subspaces of $\mathbb{P}(V)$, where ω is the associated Nochka weight of $\mathscr{A}$.

Proof. Let Γ be the gauge of $\{\mathscr{A}_\rho\}_{\rho \in S}$ defined by

$$\Gamma = \inf_{\rho \in S} \inf_{\lambda \in J_n(\mathscr{A}_\rho)} \{\|a_{\rho,\lambda(0)} \wedge \cdots \wedge a_{\rho,\lambda(n)}\|_\rho\}$$

with $0 < \Gamma \leq 1$. For $x \in \mathbb{P}(V)$, $0 < r \in \mathbb{R}$, define

$$\mathscr{A}_\rho(x, r) = \{j \in \mathbb{Z}[0, q] \mid \|x, a_{\rho,j}\|_\rho < r\}.$$

We claim that if $0 < r \leq \Gamma/\varsigma_\rho$, then $\#\mathscr{A}_\rho(x, r) \leq u$ for each $\rho \in S$. In fact, if $\#\mathscr{A}_\rho(x, r) \geq u + 1$, then $\lambda_\rho \in J_n(\mathscr{A}_\rho)$ exists such that $\mathrm{Im}\lambda_\rho \subseteq \mathscr{A}_\rho(x, r)$ because $\mathscr{A}_\rho$ is in u-subgeneral position. Hence

$$\|x, a_{\rho,\lambda_\rho(j)}\|_\rho < r, \quad j = 0, ..., n.$$

Then Lemma 6.4 implies

$$
\begin{aligned}
0 < \Gamma \ &\leq \ \|a_{\rho,\lambda_\rho(0)} \wedge \cdots \wedge a_{\rho,\lambda_\rho(n)}\|_\rho \\
&\leq \ \varsigma_\rho \max_{0 \leq j \leq n} \|x, a_{\rho,\lambda_\rho(j)}\|_\rho < \varsigma_\rho r \leq \Gamma,
\end{aligned}
$$

which is impossible, and so the claim is proved. Take $r = \Gamma/(n + 1)$. Then $\#\mathscr{A}_\rho(x, r) \leq u$ for each $\rho \in S$. Thus $\sigma_\rho \in J_u^q$ and $\lambda_\rho \in J_n(\mathscr{A}_\rho)$ exist such that $\mathscr{A}_\rho(x, r) \subset \mathrm{Im}\sigma_\rho$ and $\lambda_\rho \subset \sigma_\rho$. Hence by Lemma 6.23 and Theorem 7.2, there exists a finite set $\{b_1, ..., b_s\}$ of $\mathbb{P}(V^*)$ such that the inequality

$$
\begin{aligned}
\prod_{\rho \in S} \prod_{j=0}^{q} \left(\frac{1}{\|x, a_{\rho,j}\|_\rho}\right)^{\omega_\rho(a_{\rho,j})} \ &\leq \ \prod_{\rho \in S} \left\{\left(\frac{n+1}{\Gamma}\right)^{\frac{q-u}{\theta}} \prod_{j=0}^{u} \left(\frac{1}{\|x, a_{\rho,\sigma_\rho(j)}\|_\rho}\right)^{\omega_\rho(a_{\rho,\sigma_\rho(j)})}\right\} \\
&\leq \ \left(\frac{n+1}{\Gamma}\right)^{\frac{q-n}{\theta}\#S} \prod_{\rho \in S} \prod_{j=0}^{n} \frac{1}{\|x, a_{\rho,\lambda_\rho(j)}\|_\rho} \\
&= \ c_1 \left(\prod_{\rho \in S} \|\xi\|_\rho\right)^{n+1} \left(\prod_{\rho \in S} \prod_{j=0}^{n} \frac{1}{\|\langle\xi, \alpha_{\rho,\lambda_\rho(j)}\rangle\|_\rho}\right) \\
&\leq \ c_1 \left(\prod_{\rho \in S} \|\xi\|_\rho\right)^{n+1} \left(\max_{\rho \in S} \|\xi\|_\rho\right)^{\varepsilon}
\end{aligned}
$$

holds for all points $x = \mathbb{P}(\xi) \in \mathbb{P}(V) - \bigcup_i \ddot{E}[b_i]$, where c_1 is constant, and $\xi \in V_*$, $\alpha_{\rho,j} \in V^* - \{0\}$ with $x = \mathbb{P}(\xi)$, $a_{\rho,j} = \mathbb{P}(\alpha_{\rho,j})$. By Lemma 7.1, there exists a constant c_2 such that

$$\prod_{\rho \in S} \prod_{j=0}^{q} \left(\frac{1}{\|x, a_{\rho,j}\|_\rho} \right)^{\omega_\rho(a_{\rho,j})} \le c_2 H(x)^{n+1+\varepsilon},$$

and hence Theorem 7.8 follows.   □

Theorem 7.9. *Take $\varepsilon > 0$. Assume that the family $\mathscr{A} = \{a_0, ..., a_q\} \subset \mathbb{P}(V^*)$ is in u-subgeneral position with $q > u \ge n \ge 1$. Then*

$$\sum_{j=0}^{q} m_x(a_j) < (2u - n + 1 + \varepsilon)h(x) + O(1)$$

holds for all $x \in \mathbb{P}(V)$ outside a finite union of proper linear subspaces of $\mathbb{P}(V)$.

By using Theorem 7.8, the proof of Theorem 7.9 can be proved according to the method used in the proof of Theorem 6.26.

Theorem 7.10. *Take $\varepsilon > 0$. Assume that a family $\mathscr{A} = \{a_0, ..., a_q\} \subset \mathbb{P}(V^*)$ is in general position. Then for $n \ge k \ge 1$, the set of points of $\mathbb{P}(V) - \bigcup_j \ddot{E}[a_j]$ satisfying*

$$\sum_{j=0}^{q} m_x(a_j) \ge (2n - k + 1 + \varepsilon)h(x) + O(1)$$

is contained in a finite union of linear subspaces of dimension $k - 1$ of $\mathbb{P}(V)$. Particularly, the set of points of $\mathbb{P}(V) - \bigcup_j \ddot{E}[a_j]$ satisfying

$$\sum_{j=0}^{q} m_x(a_j) \ge (2n + \varepsilon)h(x) + O(1)$$

is finite.

Here we pose the following problem:

Conjecture 7.11. *For $\rho \in S$, $i = 0, ..., n$, take $\alpha_{\rho,i} \in \amalg_m V^* - \{0\}$ such that the system*

$$\langle \xi^{\amalg m}, \alpha_{\rho,i} \rangle = 0, \quad i = 0, ..., n$$

has only the trivial solution $\xi = 0$ in V. Then for any $\varepsilon > 0$ there exists a finite set $b_1, ..., b_s$ of $\mathbb{P}(\amalg_m V^*)$ such that the inequality

$$\prod_{\rho \in S} \prod_{i=0}^{n} \frac{1}{\|\langle \xi^{\amalg m}, \alpha_{\rho,i} \rangle\|_\rho} \le \left\{ \max_{\rho \in S} \|\xi\|_\rho \right\}^{\varepsilon}$$

holds for all S-integral points $\xi \in \mathcal{O}_{V,S} - \bigcup_i E[b_i]$.

According to the proof of Corollary 6.42, Conjecture 7.11 implies the following conjecture:

Conjecture 7.12. *Take $\varepsilon > 0$. Assume that for $\rho \in S$, the family $\mathscr{A}_\rho = \{a_{\rho,0}, ..., a_{\rho,q}\} \subset \mathbb{P}(\amalg_m V^*)$ $(q \geq n)$ is admissible. Then there exists a finite set $b_1, ..., b_s$ of $\mathbb{P}(\amalg_m V^*)$ such that the inequality*

$$\frac{1}{[\kappa:\mathbb{Q}]} \sum_{\rho \in S} \sum_{j=0}^{q} \log \frac{1}{\|x^{\amalg m}, a_{\rho,j}\|_\rho} < m(n+1+\varepsilon)h(x) + O(1)$$

holds for all $x \in \mathbb{P}(V) - \bigcup_i \ddot{E}[b_i]$.

Based on Theorem 7.10, we suggest the following problem:

Conjecture 7.13. *Take $\varepsilon > 0$. Assume that the family $\mathscr{A} = \{a_0, ..., a_q\} \subset \mathbb{P}(\amalg_m V^*)$ $(q \geq n)$ is admissible. Then the set of points of $\mathbb{P}(V) - \bigcup_j \ddot{E}[a_j]$ satisfying*

$$\sum_{j=0}^{q} m_{x^{\amalg m}}(a_j) \geq m(2n+\varepsilon)h(x) + O(1)$$

is finite.

7.4 Ru-Vojta's subspace theorem for moving targets

In this section, we will introduce the Ru-Vojta's subspace theorem for moving targets [114]. Let κ be a number field. Let V be a $(n+1)$-dimensional vector space over κ. Fix a base $e = (e_0, ..., e_n)$ of V and a infinite set $D \subset \kappa$. We will study a general mapping

$$f : D \longrightarrow \mathbb{P}(V).$$

Then there exists a mapping

$$\tilde{f} = \tilde{f}_0 e_0 + \cdots + \tilde{f}_n e_n : D \longrightarrow V_*,$$

such that $f = \mathbb{P} \circ \tilde{f}$, which will be called a reduced representation of f. The *height function* $h(f)$ of f is defined as

$$h(f)(z) = h(f(z)) = \frac{1}{[\kappa:\mathbb{Q}]} \sum_{\rho \in M_\kappa} \log \|\tilde{f}(z)\|_\rho.$$

By the product formula, this does not depend on the choice of reduced representations $\tilde{f}$ of f. If $a : D \longrightarrow \mathbb{R} \cup \{\infty\}$ and $b : D \longrightarrow \mathbb{R} \cup \{\infty\}$ are real valued functions, then

$$a \leq (\neq) b$$

means that there is a finite set E in D such that a and b are well defined in $D - E$, and

$$a(z) \leq (\neq) b(z), \quad z \in D - E.$$

Definition 7.14. *Given another mapping* $g : D \longrightarrow \mathbb{P}(V^*)$ *with a reduced representations* $\tilde{g}$, *the pair* (f, g) *is said to be free if* $\langle \tilde{f}(z), \tilde{g}(z) \rangle$ *vanishes for only finitely many* $z \in D$.

Assume that the pair (f, g) is free. The *Weil function* of f with respect to g with $f(z) \notin \ddot{E}[g(z)]$ at a place $\rho \in M_\kappa$ is $- \log \|f(z), g(z)\|_\rho$. For a finite set $S \subset M_\kappa$ of places of κ containing S_∞, the *proximity function* $m_f(g) = m_{f,S}(g)$ is defined by

$$m_f(g)(z) = m_{f(z)}(g(z)) = \frac{1}{[\kappa : \mathbb{Q}]} \sum_{\rho \in S} \log \frac{1}{\|f(z), g(z)\|_\rho}.$$

Set $S^c = M_\kappa - S$. Then

$$m_{f,S^c}(g)(z) = m_{f(z),S^c}(g(z)) = \frac{1}{[\kappa : \mathbb{Q}]} \sum_{\rho \in M_\kappa - S} \log \frac{1}{\|f(z), g(z)\|_\rho}$$

is the *valence function*. By the product formula, we have

$$| \quad m_f(g) + m_{f,S^c}(g) = h(f) + h(g), \tag{7.4}$$

and therefore

$$| \quad m_f(g) \leq h(f) + h(g).$$

Let $\mathcal{G} = \{g_j\}_{j=0}^q$ be a finite family of mappings $g_j : D \longrightarrow \mathbb{P}(V^*)$ with $n \leq q < \infty$. For each $z \in \kappa$, define the orbit $\mathcal{G}(z) = \{g_j(z)\}_{j=0}^q$. Let

$$\tilde{g}_j = \tilde{g}_{j,0}\epsilon_0 + \cdots + \tilde{g}_{j,n}\epsilon_n : D \longrightarrow V^* - \{0\}$$

be a reduced representation of g_j for $j = 0, ..., q$, where $\epsilon_0, ..., \epsilon_n$ is the dual base of $e_0, ..., e_n$.

Definition 7.15. *An infinite subset* $A \subset D$ *is said to be coherent with respect to* $\mathcal{G}$ *if, for every polynomial* $P(\xi_0, ..., \xi_q)$ *on* V^{q+1} *which is homogeneous in* ξ_j *for each* $j = 0, ..., q$, *either* $P(\tilde{g}_0(z), ..., \tilde{g}_q(z))$ *vanishes for all* $z \in A$, *or it vanishes for only finitely many* $z \in A$.

Lemma 7.16 ([114]). *There exists an infinite subset* $A \subset D$ *which is coherent with respect to* $\mathcal{G}$.

Proof. Assume, on the contrary, that for any infinite subset $A \subset D$, there is a multihomogeneous polynomial $P(\xi_0, ..., \xi_q)$ on V^{q+1} such that $P(\tilde{g}_0(z), ..., \tilde{g}_q(z)) = 0$ for infinitely many $z \in A$ but $P(\tilde{g}_0(z), ..., \tilde{g}_q(z)) \neq 0$ for some $z \in A$. Let $\mathbf{I}(A)$ denote the ideal generated by multihomogeneous polynomial $Q(\xi_0, ..., \xi_q)$ on V^{q+1} such that $Q(\tilde{g}_0(z), ..., \tilde{g}_q(z)) = 0$ for all $z \in A$. Set $A_1 = D$. Then A_1 is infinite, and not coherent, by assumption. Hence there is a multihomogeneous polynomial $P(\xi_0, ..., \xi_q)$ on V^{q+1} and an infinite subset $A_2 \subset A_1$ such that $P(\tilde{g}_0(z), ..., \tilde{g}_q(z)) = 0$ for all $z \in A_2$, but $P(\tilde{g}_0(z), ..., \tilde{g}_q(z)) \neq 0$ for all $z \in A_1 - A_2$. So $\mathbf{I}(A_1) \subset \mathbf{I}(A_2)$, and $\mathbf{I}(A_1) \neq \mathbf{I}(A_2)$. Continuing this procedure, an infinite chain of ideas

$$\mathbf{I}(A_1) \subset \mathbf{I}(A_2) \subset \cdots \subset \kappa[\xi_0, ..., \xi_q]$$

is obtained, which contradicts the fact that $\kappa[\xi_0, ..., \xi_q]$ is Noetherian. $\qquad \square$

From now on, we assume that D is coherent with respect to $\mathcal{G}$. By coherence, for each $j = 0, ..., q$, there is j_0 such that $0 \leq j_0 \leq n$ and $\tilde{g}_{j,j_0}(z) \neq 0$ for all but finitely many $z \in D$. Therefore, for $j = 0, ..., q$ and $k = 0, ..., n$, we may set

$$g_{jk} = \frac{\tilde{g}_{j,k}}{\tilde{g}_{j,j_0}}.$$

By using the product formula, we prove easily

$$0 \leq \frac{1}{[\kappa : \mathbb{Q}]} \log \max_{0 \leq k \leq n} \|g_{jk}\|_\rho \leq h(g_j).$$

Let $\mathcal{L}_m$ be the vector space generated over κ by

$$\left\{ \prod_{j,k} g_{jk}^{m_{jk}} \mid m_{jk} \in \mathbb{Z}_+ (0 \leq j \leq q, 0 \leq k \leq n), \sum_{j,k} m_{jk} = m \right\}.$$

Since $g_{j,j_0} = 1$, we have $\mathcal{L}_m \subset \mathcal{L}_{m+1}$. Write

$$1 \leq q(m) = \dim \mathcal{L}_m, \quad m = 1, 2, ... \ .$$

Let $\mathcal{R} = \mathcal{R}_\mathcal{G}$ be the field generated by $\mathcal{L}_1$ over κ. Then

$$\kappa \subset \mathcal{L}_m \subset \mathcal{R}.$$

Definition 7.17. *A mapping $f : D \longrightarrow \mathbb{P}(V)$ is said to be non-degenerate with respect to $\mathcal{G}$ (or over $\mathcal{R}$) if*

$$\varphi_0 \tilde{f}_0 + ... + \varphi_n \tilde{f}_n \neq 0$$

for all $\varphi_i \in \mathcal{R}$ with $(\varphi_0, ..., \varphi_n) \not\equiv 0$.

By the definition, if $f : D \longrightarrow \mathbb{P}(V)$ is non-degenerate with respect to $\mathcal{G}$, then $\tilde{f}_0, ..., \tilde{f}_n$ are linearly independent over $\mathcal{R}$. Further, if D is coherent with respect to $\mathcal{G}$, then (f, g_j) is free for each $j = 0, ..., q$.

Theorem 7.18 (Ru-Vojta [114]). *Let κ be a number field. Given a finite set $S \subset M_\kappa$ of places of κ containing S_∞. Let $\mathcal{G} = \{g_j\}_{j=0}^q$ be a finite family of mappings $g_j : D \longrightarrow \mathbb{P}(V^*)$ such that the orbit $\mathcal{G}(z)$ is in general position for each $z \in D$. Assume that D is coherent with respect to $\mathcal{G}$ and let $f : D \longrightarrow \mathbb{P}(V)$ be a non-degenerate mapping with respect to $\mathcal{G}$. Assume that*

$$h(g_j) = o(h(f)), \quad j = 0, ..., q.$$

Take $\varepsilon \in \mathbb{R}^+$. Then

$$\sum_{j=0}^q m_f(g_j) \leq (n + 1 + \varepsilon) h(f).$$

Proof. The theorem trivially holds when $q \leq n$ by (7.4), and so we can assume $q > n$. Define the *gauge* $\Gamma(\mathcal{G})$ of $\mathcal{G}$ by

$$\Gamma(\mathcal{G})(z) = \inf_{\rho \in S} \inf_{\lambda \in J_n^q} \{\|g_{\lambda(0)}(z) \wedge \cdots \wedge g_{\lambda(n)}(z)\|_\rho\}.$$

Then $0 < \Gamma(\mathcal{G}) \leq 1$. For $x \in \mathbb{P}(V)$, $0 < b \in \mathbb{R}$, define

$$\mathcal{G}(z)(x, b; \rho) = \{j \in \mathbb{Z}[0, q] \mid \|x, g_j(z)\|_\rho < b\}.$$

According to the proof of Lemma A.22 and Theorem 7.3, we have

$$\#\mathcal{G}(z)(x, b; \rho) \leq n, \quad \text{if } 0 < b \leq \Gamma(\mathcal{G})(z)/\varsigma_\rho. \tag{7.5}$$

Thus

$$\prod_{\rho \in S} \prod_{j=0}^{q} \frac{1}{\|x, g_j(z)\|_\rho} \leq \left(\frac{n+1}{\Gamma(\mathcal{G})(z)}\right)^{(q-n)\#S} \prod_{\rho \in S} \prod_{j=0}^{n} \frac{1}{\|x, g_{\sigma_\rho(j)}(z)\|_\rho}, \tag{7.6}$$

where $\sigma_\rho \in J_n^q$ with $\mathcal{G}(z)(x, b; \rho) \subset \mathrm{Im}\sigma_\rho$. Define

$$m_{g_0 \wedge \cdots \wedge g_n}(z) = \frac{1}{[\kappa : \mathbb{Q}]} \sum_{\rho \in S} \log \frac{1}{\|g_0(z) \wedge \cdots \wedge g_n(z)\|_\rho}$$

and

$$N_{g_0 \wedge \cdots \wedge g_n}(z) = \frac{1}{[\kappa : \mathbb{Q}]} \sum_{\rho \in M_\kappa - S} \log \frac{1}{\|g_0(z) \wedge \cdots \wedge g_n(z)\|_\rho}.$$

By the product formula, we have

$$m_{g_0 \wedge \cdots \wedge g_n} + N_{g_0 \wedge \cdots \wedge g_n} = \sum_{k=0}^{n} h(g_k). \tag{7.7}$$

Because

$$\prod_{\rho \in S} \prod_{\lambda \in J_n^q} \|g_{\lambda(0)}(z) \wedge \cdots \wedge g_{\lambda(n)}(z)\|_\rho \leq \Gamma(\mathcal{G})(z),$$

we obtain

$$0 \leq \frac{1}{[\kappa : \mathbb{Q}]} \log \frac{1}{\Gamma(\mathcal{G})} \leq \sum_{\lambda \in J_n^q} m_{g_{\lambda(0)} \wedge \cdots \wedge g_{\lambda(n)}} \leq \sum_{\lambda \in J_n^q} \sum_{k=0}^{n} h(g_{\lambda(k)}). \tag{7.8}$$

From (7.6) and (7.8), it follows that

$$\sum_{j=0}^{q} m_f(g_j) \leq \frac{1}{[\kappa : \mathbb{Q}]} \sum_{\rho \in S} \sum_{k=0}^{n} \log \frac{1}{\|f, g_{\sigma_\rho(k)}\|_\rho} + o(h(f)). \tag{7.9}$$

Steinmetz's lemma (Lemma A.12) yields

$$\liminf_{m \to \infty} \frac{q(m+1)}{q(m)} = 1.$$

Thus, given any $\delta > 0$ we may find $m \in \mathbb{Z}^+$ such that

$$q(m+1) \leq (1+\delta)q(m).$$

Fix such an m. We can take a basis $\{b_1, ..., b_{q(m+1)}\}$ of $\mathcal{L}_{m+1}$ such that $\{b_1, ..., b_{q(m)}\}$ is a basis of $\mathcal{L}_m$. Since f is non-degenerate, we deduce that

$$\varphi = \left(b_j \tilde{f}_k \mid 1 \leq j \leq q(m+1), 0 \leq k \leq n \right)$$

are linearly independent over κ. Put

$$F_j = \frac{\langle \tilde{f}, \tilde{g}_j \rangle}{\tilde{g}_{j,j_0}} = \sum_{k=0}^{n} g_{jk} \tilde{f}_k, \quad 0 \leq j \leq q.$$

According to the proof of Theorem A.15, we can prove that for any $\lambda \in J_n^q$,

$$\Phi_\lambda = \left(b_j F_{\lambda(k)} \mid 1 \leq j \leq q(m), 0 \leq k \leq n \right)$$

are linearly independent over κ.

Since $b_j F_{\lambda(k)}(1 \leq j \leq q(m), 0 \leq k \leq n)$ are linear combinations of $b_j \tilde{f}_k(1 \leq j \leq q(m+1), 0 \leq k \leq n)$ over κ, we can choose

$$\Psi_\lambda = \left(\sum_{1 \leq u \leq q(m+1), 0 \leq v \leq n} c_{jk}^{uv} b_u \tilde{f}_v \mid c_{jk}^{uv} \in \kappa, q(m) + 1 \leq j \leq q(m+1), 0 \leq k \leq n \right)$$

such that

$$(\Phi_\lambda, \Psi_\lambda) = \varphi C_\lambda \tag{7.10}$$

holds for some $C_\lambda \in GL((n+1)q(m+1); \kappa)$. Note that except for finitely many elements from D, we have $\tilde{g}_{j,j_0}(z) \neq 0$. Since $\mathcal{L}_m$ is generated by monomials in the g_{jk}, the b_j are all polynomials in the g_{jk}. Then, except for finitely many elements from D, (7.10) is well defined.

If c_λ^{ij} denote the elements of C_λ, then the corresponding $(j + kq(m))$-th column in C_λ defines an element

$$\alpha_{j,k}(\lambda) = \left(\alpha_{j,k}^1(\lambda), ..., \alpha_{j,k}^{q(m+1)}(\lambda) \right) \in V^{*q(m+1)} - \{0\},$$

where

$$\alpha_{j,k}^l(\lambda) = \sum_{i=0}^{n} c_\lambda^{(l-1)(n+1)+i,j+kq(m)} \epsilon_i, \quad 1 \leq j \leq q(m), 0 \leq k \leq n.$$

For each but finitely many $z \in D$, let $\xi(z) \in V^{q(m+1)}$ be the point defined by

$$\xi(z) = \left(b_1(z) \tilde{f}(z), ..., b_{q(m+1)}(z) \tilde{f}(z) \right).$$

Then by (7.10), we have

$$\langle \xi(z), \alpha_{j,k}(\lambda) \rangle = \sum_{l=1}^{q(m+1)} b_l(z) \langle \tilde{f}(z), \alpha_{j,k}^l(\lambda) \rangle = b_j(z) F_{\lambda(k)}(z).$$

Note that, by our assumptions, for each $a \in \mathbb{P}(V^{*q(m+1)})$, there is only finitely many $z \in D$ such that $x(z) \in \ddot{E}[a]$. Thus by applying Schmidt' subspace theorem (Theorem 7.3) to the points $x(z) = \mathbb{P}(\xi(z))$ with the family $\mathscr{A} = \{a_{j,k}(\lambda)\} \subset \mathbb{P}(V^{*q(m+1)})$, where $a_{j,k}(\lambda) = \mathbb{P}(\alpha_{j,k}(\lambda))$, we have

$$\frac{1}{[\kappa : \mathbb{Q}]} \sum_{\rho \in S} \sum_{k=0}^{n} \sum_{j=1}^{q(m)} \log \frac{1}{\|x(z), a_{j,k}(\lambda)\|_\rho} < ((n+1)q(m+1) + \delta)h(x(z)) + O(1). \quad (7.11)$$

We now translate the various terms above. First consider the height. Since

$$\|\xi(z)\|_\rho = \max_{j,k} \|b_j(z)\tilde{f}_k(z)\|_\rho = \max_j \|b_j(z)\|_\rho \cdot \max_k \|\tilde{f}_k(z)\|_\rho,$$

we have

$$h(x(z)) = h(f(z)) + o(h(f(z))).$$

On other hand, we have

$$\begin{aligned}
\|x(z), a_{j,k}(\lambda)\|_\rho &= \frac{\|\langle \xi(z), \alpha_{j,k}(\lambda)\rangle\|_\rho}{\|\xi(z)\|_\rho \cdot \|\alpha_{j,k}(\lambda)\|_\rho} \\
&= \frac{\|b_j(z)F_{\lambda(k)}(z)\|_\rho}{\|\tilde{f}(z)\|_\rho \max_j \|b_j(z)\|_\rho \cdot \max_l \|c_\lambda^{l,j+kq(m)}\|_\rho} \\
&= c\frac{\|b_j(z)\|_\rho \max_i \|g_{\lambda(k),i}(z)\|_\rho}{\max_j \|b_j(z)\|_\rho} \|f(z), g_{\lambda(k)}(z)\|_\rho,
\end{aligned}$$

where c is a positive constant. But

$$-o(h(f(z))) \leq -\log \max_{0 \leq i \leq n} \|g_{\lambda(k),i}(z)\|_\rho \leq 0$$

by the assumption of the theorem, and likewise

$$0 \leq -\log \frac{\|b_j(z)\|_\rho}{\max_{1 \leq j \leq q(m+1)} \|b_j(z)\|_\rho} \leq o(h(f(z)))$$

since the b_j are polynomials in the g_{jk}. Thus

$$\log \frac{1}{\|x(z), a_{j,k}(\lambda)\|_\rho} = \log \frac{1}{\|f(z), g_{\lambda(k)}(z)\|_\rho} + o(h(f(z))).$$

Combining this with (7.11), we have

$$\frac{q(m)}{[\kappa : \mathbb{Q}]} \sum_{\rho \in S} \sum_{k=0}^{n} \log \frac{1}{\|f(z), g_{\lambda(k)}(z)\|_\rho} < ((n+1)q(m+1) + \delta)h(f(z))$$
$$+ o(h(f(z))) \quad (7.12)$$

for all $\lambda \in J_n^q$. This and (7.9) imply

$$q(m) \sum_{j=0}^{q} m_f(g_j) \leq ((n+1)q(m+1) + \delta)h(f) + o(h(f)),$$

and hence

$$\sum_{j=0}^{q} m_f(g_j) \le \left((n+1)\frac{q(m+1)}{q(m)} + \frac{\delta}{q(m)}\right)h(f) + o(h(f)) < (n+1+\varepsilon)h(f).$$

Theorem 7.18 is thus proved. □

7.5 Subspace theorem for degenerate mappings

We will keep the notations and terminologies of §7.4, and continue to introduce Ru-Vojta's results [114].

Theorem 7.19 (Ru-Vojta [114]). *Let κ be a number field. Given a finite set $S \subset M_\kappa$ of places of κ containing S_∞. Let u be an integer with $n \le u < q$. Let $\mathcal{G} = \{g_j\}_{j=0}^{q}$ be a finite family of mappings $g_j : D \longrightarrow \mathbb{P}(V^*)$ such that the orbit $\mathcal{G}(z)$ is in u-subgeneral position for each $z \in D$. Assume that D is coherent with respect to $\mathcal{G}$ and let $f : D \longrightarrow \mathbb{P}(V)$ be a non-degenerate mapping with respect to $\mathcal{G}$ satisfying*

$$h(g_j) = o(h(f)), \quad j = 0, ..., q.$$

Take $\varepsilon \in \mathbb{R}^+$. Then

$$\sum_{j=0}^{q} m_f(g_j) \le (2u - n + 1 + \varepsilon)h(f).$$

Proof. By (7.4), the theorem is trivial when $q \le 2u - n + 1$, therefore we may assume that $q > 2u - n + 1$. For every point $z \in D$, we have the Nochka weight function ω and the Nochka constant θ for the orbit $\mathcal{G}(z)$. Define the *gauge* $\Gamma(\mathcal{G})$ of $\mathcal{G}$ by

$$\Gamma(\mathcal{G})(z) = \inf_{\rho \in S} \inf_{\lambda \in J_n(\mathcal{G}(z))} \{\|g_{\lambda(0)}(z) \wedge \cdots \wedge g_{\lambda(n)}(z)\|_\rho\}.$$

Then $0 < \Gamma(\mathcal{G}) \le 1$. For $x \in \mathbb{P}(V)$, $0 < b \in \mathbb{R}$, define

$$\mathcal{G}(z)(x, b; \rho) = \{j \in \mathbb{Z}[0, q] \mid \|x, g_j(z)\|_\rho < b\}.$$

According to the proof of Lemma A.22 and Theorem 7.3, we have

$$\#\mathcal{G}(z)(x, b; \rho) \le u, \quad \text{if } 0 < b \le \Gamma(\mathcal{G})(z)/\varsigma_\rho. \tag{7.13}$$

Now, according to the proof of Lemma A.23, we have

$$\prod_{\rho \in S} \prod_{j=0}^{q} \left(\frac{1}{\|x, g_j(z)\|_\rho}\right)^{\omega(g_j(z))} \le \left(\frac{n+1}{\Gamma(\mathcal{G})(z)}\right)^{(q-u)\#S} \prod_{\rho \in S} \prod_{j=0}^{n} \frac{1}{\|x, g_{\sigma_\rho(j)}(z)\|_\rho}, \tag{7.14}$$

where $\sigma_\rho \in J_n(\mathcal{G}(z))$. Since, by (7.7),

$$\prod_{\rho \in S} \prod_{\lambda \in J_n(\mathcal{G}(z))} \|g_{\lambda(0)}(z) \wedge \cdots \wedge g_{\lambda(n)}(z)\|_\rho \le \Gamma(\mathcal{G})(z),$$

we have

$$0 \le \frac{1}{[\kappa : \mathbb{Q}]} \log \frac{1}{\Gamma(\mathcal{G})(z)} \le \sum_{\lambda \in J_n(\mathcal{G}(z))} m_{g_{\lambda(0)} \wedge \cdots \wedge g_{\lambda(n)}}(z)$$

$$\le \sum_{\lambda \in J_n(\mathcal{G}(z))} \sum_{k=0}^{n} h(g_{\lambda(k)}(z)). \tag{7.15}$$

From (7.14), (7.15) and (7.12), it follow that

$$\sum_{j=0}^{q} \omega(g_j) m_f(g_j) \le \frac{1}{[\kappa : \mathbb{Q}]} \sum_{\rho \in S} \sum_{k=0}^{n} \log \frac{1}{\|f, g_{\sigma_\rho(k)}\|_\rho} + o(h(f))$$

$$\le (n + 1 + \varepsilon) h(f) + o(h(f)).$$

The theorem follows from this and the proof of Theorem A.25. □

Definition 7.20. *A mapping $f : D \longrightarrow \mathbb{P}(V)$ is said to be k-flat with respect to $\mathcal{G}$ (or over $\mathcal{R}$) if k is the largest integer with the property that $\tilde{f}_{\lambda(0)}, \ldots, \tilde{f}_{\lambda(k+1)}$ are linearly dependent over $\mathcal{R}$ for all $\lambda \in J_{k+1}^n$, but there is an injection $\eta \in J_k^n$ such that*

$$\varphi_0 \tilde{f}_{\eta(0)} + \ldots + \varphi_k \tilde{f}_{\eta(k)} \ne 0$$

for all $\varphi_i \in \mathcal{R}$ with $(\varphi_0, \ldots, \varphi_k) \not\equiv 0$.

Theorem 7.21 (Ru-Vojta [114]). *Let κ be a number field. Given a finite set $S \subset M_\kappa$ of places of κ containing S_∞. Let $\mathcal{G} = \{g_j\}_{j=0}^q$ be a finite family of mappings $g_j : D \longrightarrow \mathbb{P}(V^*)$ such that the orbit $\mathcal{G}(z)$ is in general position for each $z \in D$. Assume that D is coherent with respect to $\mathcal{G}$ and let $f : D \longrightarrow \mathbb{P}(V)$ be a k-flat mapping with respect to $\mathcal{G}$ such that each pair (f, g_j) is free for $j = 0, \ldots, q$. Assume that*

$$h(g_j) = o(h(f)), \quad j = 0, \ldots, q.$$

Take $\varepsilon \in \mathbb{R}^+$. Then

$$\sum_{j=0}^{q} m_f(g_j) \le (2n - k + 1 + \varepsilon) h(f).$$

    Proof. By (7.4), the theorem is trivial when $q \le 2n - k + 1$, hence we may assume that $q > 2n - k + 1$. W. l. o. g., we may assume that $\eta(i) = i$ for all $i = 0, \ldots, k$ in Definition 7.20, and thus for each $r = k + 1, \ldots, n$, there are $\varphi_{r,0}, \ldots, \varphi_{r,k} \in \mathcal{R}$ such that

$$\tilde{f}_r = \varphi_{r,0} \tilde{f}_0 + \ldots + \varphi_{r,k} \tilde{f}_k.$$

Let W be the vector space spanned by $e_0, \ldots, e_k$ over κ. Then $\epsilon_0, \ldots, \epsilon_k$ span W^*. Except for finitely many points from D, a mapping $\tilde{f} : \kappa \longrightarrow \mathbb{P}(W)$ is defined with a representation

$$\tilde{f} = \sum_{j=0}^{k} \tilde{f}_j e_j : D \longrightarrow W.$$

The mapping $\hat{f}$ is nondegenerate over $\mathcal{R}$. We have

$$\left|\tilde{\hat{f}}\right| = \max_{0 \le j \le k} |\tilde{f}_j| \le \max_{0 \le j \le n} |\tilde{f}_j| = |\tilde{f}|$$

$$\le |\tilde{\hat{f}}| \max \left\{ 1, (k+1) \max_{0 \le j \le k} \max_{k+1 \le r \le n} |\varphi_{r,j}| \right\}.$$

Hence

$$h(f) = h(\hat{f}) + o(h(f)).$$

Also mappings $\hat{g}_j : D \longrightarrow \mathbb{P}(W^*)$ are defined with representations

$$\tilde{\hat{g}}_j = \sum_{i=0}^{k} \left(\tilde{g}_{j,i} + \sum_{r=k+1}^{n} \varphi_{r,i} \tilde{g}_{j,r} \right) \epsilon_i : D \longrightarrow W^*, \quad j = 0, ..., q.$$

We obtain

$$\langle \tilde{f}, \tilde{g}_j \rangle = \left\langle \tilde{\hat{f}}, \tilde{\hat{g}}_j \right\rangle .$$

Since (f, g_j) is free, then $\tilde{\hat{g}}_j(z)$ vanishes for only finitely many $z \in D$. Hence $\hat{g}_j$ is well defined except for finitely many points from D. Thus

$$\log \frac{1}{\|f, g_j\|_\rho} = \log \frac{1}{\|\hat{f}, \hat{g}_j\|_\rho} + o(h(f)), \quad 0 \le j \le q, \ \rho \in S. \tag{7.16}$$

It is easy to show that the set $\{\hat{g}_0(z), ..., \hat{g}_q(z)\}$ is in n-subgeneral position for each $z \in D$. By applying Theorem 7.19 to $\hat{f}$ and to the family $\{\hat{g}_0, ..., \hat{g}_q\}$, we obtain Theorem 7.21. $\square$

Appendix A

The Cartan conjecture for moving targets

In this appendix, we will introduce the Ru-Stoll's methods for proving the second main theorem of moving targets. In proving Roth's theorem with moving targets in number theory, Vojta [134] observed that the general form of the second main theorem with fixed targets implies the second main theorem with moving targets (also see [110], [114]). For non-Archimedean case, the moving target version of the second main theorem follows trivially (see Theorem A.27). However, we think that the Ru-Stoll's methods have independent interesting.

A.1 Non-degenerate holomorphic curves

Let κ be an algebraically closed field of characteristic zero, complete for a non-trivial non-Archimedean absolute value $|\cdot|$.

Definition A.1. *Take $\Phi \subset \mathcal{M}(\kappa)$. A meromorphic function $G \not\equiv 0$ is said to be a universal denominator of Φ if $G\varphi$ is an entire function for every $\varphi \in \Phi$, and such that for every $z \in \kappa$, there exists $\phi \in \Phi$ with $(G\varphi)(z) \neq 0$.*

Lemma A.2. *If Φ is a finite set of $\mathcal{M}(\kappa)$ with $\Phi - \{0\} \neq \emptyset$, then there exists a universal denominator of Φ. Further if $\Phi \cap \kappa_* \neq \emptyset$, universal denominators of Φ are entire functions.*

Proof. Suppose $\Phi = \{\varphi_1, ..., \varphi_q\}$. Since greatest common divisors of any two elements in $\mathcal{A}(\kappa)$ exists, then there are $g_i, h_i \in \mathcal{A}(\kappa)$ with $\varphi_i = \frac{g_i}{h_i}$ such that g_i and h_i have no common factors in the ring $\mathcal{A}(\kappa)$. Define

$$\delta(z) = \max\{\mu_{h_i}^\infty(z) \mid 1 \leq i \leq q\}$$

for all $z \in \kappa$. By Corollary 1.25, there exists an entire function $h \not\equiv 0$ with $\mu_h^0 = \delta$. Let g be a greatest common divisor of $g_1, ..., g_q$ and set $G = h/g$. Obviously, $G\varphi_i$ is entire for $i = 1, ..., q$. Take $z \in \kappa$. Then $\delta(z) = \mu_{h_i}^\infty(z)$ for some i. If $\delta(z) > 0$, then $\mu_{g_i}^0(z) = 0$ by the assumption, and hence $(G\varphi_i)(z) \neq 0$. Suppose $\delta(z) = 0$. Since g is a greatest common divisor of $g_1, ..., g_q$, then there is some i with $(g_i/g)(z) \neq 0$, and so $(G\varphi_i)(z) \neq 0$. Thus G

is a universal denominator of Φ. Further if $\Phi \cap \kappa_* \neq \emptyset$, then the greatest common divisor g of $g_1, ..., g_q$ is a nonzero constant, and so $G = h/g$ is an entire function. $\qquad\square$

Definition A.3. *Let W be a vector space of dimension $m + 1 > 0$ over κ with a base $s_0, ..., s_m$. Let $h : \kappa \longrightarrow \mathbb{P}(W)$ be a holomorphic curve with a representation*

$$\tilde{h} = \tilde{h}_0 s_0 + \cdots + \tilde{h}_m s_m : \kappa \longrightarrow W.$$

The base is said to be admissible for h if $\tilde{h}_0 \not\equiv 0$. If the base is admissible for h, the k-th coordinate function of h is defined by

$$h_k = \frac{\tilde{h}_k}{\tilde{h}_0}, \quad k = 0, ..., m.$$

Further, let W^* be the dual vector space of W. The proper linear subspace

$$B(h) = \{\beta \in W^* \mid \langle \tilde{h}, \beta \rangle \equiv 0\}$$

of W^* does not depend on the choice of the representation $\tilde{h}$. If $\beta \in W^* - \{0\}$ and $b = \mathbb{P}(\beta)$, then (h, b) is free, that is, $h(\kappa) \not\subset \ddot{E}[b]$, if and only if $\beta \in W^* - B(h)$. If $\alpha \in W^*$ and $\beta \in W^* - B(h)$, the meromorphic function

$$h_{\alpha,\beta} = \frac{\langle \tilde{h}, \alpha \rangle}{\langle \tilde{h}, \beta \rangle}$$

does not depend on the choice of the representation $\tilde{h}$ and is called the (α, β)-*coordinate function of h.*

Definition A.4. *Let $\mathcal{R}$ be a subfield of $\mathcal{M}(\kappa)$ with $\kappa \subset \mathcal{R}$. Then h is said to be defined over $\mathcal{R}$ if $h_{\alpha,\beta} \in \mathcal{R}$ for all $\alpha \in W^*$ and $\beta \in W^* - B(h)$.*

If the base is admissible for h, it is easy to show that h is defined over $\mathcal{R}$ if and only if $h_k \in \mathcal{R}$ for $k = 1, ..., m$. Note that $\mu(r, \tilde{h}_k) = \mu(r, h_k)\mu(r, \tilde{h}_0)$. For $k = 1, ..., m$, we have

$$
\begin{aligned}
T(r, h_k) &= \max\{\log \mu(r, \tilde{h}_0), \log \mu(r, \tilde{h}_k)\} + O(1) \\
&\leq T(r, h) + O(1) \\
&\leq \sum_{j=1}^{m} T(r, h_j) + O(1).
\end{aligned}
\tag{A.1}
$$

Let V be a normed vector space of dimension $n + 1 > 0$ over κ and let $|\ |$ be a norm defined over a base $e = (e_0, ..., e_n)$ of V. Let $f : \kappa \longrightarrow \mathbb{P}(V)$ be a non-constant holomorphic curve. Then the holomorphic curve $h : \kappa \longrightarrow \mathbb{P}(W)$ is said to *grow slower* than f if

$$\lim_{r \to \infty} \frac{T(r, h)}{T(r, f)} = 0. \tag{A.2}$$

Let $\mathcal{M}_f(\kappa)$ be the set of all meromorphic functions on κ that grow slower than f. Note that for $a_1, a_2 \in \mathcal{M}_f(\kappa)$, we have the estimates

$$T(r, a_1 + a_2) \leq T(r, a_1) + T(r, a_2),$$

$$T(r, a_1 a_2) \leq T(r, a_1) + T(r, a_2),$$

$$T(r, -a_1) = T(r, a_1),$$

$$T\left(r, \frac{1}{a_1}\right) = T(r, a_1) - \log \mu(\rho_0, a_1).$$

Hence $\mathcal{M}_f(\kappa)$ is a field with $\kappa \subset \mathcal{M}_f(\kappa)$. Then (A.1) implies that h is defined over $\mathcal{M}_f(\kappa)$ if and only if h grow slower than f. Thus, if the subfield $\mathcal{R}$ of $\mathcal{M}(\kappa)$ is contained in $\mathcal{M}_f(\kappa)$, then a holomorphic curve $h : \kappa \longrightarrow \mathbb{P}(W)$ defined over $\mathcal{R}$ must grow slower than f.

Definition A.5. *A holomorphic curve $f : \kappa \longrightarrow \mathbb{P}(V)$ is said to be linearly non-degenerate over $\mathcal{R}$ if (f, g) is free for each holomorphic curve $g : \kappa \longrightarrow \mathbb{P}(V^*)$ defined over $\mathcal{R}$.*

Lemma A.6. *Let $f : \kappa \longrightarrow \mathbb{P}(V)$ be a holomorphic curve with a representation $\tilde{f} = \tilde{f}_0 e_0 + \cdots + \tilde{f}_n e_n : \kappa \longrightarrow V$. Then f is linearly non-degenerate over $\mathcal{R}$ if and only if $\tilde{f}_0, ..., \tilde{f}_n$ are linearly independent over $\mathcal{R}$.*

Proof. W. l. o. g. we can renumber the base such that $\tilde{f}_0 \not\equiv 0$. Then, the base $e_0, e_1, ..., e_n$ is admissible for f, and so the k-th coordinate function of f is given by

$$f_k = \frac{\tilde{f}_k}{\tilde{f}_0}, \quad k = 0, 1, ..., n.$$

Let $\epsilon_0, ..., \epsilon_n$ be the dual base of $e_0, ..., e_n$.

Assume that f is linearly non-degenerate over $\mathcal{R}$. Take $g_k \in \mathcal{R}$ for $k = 0, ..., n$ such that

$$g_0 \tilde{f}_0 + g_1 \tilde{f}_1 + \cdots + g_n \tilde{f}_n \equiv 0.$$

Assume $(g_1, ..., g_n) \not\equiv 0$. Let G be an universal denominator of $g_0, g_1, ..., g_n$. Then $\tilde{g}_k = G g_k$ is an entire function. Then

$$\tilde{g} = \tilde{g}_0 \epsilon_0 + \cdots + \tilde{g}_n \epsilon_n : \kappa \longrightarrow V^*$$

defines a holomorphic curve $g = \mathbb{P} \circ \tilde{g} : \kappa \longrightarrow \mathbb{P}(V^*)$ such that $\tilde{g}$ is a reduced representation of g. Since $(g_1, ..., g_n) \not\equiv 0$, then $\tilde{g}_i = G g_i \not\equiv 0$ for some $i \in \mathbb{Z}[1, n]$. Note that

$$\frac{\tilde{g}_k}{\tilde{g}_i} = \frac{g_k}{g_i} \in \mathcal{R}, \quad k = 0, ..., n.$$

Hence g is defined over $\mathcal{R}$, however

$$\langle \tilde{f}, \tilde{g} \rangle = \sum_{k=0}^{n} \tilde{g}_k \tilde{f}_k = G \sum_{k=0}^{n} g_k \tilde{f}_k \equiv 0,$$

that is, (f, g) is not free, which contradicts with the assumption that f is linearly non-degenerate over $\mathcal{R}$. Therefore, $(g_1, ..., g_n) \equiv 0$, and so $g_0 \equiv 0$ since $\tilde{f}_0 \not\equiv 0$. Thus $\tilde{f}_0, ..., \tilde{f}_n$ are linearly independent over $\mathcal{R}$.

Assume that $\tilde{f}_0, ..., \tilde{f}_n$ are linearly independent over $\mathcal{R}$. Let $g : \kappa \longrightarrow \mathbb{P}(V^*)$ be a holomorphic curve defined over $\mathcal{R}$ with a representation $\tilde{g} = \tilde{g}_0 \epsilon_0 + \cdots + \tilde{g}_n \epsilon_n$. We have to show that $\langle \tilde{f}, \tilde{g} \rangle \not\equiv 0$. Note that $i \in \mathbb{Z}[0, n]$ exists such that $\tilde{g}_i \not\equiv 0$. Thus

$$g_k = \frac{\tilde{g}_k}{\tilde{g}_i} \in \mathcal{R}, \quad k = 0, 1, ..., n,$$

with $(g_0, ..., g_n) \not\equiv 0$, and hence

$$\langle \tilde{f}, \tilde{g} \rangle = \sum_{k=0}^{n} \tilde{g}_k \tilde{f}_k = \tilde{g}_i \sum_{k=0}^{n} g_k \tilde{f}_k \not\equiv 0,$$

since $\tilde{f}_0, ..., \tilde{f}_n$ are linearly independent over $\mathcal{R}$. Thus (f, g) is free, and then f is linearly non-degenerate over $\mathcal{R}$. $\square$

If f is linearly degenerate over $\mathcal{R}$, let $\mathcal{G}^*$ be the set of all non-Archimedean holomorphic curves $g : \kappa \longrightarrow \mathbb{P}(V^*)$ defined over $\mathcal{R}$ such that (f, g) is not free. Take a subset $\mathcal{G} = \{g_j\}_{j=0}^{q} \subseteq \mathcal{G}^*$ with $q + 1 = \#\mathcal{G}^*$ if $\mathcal{G}^*$ is finite, otherwise $q = \infty$. Take a reduced representation of g_j

$$\tilde{g}_j = \tilde{g}_{j0} \epsilon_0 + \cdots + \tilde{g}_{jn} \epsilon_n : \kappa \longrightarrow V^*.$$

For $\lambda \in J_l^q$, write $\mathcal{G}_\lambda = \{g_{\lambda(j)}\}_{j=0}^{l}$ and abbreviate

$$\tilde{g}_\lambda = \tilde{g}_{\lambda(0)} \wedge \cdots \wedge \tilde{g}_{\lambda(l)} : \kappa \longrightarrow \bigwedge_{l+1} V^*.$$

If $\tilde{g}_\lambda \not\equiv 0$, then

$$g_\lambda = g_{\lambda(0)} \wedge \cdots \wedge g_{\lambda(l)} = \mathbb{P} \circ \tilde{g}_\lambda : \kappa \longrightarrow \mathbb{P}\left(\bigwedge_{l+1} V^* \right)$$

is a non-Archimedean holomorphic curve, and the family $\mathcal{G}_\lambda$ is said to be *linearly independent*. Define

$$J_l(\mathcal{G}) = \{\lambda \in J_l^q \mid \tilde{g}_\lambda \not\equiv 0\}$$

and

$$\ell_f(\mathcal{R}) = \max\{l + 1 \mid J_l(\mathcal{G}) \neq \emptyset, \mathcal{G} \subseteq \mathcal{G}^*\}$$

with $1 \leq \ell_f(\mathcal{R}) \leq n + 1$. Write

$$V_\mathcal{R} = V \otimes_\kappa \mathcal{R} = \mathcal{R}\{e_0, ..., e_n\},$$

$$V_\mathcal{R}^* = V^* \otimes_\kappa \mathcal{R} = \mathcal{R}\{\epsilon_0, ..., \epsilon_n\},$$

and define a linear subspace of $V_\mathcal{R}^*$

$$E[f] = \{w \in V_\mathcal{R}^* \mid \langle \tilde{f}, w \rangle \equiv 0\}.$$

It is easy to prove $\ell_f(\mathcal{R}) = \dim_\mathcal{R} E[f]$. Put

$$k = n - \ell_f(\mathcal{R}).$$

Then f is said to be *k-flat over $\mathcal{R}$*. In order to simplify our notation, we define $\ell_f(\mathcal{R}) = 0$ if f is linearly non-degenerate over $\mathcal{R}$, that is, $E[f] = \emptyset$, and say that f is *n-flat over $\mathcal{R}$*.

Lemma A.7. *The number k is nonnegative, i.e., $0 \leq \ell_f(\mathcal{R}) \leq n$.*

Proof. If $k < 0$, that is, $\ell_f(\mathcal{R}) = n + 1$, then there is a $\lambda \in J_n(\mathcal{G})$ such that

$$\tilde{g}_{\lambda(0)} \wedge \cdots \wedge \tilde{g}_{\lambda(n)} \not\equiv 0; \quad \langle \tilde{f}, \tilde{g}_{\lambda(j)} \rangle \equiv 0 \ (0 \leq j \leq n).$$

By Cramer's rule, $\tilde{f} \equiv 0$, which is impossible. □

Lemma A.8. *Let $f : \kappa \longrightarrow \mathbb{P}(V)$ be a non-Archimedean holomorphic curve which is linearly degenerate over $\mathcal{R}$. Then f is k-flat over $\mathcal{R}$ if and only if there is a base $\xi = (\xi_0, ..., \xi_n)$ of $V_{\mathcal{R}}$ over $\mathcal{R}$ with dual base $\alpha = (\alpha_0, ..., \alpha_n)$ such that for any representation $\tilde{f} : \kappa \longrightarrow V$,*

$$\tilde{f} = \sum_{j=0}^{k} \langle \tilde{f}, \alpha_j \rangle \xi_j.$$

Moreover, the base can be chosen such that $\alpha_j \in V^*$ for $j = 0, ..., k$ and

$$\langle \tilde{f}, \alpha_0 \rangle, ..., \langle \tilde{f}, \alpha_k \rangle$$

are holomorphic and linearly independent over $\mathcal{R}$.

Proof. We follow the proof by Ru and Stoll in Theorem 6.9 of [113]. Assume that f is k-flat over $\mathcal{R}$. Let $\alpha_{k+1}, ..., \alpha_n \in V^*$ be a base of $E[f]$ over $\mathcal{R}$. Then $\alpha_0, ..., \alpha_k \in V^*$ exist such that $\alpha_0, ..., \alpha_n$ is a base of $V_{\mathcal{R}}^*$ over $\mathcal{R}$. Let $\xi_0, ..., \xi_n$ be the dual base. Since

$$\langle \tilde{f}, \alpha_j \rangle \equiv 0, \quad \alpha_j \in E[f], \quad j = k+1, ..., n,$$

we have

$$\tilde{f} = \sum_{j=0}^{n} \langle \tilde{f}, \alpha_j \rangle \xi_j = \sum_{j=0}^{k} \langle \tilde{f}, \alpha_j \rangle \xi_j.$$

Assume that there exists $w_j \in \mathcal{R}$ for $j = 0, ..., k$ such that

$$w_0 \langle \tilde{f}, \alpha_0 \rangle + \cdots + w_k \langle \tilde{f}, \alpha_k \rangle \equiv 0.$$

Then $w = w_0 \alpha_0 + \cdots + w_k \alpha_k \in E[f]$, and so $w_j \in \mathcal{R}$ for $j = k+1, ..., n$ exist such that

$$w_0 \alpha_0 + \cdots + w_k \alpha_k = w = -w_{k+1} \alpha_{k+1} - \cdots - w_n \alpha_n.$$

Thus $w_j = 0$ for $j = 0, ..., n$. Hence $\langle \tilde{f}, \alpha_0 \rangle, ..., \langle \tilde{f}, \alpha_k \rangle$ are linearly independent over $\mathcal{R}$.

Conversely, since $\xi = (\xi_0, ..., \xi_n)$ is a base of $V_{\mathcal{R}}$ over $\mathcal{R}$ with dual base $\alpha = (\alpha_0, ..., \alpha_n)$, we have

$$\sum_{j=0}^{k} \langle \tilde{f}, \alpha_j \rangle \xi_j = \tilde{f} = \sum_{j=0}^{n} \langle \tilde{f}, \alpha_j \rangle \xi_j,$$

which implies

$$\langle \tilde{f}, \alpha_j \rangle \equiv 0, \quad j = k+1, ..., n.$$

Therefore, $\alpha_j \in E[f]$ for $j = k+1, ..., n$, and $\dim_{\mathcal{R}} E[f] \geq n - k$. If $\dim_{\mathcal{R}} E[f] > n - k$, there is

$$w = \sum_{j=0}^{n} w_j \alpha_j \in E[f], \quad w_j \in \mathcal{R} \ (j = 0, ..., n),$$

such that $w_{j_0} \not\equiv 0$ for some $j_0 \in \mathbb{Z}[0, k]$. However,

$$0 = \langle \tilde{f}, w \rangle = \sum_{j=0}^{n} w_j \langle \tilde{f}, \alpha_j \rangle = \sum_{j=0}^{k} w_j \langle \tilde{f}, \alpha_j \rangle,$$

which contradicts the assumption that $\langle \tilde{f}, \alpha_0 \rangle, ..., \langle \tilde{f}, \alpha_k \rangle$ are linearly independent over $\mathcal{R}$. Thus $\dim_{\mathcal{R}} E[f] = n - k$, that is, f is k-flat over $\mathcal{R}$. $\qquad\square$

A.2 The Steinmetz lemma

Let κ be an algebraically closed field of characteristic zero, complete for a non-trivial non-Archimedean absolute value $|\cdot|$. Let V be a vector space of dimension $n+1 > 0$ over κ. Let $\mathcal{G} = \{g_j\}_{j=0}^{q}$ be a finite family of holomorphic curves $g_j : \kappa \longrightarrow \mathbb{P}(V^*)$ with $n \leq q < \infty$ and $J_n(\mathcal{G}) \neq \emptyset$. For each $z \in \kappa$, define the orbit $\mathcal{G}(z) = \{g_j(z)\}_{j=0}^{q}$. Let $\tilde{g}_j : \kappa \longrightarrow V^*$ be a reduced representation of g_j for $j = 0, ..., q$.

Definition A.9. *A base* $e = (e_0, ..., e_n)$ *of* V *is said to be perfect for* $\mathcal{G}$ *if for every* $\lambda \in J_n(\mathcal{G})$

$$\langle e_0 \wedge \cdots \wedge e_j, \tilde{g}_{\pi(0)} \wedge \cdots \wedge \tilde{g}_{\pi(j)} \rangle \not\equiv 0, \quad \pi \in \mathcal{J}_\lambda, \quad j = 0, 1, ..., n,$$

where $\mathcal{J}_\lambda$ *is the permutation group of* $\{\lambda(0), ..., \lambda(n)\}$.

Define

$$I = I(\mathcal{G}) = \bigcup_{l=0}^{n} \bigcup_{\lambda \in J_l(\mathcal{G})} \tilde{g}_\lambda^{-1}(0).$$

Obviously, $J_l(\mathcal{G}) = J_l(\mathcal{G}(z))$ if $z \in \kappa - I$. Thus if $z \in \kappa - I$, a perfect base of V for $\mathcal{G}(z)$ is a perfect base of V for $\mathcal{G}$. In particular, perfect bases of V for $\mathcal{G}$ exist.

Let $|\ |$ be a norm defined over a perfect base $e = (e_0, ..., e_n)$ of V for $\mathcal{G}$. Write

$$\tilde{g}_j = \tilde{g}_{j0}\epsilon_0 + \cdots + \tilde{g}_{jn}\epsilon_n : \kappa \longrightarrow V^*, \quad j = 0, ..., q,$$

where $\epsilon = (\epsilon_0, ..., \epsilon_n)$ is the dual of e. Note that for $\lambda \in J_n(\mathcal{G})$, the base ϵ is admissible for each $g_{\lambda(j)}$ since e is perfect. The coordinate functions

$$g_{\lambda(j)k} = \frac{\tilde{g}_{\lambda(j)k}}{\tilde{g}_{\lambda(j)0}}, \quad \lambda \in J_n(\mathcal{G}), \quad j = 0, ..., n, \quad k = 0, ..., n,$$

are defined. Define

$$\tilde{\mathcal{G}} = \{g_{\lambda(j)k} \mid \lambda \in J_n(\mathcal{G}), \ 0 \leq j \leq n, \ 0 \leq k \leq n\}.$$

Let $\mathcal{L} = \mathcal{L}(\tilde{\mathcal{G}})$ be the finite dimensional vector space generated by $\tilde{\mathcal{G}}$ over κ and let $\mathcal{R} = \mathcal{R}(\tilde{\mathcal{G}})$ be the field generated by $\tilde{\mathcal{G}}$ over κ. Then

$$\kappa \subseteq \mathcal{L} \subseteq \mathcal{R} \subseteq \mathcal{M}(\kappa)$$

and $\mathcal{R} = \kappa$ if each element of $\mathcal{G}_\lambda$ is constant for $\lambda \in J_n(\mathcal{G})$. Also $\mathcal{R}$ is the smallest field such that each element of $\mathcal{G}_\lambda$ is defined over $\mathcal{R}$ for $\lambda \in J_n(\mathcal{G})$.

Let G_0 be a universal denominator of $\tilde{\mathcal{G}}$. Since $1 \in \tilde{\mathcal{G}}$, then $G_0 \not\equiv 0$, and G_0 is an entire function. Let $\mathcal{L}^*$ be the dual vector space of $\mathcal{L}$ and set

$$\dim \mathcal{L}^* = \dim \mathcal{L} = l + 1.$$

Let $\theta_0, ..., \theta_l$ be a base of $\mathcal{L}$ over κ with $\theta_0 = 1$. Then $\tilde{G}_j = G_0\theta_j$ is an entire function on κ for $j = 0, ..., l$. Let $\tau_0, ..., \tau_l$ be the dual base of $\theta_0, ..., \theta_l$. Then a holomorphic vector function

$$\tilde{G} : \kappa \longrightarrow \mathcal{L}^* - \{0\}$$

is uniquely defined by

$$\tilde{G}(z) = \sum_{j=0}^{l} \tilde{G}_j(z)\tau_j, \quad z \in \kappa.$$

For any $\phi \in \mathcal{L}$, write $\phi = a_0\theta_0 + \cdots + a_l\theta_l$. Then

$$\langle \phi, \tilde{G}(z) \rangle = \sum_{j=0}^{l} a_j \tilde{G}_j(z) = \sum_{j=0}^{l} a_j \left(G_0\theta_j \right)(z),$$

that is,

$$\langle \phi, \tilde{G}(z) \rangle = (G_0\phi)(z),$$

for all $z \in \kappa$. The holomorphic curve

$$G = \mathbb{P} \circ \tilde{G} : \kappa \longrightarrow \mathbb{P}(\mathcal{L}^*)$$

is called the *adjunct mapping* of $\tilde{\mathcal{G}}$.

Lemma A.10. *The adjunct mapping* $G : \kappa \longrightarrow \mathbb{P}(\mathcal{L}^*)$ *is linearly non-degenerate, i.e., its image can not be contained in any hyperplane.*

Proof. Assume that G is linearly degenerate. Then $\phi \in \mathcal{L}$ with $\phi \not\equiv 0$ exists such that $\langle \phi, \tilde{G}(z) \rangle = 0$ for all $z \in \kappa$. Hence $G_0\phi \equiv 0$, and so $\phi \equiv 0$ since $G_0 \not\equiv 0$, against the choice of ϕ. Therefore, G is linearly non-degenerate. $\square$

Further for any $m \in \mathbb{Z}^+$, we obtain the non-Archimedean holomorphic curve

$$G^{\amalg m} : \kappa \longrightarrow \mathbb{P}(\amalg_m \mathcal{L}^*)$$

with a reduced representation

$$\tilde{G}^{\amalg m} : \kappa \longrightarrow \amalg_m \mathcal{L}^* - \{0\}.$$

Let $\mathcal{L}^*(m)$ be the smallest linear subspace of $\amalg_m\mathcal{L}^*$ containing the image $\tilde{G}^{\amalg m}(\kappa)$ and define

$$q(m) = \dim \mathcal{L}^*(m).$$

Then $\mathcal{L}^*(1) = \mathcal{L}^*$, and

$$G^{\amalg m} : \kappa \longrightarrow \mathbb{P}(\mathcal{L}^*(m))$$

is linearly non-degenerate.

Define

$$\tilde{\mathcal{G}}_m = \{\phi_1\phi_2\cdots\phi_m \mid \phi_j \in \tilde{\mathcal{G}}, \; j = 1, ..., m\}$$

and write

$$\mathcal{L}_m = \mathcal{L}(\tilde{\mathcal{G}}_m), \quad \mathcal{L}_0 = \kappa.$$

Observe that

$$\kappa \subseteq \mathcal{L}_m \subseteq \mathcal{L}_{m+1} \subseteq \mathcal{R}(\tilde{\mathcal{G}}_{m+1}) \subseteq \mathcal{R}(\tilde{\mathcal{G}}) = \mathcal{R}.$$

There exists one and only one surjective linear mapping $\sigma : \amalg_m\mathcal{L} \longrightarrow \mathcal{L}_m$ such that

$$\sigma(\phi_1 \amalg \cdots \amalg \phi_m) = \phi_1 \cdots \phi_m$$

whenever $\phi_j \in \mathcal{L}$ for $j = 1, ..., m$. Let $\theta = (\theta_0, ..., \theta_l)$ be a base of $\mathcal{L} = \mathcal{L}_1$ with $\theta_0 = 1$, $q(1) = l + 1$ and let $\tau = (\tau_0, ..., \tau_l)$ be the dual base. Write

$$\tilde{G} = \sum_{j=0}^{l} \tilde{G}_j \tau_j,$$

with $\tilde{G}_j = G_0\theta_j$ for $j = 0, ..., l$. Then

$$\tilde{G}^{\amalg m} = \sum_{\lambda \in J_{l,m}} \frac{m!}{\lambda!} \tilde{G}_0^{\lambda(0)} \cdots \tilde{G}_l^{\lambda(l)} \tau^{\amalg\lambda} = G_0^m \sum_{\lambda \in J_{l,m}} \frac{m!}{\lambda!} \theta_0^{\lambda(0)} \cdots \theta_l^{\lambda(l)} \tau^{\amalg\lambda}.$$

We obtain

$$\langle \phi, \tilde{G}^{\amalg m} \rangle = G_0^m \sigma(\phi),$$

for all $\phi \in \amalg_m\mathcal{L}$.

Lemma A.11. *Let $\tau_{m,1}, ..., \tau_{m,q(m)}$ be a base of $\mathcal{L}^*(m)$ over κ. Then*

$$\tilde{G}^{\amalg m} = \sum_{j=1}^{q(m)} B_{m,j}\tau_{m,j}, \tag{A.3}$$

where each $B_{m,j}$ is an entire function with $G_0^{-m}B_{m,j} \in \mathcal{L}_m$. Moreover,

$$G_0^{-m}B_{m,1}, ..., G_0^{-m}B_{m,q(m)}$$

is a base of $\mathcal{L}_m$.

Proof. Here we follow Stoll's proof in Theorem 3.9 of [127] and abbreviate $s = q(m)$. Extend the base $\tau_{m,1}, ..., \tau_{m,s}$ of $\mathcal{L}^*(m)$ over κ to a base $\tau_{m,1}, ..., \tau_{m,r}$ of $\amalg_m \mathcal{L}^*$ and let $\tau_{m,1}^*, ..., \tau_{m,r}^*$ be the dual base in $\amalg_m \mathcal{L}$. Then

$$\tilde{G}^{\amalg m} = \sum_{j=1}^{r} \langle \tau_{m,j}^*, \tilde{G}^{\amalg m} \rangle \tau_{m,j} = \sum_{j=1}^{r} G_0^m \sigma \left(\tau_{m,j}^* \right) \tau_{m,j}. \tag{A.4}$$

Comparing (A.3) and (A.4) we have

$$G_0^m \sigma \left(\tau_{m,j}^* \right) = \begin{cases} B_{m,j} & : \quad 1 \le j \le s \\ 0 & : \quad s+1 \le j \le r. \end{cases}$$

Hence $B_{m,j}$ are entire functions such that $G_0^{-m} B_{m,j} \in \mathcal{L}_m$ for $j = 1, ..., s$ and $\sigma \left(\tau_{m,j}^* \right) = 0$ for $j = s+1, ..., r$.

Let $\theta = (\theta_0, ..., \theta_l)$ be a base of $\mathcal{L}$. Then $\{\theta^{\amalg \lambda}\}_{\lambda \in J_{l,m}}$ is a base of $\amalg_m \mathcal{L}$. There are constants $b_{\lambda j} \in \kappa$ such that

$$\theta^{\amalg \lambda} = \sum_{j=1}^{r} b_{\lambda j} \tau_{m,j}^*, \quad \lambda \in J_{l,m}.$$

Take any $\psi \in \mathcal{L}_m$. Pick $\phi \in \amalg_m \mathcal{L}$ with $\sigma(\phi) = \psi$. Then $a_\lambda \in \kappa$ exist such that

$$\phi = \sum_{\lambda \in J_{l,m}} a_\lambda \theta^{\amalg \lambda}.$$

Then

$$\begin{aligned} G_0^m \psi &= G_0^m \sigma(\phi) = \langle \phi, \tilde{G}^{\amalg m} \rangle = \sum_{\lambda \in J_{l,m}} a_\lambda \langle \theta^{\amalg \lambda}, \tilde{G}^{\amalg m} \rangle \\ &= \sum_{\lambda \in J_{l,m}} a_\lambda \sum_{j=1}^{r} b_{\lambda j} \langle \tau_{m,j}^*, \tilde{G}^{\amalg m} \rangle \\ &= \sum_{j=1}^{r} c_j G_0^m \sigma \left(\tau_{m,j}^* \right) = \sum_{j=1}^{s} c_j B_{m,j}, \end{aligned}$$

where

$$c_j = \sum_{\lambda \in J_{l,m}} a_\lambda b_{\lambda j} \in \kappa.$$

Therefore, $G_0^{-m} B_{m,1}, ..., G_0^{-m} B_{m,q(m)}$ generate $\mathcal{L}_m$ over κ.

Assume that there are $d_j \in \kappa$ such that

$$\sum_{j=1}^{s} d_j G_0^{-m} B_{m,j} \equiv 0.$$

A linear function $h : \mathcal{L}^*(m) \longrightarrow \kappa$ is uniquely given by

$$h \left(\tau_{m,j} \right) = d_j, \quad j = 1, ..., s.$$

Then

$$h \circ \tilde{G}^{\mathrm{II}m} = \sum_{j=1}^{s} B_{m,j} h\left(\tau_{m,j}\right) = \sum_{j=1}^{s} B_{m,j} d_j \equiv 0,$$

and hence $h \equiv 0$ since $G^{\mathrm{II}m} : \kappa \longrightarrow \mathbb{P}(\mathcal{L}^*(m))$ is linearly non-degenerate. Thus $d_j = 0$ for $j = 1, ..., s$. Therefore, the functions $G_0^{-m} B_{m,1}, ..., G_0^{-m} B_{m,q(m)}$ are linearly independent over κ and, therefore, constitute a base of $\mathcal{L}_m$ over κ. $\square$

In particular, we have

$$1 \le q(m) = \dim \mathcal{L}_m \le \dim \mathcal{L}_{m+1} = q(m+1), \quad m = 1, ...,$$

and can easily prove the Steinmetz's lemma (cf. Stoll [127], Lemma 3.12):

Lemma A.12. *For any $u \in \mathbb{Z}^+$, the sequence $q(m)$ satisfies the following relation:*

$$\liminf_{m \to \infty} \frac{q(m+1)}{q(m+1-u)} = 1.$$

Proof. Set

$$a = \liminf_{m \to \infty} \frac{q(m+1)}{q(m+1-u)} \ge 1.$$

Assume that $a > 1$. Take $b \in \mathbb{R}$ with $a > b > 1$. Then $u < m_0 \in \mathbb{Z}^+$ exists such that

$$q(m+1) \ge b q(m+1-u), \quad m \ge m_0.$$

Take $m \ge m_0 + u$. One and only one integer k exists such that $(k-1)u \le m+1-m_0 < ku$. Hence, by induction,

$$q(m+1) \ge q(m+1-ku)b^k \ge b^k, \quad m \ge m_0 + u.$$

Note that

$$q(m+1) \le \binom{m+q(1)}{m+1} = \prod_{j=1}^{q(1)-1} \frac{m+j+1}{j} \le (m+q(1))^{q(1)-1}.$$

Therefore

$$0 < \log b \le \frac{\log q(m+1)}{k} \le \frac{u(q(1)-1)\log(m+q(1))}{m+1-m_0} \to 0$$

as $m \to \infty$, which is impossible. Hence $a = 1$. $\square$

Since the linear mapping $\sigma : \mathrm{II}_m \mathcal{L} \longrightarrow \mathcal{L}_m$ is surjective, the dual linear mapping $\sigma^* : \mathcal{L}_m^* \longrightarrow \mathrm{II}_m \mathcal{L}^*$ is injective. Hence $\sigma^*(\mathcal{L}_m^*)$ is a $q(m)$-dimensional linear subspace of $\mathrm{II}_m \mathcal{L}^*$.

Lemma A.13. *$\sigma^*(\mathcal{L}_m^*) = \mathcal{L}^*(m)$ and the restriction $\rho = \sigma^* : \mathcal{L}_m^* \longrightarrow \mathcal{L}^*(m)$ is an isomorphism.*

Proof. Any $z \in \kappa$ induces a linear mapping $\alpha_z \in \mathcal{L}_m^*$ defined by

$$\alpha_z(\psi) = (G_0^m \psi)(z), \quad \psi \in \mathcal{L}_m.$$

Then, for any $\xi \in \mathrm{II}_m \mathcal{L}$,

$$\langle \xi, \sigma^*(\alpha_z) \rangle = \langle \sigma(\xi), \alpha_z \rangle = \alpha_z(\sigma(\xi)) = (G_0^m \sigma(\xi))(z) = \langle \xi, \tilde{G}^{\mathrm{II}m}(z) \rangle,$$

and $\sigma^*(\alpha_z) = \tilde{G}^{\mathrm{II}m}(z) \in \mathcal{L}^*(m)$. Therefore

$$\tilde{G}^{\mathrm{II}m}(z) = \sigma^*(\alpha_z) \in \sigma^*(\mathcal{L}_m^*), \quad z \in \kappa,$$

which means $\mathcal{L}^*(m) \subset \sigma^*(\mathcal{L}_m^*)$ with

$$\dim \mathcal{L}^*(m) = q(m) = \dim \sigma^*(\mathcal{L}_m^*).$$

Hence $\sigma^*(\mathcal{L}_m^*) = \mathcal{L}^*(m)$. $\qquad\square$

Take $m, s \in \mathbb{Z}^+$ and $\psi \in \mathcal{L}_s$. A linear "action" mapping

$$H_\psi = H_{\psi,m} : \mathcal{L}_m \longrightarrow \mathcal{L}_{m+s} \tag{A.5}$$

is defined by

$$H_\psi(\Psi) = \psi \Psi, \quad \Psi \in \mathcal{L}_m. \tag{A.6}$$

Take $\tilde{\psi} \in \mathrm{II}_s \mathcal{L}$ with $\sigma(\tilde{\psi}) = \psi$. Then a lifting linear mapping

$$\tilde{H}_\psi = \tilde{H}_{\psi,m} : \mathrm{II}_m \mathcal{L} \longrightarrow \mathrm{II}_{m+s} \mathcal{L}$$

of H_ψ is induced by

$$\tilde{H}_\psi(\Psi) = \tilde{\psi} \mathrm{II} \Psi, \quad \Psi \in \mathrm{II}_m \mathcal{L}.$$

Obviously, the lifting mapping of H_ψ satisfies

$$\sigma \circ \tilde{H}_\psi = H_\psi \circ \sigma.$$

When $\psi \neq 0$, $\tilde{H}_\psi$ and H_ψ are injective and, hence, a linear surjective mapping

$$\mathbf{H}_\psi = \mathbf{H}_{\psi,m} = \rho \circ H_\psi^* \circ \rho^{-1} : \mathcal{L}^*(m+s) \longrightarrow \mathcal{L}^*(m)$$

is defined. Since $\tilde{H}_\psi^* \circ \sigma^* = \sigma^* \circ H_\psi^*$, one has

$$\mathbf{H}_\psi = \tilde{H}_\psi^*|_{\mathcal{L}^*(m+s)}. \tag{A.7}$$

For any $z \in \kappa$, $\Psi \in \mathrm{II}_m \mathcal{L}$,

$$
\begin{aligned}
\left\langle \Psi, \mathbf{H}_\psi \left(\tilde{G}^{\mathrm{II}m+s}(z) \right) \right\rangle
&= \left\langle \Psi, \tilde{H}_\psi^* \left(\tilde{G}^{\mathrm{II}m+s}(z) \right) \right\rangle \\
&= \left\langle \tilde{H}_\psi(\Psi), \tilde{G}^{\mathrm{II}m+s}(z) \right\rangle = \left\langle \tilde{\psi} \mathrm{II} \, \Psi, \tilde{G}^{\mathrm{II}m+s}(z) \right\rangle \\
&= \left(G_0^{m+s} \psi \sigma(\Psi) \right)(z) = (G_0^s \psi)(z) \left(G_0^m \sigma(\Psi) \right)(z) \\
&= (G_0^s \psi)(z) \left\langle \Psi, \tilde{G}^{\mathrm{II}m}(z) \right\rangle = \left\langle \Psi, (G_0^s \psi)(z) \tilde{G}^{\mathrm{II}m}(z) \right\rangle,
\end{aligned}
$$

and, hence,

$$\mathbf{H}_\psi \circ \tilde{G}^{\amalg m+s} = G_0^s \psi \tilde{G}^{\amalg m}. \tag{A.8}$$

In particular, if $s = 1 = \psi$, the linear mapping

$$H_{1,m} : \mathcal{L}_m \longrightarrow \mathcal{L}_{m+1}$$

is the inclusion mapping. Thus we can iterate s-times

$$H_m^s = H_{1,m+s-1} \circ \cdots \circ H_{1,m+1} \circ H_{1,m}$$

and obtain the inclusion mapping

$$H_m^s : \mathcal{L}_m \longrightarrow \mathcal{L}_{m+s}.$$

Similarly, the lifting linear mapping

$$\tilde{H}_{1,m} : \amalg_m \mathcal{L} \longrightarrow \amalg_{m+1} \mathcal{L}$$

of $H_{1,m}$ is also an inclusion, as is the s-th iteration

$$\tilde{H}_m^s : \amalg_m \mathcal{L} \longrightarrow \amalg_{m+s} \mathcal{L}.$$

Hence, the linear surjective mapping

$$\mathbf{H}_{1,m} = \tilde{H}_{1,m}^* |_{\mathcal{L}^*(m+1)} : \mathcal{L}^*(m+1) \longrightarrow \mathcal{L}^*(m)$$

and its s-th iteration

$$\mathbf{H}_m^s = \left(\tilde{H}_m^s \right)^* |_{\mathcal{L}^*(m+s)} : \mathcal{L}^*(m+s) \longrightarrow \mathcal{L}^*(m)$$

are projections.

Assume $\psi \in \mathcal{L}_s - \{0\}$. Since $\mathbf{H}_\psi$ is surjective, we have

$$\dim \mathrm{Ker} \mathbf{H}_\psi = q(m+s) - q(m),$$

and so the linear isomorphism

$$\eta_\psi = \eta_{\psi,m} : \mathcal{L}^*(m) \longrightarrow \{\mathcal{L}^*(m+s) - \mathrm{Ker} \mathbf{H}_\psi\} \cup \{0\}$$

is well defined. The direct sum decomposition

$$\mathcal{L}^*(m+s) = \eta_\psi \left(\mathcal{L}^*(m) \right) \oplus \mathrm{Ker} \mathbf{H}_\psi$$

gives us the respective projection $Q_\psi = Q_{\psi,m}$ and inclusions $\zeta_\psi = \zeta_{\psi,m}$

$$0 \longrightarrow \mathrm{Ker} \mathbf{H}_\psi \overset{\zeta_\psi}{\longrightarrow} \mathcal{L}^*(m+s) \overset{\mathbf{H}_\psi}{\longrightarrow} \mathcal{L}^*(m) \longrightarrow 0, \tag{A.9}$$

$$0 \longrightarrow \mathcal{L}^*(m) \xrightarrow{\eta_\psi} \mathcal{L}^*(m+s) \xrightarrow{Q_\psi} \mathrm{Ker}\mathbf{H}_\psi \longrightarrow 0, \tag{A.10}$$

such that both sequences are exact. In particular, we have two exact sequences

$$0 \longrightarrow \mathrm{Ker}\mathbf{H}_m^s \xrightarrow{\zeta_m^s} \mathcal{L}^*(m+s) \xrightarrow{\mathbf{H}_m^s} \mathcal{L}^*(m) \longrightarrow 0, \tag{A.11}$$

$$0 \longrightarrow \mathcal{L}^*(m) \xrightarrow{\eta_m^s} \mathcal{L}^*(m+s) \xrightarrow{Q_m^s} \mathrm{Ker}\mathbf{H}_m^s \longrightarrow 0, \tag{A.12}$$

with the inclusions ζ_m^s, η_m^s, and the projection Q_m^s, $\mathbf{H}_m^s$. Thus, there exist the direct sum decomposition

$$\mathcal{L}^*(m+s) = \mathcal{L}^*(m) \oplus \mathrm{Ker}\mathbf{H}_m^s,$$

and an isomorphism

$$\xi_\psi : \mathrm{Ker}\mathbf{H}_\psi \longrightarrow \mathrm{Ker}\mathbf{H}_m^s.$$

Noting that

$$\eta_\psi \circ \mathbf{H}_\psi + Q_\psi = id : \mathcal{L}^*(m+s) \longrightarrow \mathcal{L}^*(m+s), \tag{A.13}$$

if we denote the isomorphism

$$\dot{\mathbf{H}}_\psi = \dot{\mathbf{H}}_{\psi,m} = \eta_\psi^{-1} \oplus \xi_\psi : \mathcal{L}^*(m+s) \longrightarrow \mathcal{L}^*(m+s),$$

then

$$\dot{\mathbf{H}}_\psi = \mathbf{H}_\psi + \xi_\psi \circ Q_\psi$$

Therefore

$$\dot{\mathbf{H}}_\psi \circ \tilde{G}^{\mathrm{II}m+s} = \mathbf{H}_\psi \circ \tilde{G}^{\mathrm{II}m+s} + Q_\psi = G_0^s \psi \tilde{G}^{\mathrm{II}m} + Q_\psi, \tag{A.14}$$

where

$$Q_\psi = Q_{\psi,m} = \xi_\psi \circ Q_\psi \circ \tilde{G}^{\mathrm{II}m+s}$$

In particular, noting that,

$$\mathbf{H}_m^s + Q_m^s = id : \mathcal{L}^*(m+s) \longrightarrow \mathcal{L}^*(m+s), \tag{A.15}$$

we have

$$\tilde{G}^{\mathrm{II}m+s} = \mathbf{H}_m^s \circ \tilde{G}^{\mathrm{II}m+s} + Q_m^s = G_0^s \tilde{G}^{\mathrm{II}m} + Q_m^s, \tag{A.16}$$

where

$$Q_m^s = Q_m^s \circ \tilde{G}^{\mathrm{II}m+s}.$$

Lemma A.14. *Let* $\alpha : \mathcal{L}^*(m+s) \longrightarrow \kappa$ *be a* κ-*linear mapping. Then one and only one function* $\Psi \in \mathcal{L}_{m+s}$ *exists such that*

$$\alpha \circ Q_\psi = G_0^{m+s} \Psi.$$

Proof. Since $\tilde{\alpha} = \alpha \circ \xi_\psi \circ Q_\psi : \mathcal{L}^*(m+s) \longrightarrow \kappa$ is κ-linear mapping, and because $\mathcal{L}^*(m+s) = \sigma^*(\mathcal{L}_{m+s}^*)$, one and only one function $\tilde{\Psi} \in \mathrm{II}_{m+s}\mathcal{L}$ exists such that $\Psi = \sigma(\tilde{\Psi}) \in \mathcal{L}_{m+s}$ and $\tilde{\alpha}(z) = \langle \tilde{\Psi}, z \rangle$ for all $z \in \mathcal{L}^*(m+s)$. Hence

$$\alpha \circ Q_\psi = \tilde{\alpha} \circ \tilde{G}^{\mathrm{II}m+s} = \left\langle \tilde{\Psi}, \tilde{G}^{\mathrm{II}m+s} \right\rangle = G_0^{m+s} \sigma(\tilde{\Psi}) = G_0^{m+s} \Psi.$$

$$\square$$

A.3   A defect relation for moving targets

We will adept the notations and terminologies used in § A.2 and will consider the holomorphic curve

$$f : \kappa \longrightarrow \mathbb{P}(V),$$

with a reduced representation

$$\tilde{f} = \tilde{f}_0 e_0 + \cdots + \tilde{f}_n e_n : \kappa \longrightarrow V_*.$$

For $q \geq n$, let

$$g_j : \kappa \longrightarrow \mathbb{P}(V^*), \quad j = 0, ..., q$$

be $q + 1$ holomorphic curves with reduced representations

$$\tilde{g} = \tilde{g}_{j0}\epsilon_0 + \cdots + \tilde{g}_{jn}\epsilon_n : \kappa \longrightarrow V_*^*.$$

The family $\{g_j\}$ is said to be *in general position* if $\det(\tilde{g}_{\lambda(k)l}) \not\equiv 0$ for any $\lambda \in J_n^q$. If so, we may assume that

$$\tilde{g}_{j0} \not\equiv 0, \quad j = 0, ..., q$$

by changing the homogeneous coordinate system of $\mathbb{P}(V^*)$ if necessary. Then put

$$g_{jk} = \frac{\tilde{g}_{jk}}{\tilde{g}_{j0}}$$

with $g_{j0} = 1$. Let $\mathcal{R}$ be the smallest subfield containing

$$\{g_{jk} \mid 0 \leq j \leq q, 0 \leq k \leq n\} \cup \kappa$$

of the meromorphic function field on κ. We have the following *defect relation* (see [62]):

Theorem A.15. *Given the holomorphic curves* $f : \kappa \longrightarrow \mathbb{P}(V)$ *and*

$$g_j : \kappa \longrightarrow \mathbb{P}(V^*), \quad j = 0, ..., q$$

with $q \geq n$. *If the family* $\{g_j\}$ *is in general position such that*

$$T(r, g_j) = o(T(r, f)), \quad r \to \infty, \quad j = 0, ..., q,$$

and if f *is non-degenerate over* $\mathcal{R}$, *then*

$$\sum_{j=0}^{q} \delta_f(g_j) \leq n + 1.$$

Proof. It is easy to check that

$$T(r, h) = o(T(r, f)), \quad r \to \infty, \quad h \in \mathcal{R}.$$

Take a positive integer m. Let $\mathcal{L}_m$ be the vector space generated over κ by

$$\left\{ \prod_{j,k} g_{jk}^{m_{jk}} \mid m_{jk} \in \mathbb{Z}_+ (0 \leq j \leq q, 0 \leq k \leq n), \sum_{j,k} m_{jk} = m \right\},$$

where $\mathbb{Z}_+$ is the set of non-negative integers. Since $g_{j0} = 1$, we have $\mathcal{L}_m \subset \mathcal{L}_{m+1}$. Thus we can take a basis $\{b_1, ..., b_t\}$ of $\mathcal{L}_{m+1}$ such that $\{b_1, ..., b_s\}$ is a basis of $\mathcal{L}_m$, where

$$t = q(m+1) = \dim \mathcal{L}_{m+1}, \quad s = q(m) = \dim \mathcal{L}_m.$$

Since f is non-degenerate over $\mathcal{R}$, we can deduce that

$$\varphi = \left(b_j \tilde{f}_k \mid 1 \leq j \leq t, 0 \leq k \leq n \right)$$

are linearly independent over κ. Put

$$F_j = \frac{\langle \tilde{f}, \tilde{g}_j \rangle}{\tilde{g}_{j0}} = \tilde{f}_0 + g_{j1}\tilde{f}_1 + ... + g_{jn}\tilde{f}_n, \quad 0 \leq j \leq q.$$

We claim that for any $\lambda \in J_n^q$,

$$\Phi_\lambda = \left(b_j F_{\lambda(k)} \mid 1 \leq j \leq s, 0 \leq k \leq n \right)$$

are linearly independent over κ. In fact, if there are $a_{jk} \in \kappa$ such that

$$\sum_{k=0}^n \sum_{j=1}^s a_{jk} b_j F_{\lambda(k)} = 0,$$

then

$$\sum_{i=0}^n \left(\sum_{k=0}^n \sum_{j=1}^s a_{jk} b_j g_{\lambda(k)i} \right) \tilde{f}_i = 0$$

and hence

$$\sum_{k=0}^n \sum_{j=1}^s a_{jk} b_j g_{\lambda(k)i} = 0, \quad i = 0, ..., n,$$

since $\tilde{f}_0, ..., \tilde{f}_n$ are linearly independent over $\mathcal{R}$. Because $\det(g_{\lambda(k)i}) \not\equiv 0$, we obtain

$$\sum_{j=1}^s a_{jk} b_j = 0, \quad k = 0, ..., n,$$

and hence $a_{jk} = 0$ ($0 \leq k \leq n$, $1 \leq j \leq s$) since $b_1, ..., b_s$ is the base of $\mathcal{L}_m$.

Since $b_j F_{\lambda(k)} (1 \leq j \leq s, 0 \leq k \leq n)$ are linear combinations of $b_j \tilde{f}_k (1 \leq j \leq t, 0 \leq k \leq n)$ over κ, we can choose

$$\Psi_\lambda = \left(\sum_{1 \leq u \leq t, 0 \leq v \leq n} c_{jk}^{uv} b_u \tilde{f}_v \mid c_{jk}^{uv} \in \kappa, s+1 \leq j \leq t, 0 \leq k \leq n \right)$$

such that

$$(\Phi_\lambda, \Psi_\lambda) = \varphi C_\lambda$$

holds for some $C_\lambda \in GL((n+1)t; \kappa)$. Then we have the equality of Wronskian determinants

$$\mathbf{W}(\Phi_\lambda, \Psi_\lambda) = \mathbf{W}(\varphi) \cdot \det(C_\lambda).$$

Put

$$\mathbf{W}_\lambda = \mathbf{W}(\Phi_\lambda, \Psi_\lambda), \quad \mathbf{W} = \mathbf{W}(\varphi).$$

Then

$$\mathbf{W}_\lambda = c_\lambda \mathbf{W}, \quad c_\lambda = \det(C_\lambda).$$

Next, we fix $z \in \kappa - \kappa[0; \rho_0]$ such that

$$\mathbf{W}(z) \neq 0, \infty, \ f_i(z) \neq 0 \ (i = 0, ..., n), \ F_j(z) \neq 0, \infty \ (j = 0, 1, ..., q).$$

We take distinct indices $\lambda_0, ..., \lambda_n = \beta_0, ..., \beta_{q-n}$ such that

$$0 < \mid F_{\lambda_0}(z) \mid \leq \cdots \leq \mid F_{\lambda_n}(z) \mid \leq \mid F_{\beta_1}(z) \mid \leq \cdots \leq \mid F_{\beta_{q-n}}(z) \mid < \infty.$$

Since $\{g_j\}$ is in general position, we have representations

$$\tilde{f}_k = \sum_{j=0}^{n} A_{kj}^\lambda F_{\lambda_j}, \quad A_{kj}^\lambda \in \mathcal{R}, \quad k = 0, ..., n.$$

Therefore

$$\mid \tilde{f}_k(z) \mid \leq \max_j \mid A_{kj}^\lambda(z) \mid \mid F_{\lambda_j}(z) \mid \leq \left\{ \max_j \mid A_{kj}^\lambda(z) \mid \right\} \mid F_{\beta_l}(z) \mid$$

for $k = 0, ..., n, l = 0, ..., q - n$. Hence we obtain

$$\mid \tilde{f}(z) \mid \leq A(z) \mid F_{\beta_l}(z) \mid, \quad l = 0, ..., q - n,$$

where

$$A(z) = \max_{k,j} \mid A_{kj}^\lambda(z) \mid.$$

Take $\lambda \in J_n^q$ with $\mathrm{Im}(\lambda) = \{\lambda_0, ..., \lambda_n\}$. By $\mathbf{W}_\lambda = c_\lambda \mathbf{W}$, we have

$$\log \frac{\mid F_0 \cdots F_q \mid^s}{\mid \mathbf{W} \mid} = \log \mid F_{\beta_1} \cdots F_{\beta_{q-n}} \mid^s - \log \frac{\mid \mathbf{W}_\lambda \mid}{\mid F_{\lambda_0} \cdots F_{\lambda_n} \mid^s} + c_1$$

$$= \log \mid F_{\beta_1} \cdots F_{\beta_{q-n}} \mid^s - \log D_\lambda - (n+1)(t-s)\log|\tilde{f}| + c_1$$

for some constant c_1, where

$$D_\lambda = \frac{\mid \mathbf{W}_\lambda \mid}{\mid F_{\lambda_0} \cdots F_{\lambda_n} \mid^s |\tilde{f}|^{(n+1)(t-s)}}.$$

Then

$$\log \mid F_{\beta_1} \cdots F_{\beta_{q-n}} \mid^s \leq \log \frac{\mid F_0 \cdots F_q \mid^s}{\mid \mathbf{W} \mid} + \log^+ D_\lambda + (n+1)(t-s)\log|\tilde{f}| + c_1.$$

Therefore

$$s(q-n)\log|\tilde{f}| \leq \log\frac{|F_0\cdots F_q|^s}{|\mathbf{W}|} + \log^+ D_\lambda$$
$$+ (n+1)(t-s)\log|\tilde{f}| + s(q-n)\log^+ A + c_1.$$

Note that

$$D_\lambda = |\mathbf{S}(\Phi_\lambda,\Psi_\lambda)|\frac{|\mathbf{K}(\Phi_\lambda,\Psi_\lambda)|}{|F_{\lambda_0}\cdots F_{\lambda_n}|^s|\tilde{f}|^{(n+1)(t-s)}}$$
$$\leq |\mathbf{S}(\Phi_\lambda,\Psi_\lambda)|\left|\prod_{j=1}^{s}b_j\right|^{n+1}\left(\prod_{j=s+1}^{t}\prod_{k=0}^{n}\max_{1\leq u\leq t, 0\leq v\leq n}|c_{jk}^{uv}b_u|\right).$$

Set $|z|=r$. By using the lemma of logarithmic derivative (see (6.3)), we can prove that

$$\log^+ D_\lambda(z) = o(T(r,f)).$$

Note that

$$\log^+ A(z) = o(T(r,f)).$$

We obtain

$$s(q-n)T(r,f) \leq s\sum_{j=0}^{q}\left\{N_f(r,g_j)-N\left(r,\frac{1}{\tilde{g}_{j0}}\right)\right\} - N\left(r,\frac{1}{\mathbf{W}}\right)$$
$$+ (n+1)(t-s)T(r,f) + o(T(r,f)),$$

which implies

$$\left\{q+1-(n+1)\frac{t}{s}\right\}T(r,f) \leq \sum_{j=0}^{q}N_f(r,g_j) + o(T(r,f)). \tag{A.17}$$

Note that the set of r for which (A.17) holds is dense in (ρ_0,∞). Thus (A.17) also holds for all $\rho_0 < r < \infty$, by continuity of the functions involved in the inequality. By Steinmetz' lemma, we have

$$\liminf\frac{t}{s} = \liminf_{m\to\infty}\frac{q(m+1)}{q(m)} = 1.$$

Thus, the theorem follows from this and (A.17). □

Here we follow the Shirosaki's simplified proof [121] of Ru-Stoll's theorem [112] by using the Cartan's method. The main idea comes from Stoll's presentation [127] of Chuang-Steinmetz's method in proving the defect relation for small functions.

A.4 The Ru-Stoll techniques

We continue the discussion in § A.2. Select a reduced representation $\tilde{f} : \kappa \longrightarrow V_*$ of f with

$$\tilde{f} = \sum_{j=0}^{n}\tilde{f}_j e_j, \quad \tilde{f}_0 \neq 0,$$

and decompose V into a direct sum

$$V = V' \oplus V'',$$

where

$$V' = \kappa e_0, \quad V'' = \kappa e_1 \oplus \cdots \oplus \kappa e_n.$$

For a positive integer m, define a vector space

$$V(m) = \left\{ \mathcal{L}^*(m) \otimes V' \right\} \oplus \left\{ \mathcal{L}^*(m+1) \otimes V'' \right\}$$

with the dimension

$$k(m) + 1 = \dim V(m) = q(m) + nq(m+1).$$

A holomorphic vector function

$$\tilde{F}_m = \tilde{f}_0 G_0 \tilde{G}^{\amalg m} \otimes e_0 + \sum_{i=1}^{n} \tilde{f}_i \tilde{G}^{\amalg m+1} \otimes e_i : \kappa \longrightarrow V(m)$$

is defined. Because $\tilde{f}_0 G_0 \not\equiv 0$, we have $\tilde{F}_m \not\equiv 0$. Hence, an non-Archimedean holomorphic curve, called the m-th Steinmetz-Stoll (non-Archimedean) mapping,

$$F_m = \mathbb{P} \circ \tilde{F}_m : \kappa \longrightarrow \mathbb{P}(V(m))$$

is defined. According to Theorem 6.5 of Stoll [127], we can prove the following:

Lemma A.16. *If f is linearly non-degenerate over $\mathcal{R}$, the m-th Steinmetz-Stoll mapping F_m is linearly non-degenerate.*

Proof. By Lemma A.11, we have

$$\tilde{F}_m = \tilde{f}_0 G_0 \sum_{j=1}^{q(m)} B_{m,j} \tau_{m,j} \otimes e_0 + \sum_{i=1}^{n} \sum_{j=1}^{q(m+1)} \tilde{f}_i B_{m+1,j} \tau_{m+1,j} \otimes e_i.$$

Take any $\alpha \in V(m)^*$ with $\langle \tilde{F}_m, \alpha \rangle \equiv 0$. We are going to show that $\alpha = 0$. Set

$$a_j = \langle \tau_{m,j} \otimes e_0, \alpha \rangle \in \kappa, \quad b_{ij} = \langle \tau_{m+1,j} \otimes e_i, \alpha \rangle \in \kappa,$$

and define

$$B = \sum_{j=1}^{q(m)} a_j B_{m,j}, \quad D_i = \sum_{j=1}^{q(m+1)} b_{ij} B_{m+1,j}.$$

Then

$$G_0^{-m} B \in \mathcal{L}_m \subset \mathcal{R}, \quad G_0^{-m-1} D_i \in \mathcal{L}_{m+1} \subset \mathcal{R} \ (i = 1, ..., n).$$

We obtain

$$0 = \langle \tilde{F}_m, \alpha \rangle = \tilde{f}_0 G_0 B + \sum_{i=1}^{n} D_i \tilde{f}_i$$

or

$$G_0^{-m} B \tilde{f}_0 + \sum_{i=1}^{n} G_0^{-m-1} D_i \tilde{f}_i \equiv 0.$$

Because f is linearly non-degenerate over $\mathcal{R}$, Lemma A.6 implies $B \equiv 0$ and $D_i \equiv 0$ for $i = 1, ..., n$, and hence

$$a_j = 0 \ (j = 1, ..., q(m)), \quad b_{ij} = 0 \ (i = 1, ..., n; \ j = 1, ..., q(m+1))$$

by Lemma A.11. Hence $\alpha = 0$. Therefore, the mapping F_m is linearly non-degenerate. $\square$

In the following, we introduce Ru-Stoll's techniques for studying Steinmetz-Stoll mappings (cf. [112]). Write

$$\bar{g}_j = \frac{\tilde{g}_j}{\tilde{g}_{j0}} = g_{j0} \epsilon_0 + \cdots + g_{jn} \epsilon_n, \quad j = 0, ..., q.$$

Take $m > s(n)$ and $\lambda \in J_n(\mathcal{G})$. Define

$$\psi_{jk} = \langle e_0 \wedge \cdots \wedge e_{j-1} \wedge e_k, \bar{g}_{\lambda(0)} \wedge \cdots \wedge \bar{g}_{\lambda(j)} \rangle \in \mathcal{L}_j$$

for $0 \leq j \leq k \leq n$, and write

$$\phi_{jk} = (-1)^{j+k} \langle e_0 \wedge \cdots \wedge e_{j-1}, \bar{g}_{\lambda(0)} \wedge \cdots \wedge \bar{g}_{\lambda(k-1)} \wedge \bar{g}_{\lambda(k+1)} \wedge \cdots \wedge \bar{g}_{\lambda(j)} \rangle \in \mathcal{L}_{j-1}$$

for $1 \leq j \leq n$, $0 \leq k \leq j$. Because $e_0, ..., e_n$ is a perfect base for $\mathcal{G}$, one has

$$\psi_{00} = \phi_{1,0} = 1, \quad \psi_{jj} \not\equiv 0, \quad \phi_{jj} = \psi_{j-1,j-1} \not\equiv 0.$$

Note that

$$\tilde{f} \angle \left(\bar{g}_{\lambda(0)} \wedge \cdots \wedge \bar{g}_{\lambda(j)} \right) = \sum_{k=0}^{j} \langle \tilde{f}, \bar{g}_{\lambda(k)} \rangle (-1)^k \bar{g}_{\lambda(0)} \wedge \cdots \wedge \bar{g}_{\lambda(k-1)} \wedge \bar{g}_{\lambda(k+1)} \wedge \cdots \wedge \bar{g}_{\lambda(j)}.$$

Hence

$$\left\langle e_0 \wedge \cdots \wedge e_{j-1} \wedge \tilde{f}, \bar{g}_{\lambda(0)} \wedge \cdots \wedge \bar{g}_{\lambda(j)} \right\rangle$$
$$= (-1)^j \left\langle \tilde{f} \wedge e_0 \wedge \cdots \wedge e_{j-1}, \bar{g}_{\lambda(0)} \wedge \cdots \wedge \bar{g}_{\lambda(j)} \right\rangle$$
$$= (-1)^j \left\langle e_0 \wedge \cdots \wedge e_{j-1}, \tilde{f} \angle \left(\bar{g}_{\lambda(0)} \wedge \cdots \wedge \bar{g}_{\lambda(j)} \right) \right\rangle$$
$$= \sum_{k=0}^{j} \langle \tilde{f}, \bar{g}_{\lambda(k)} \rangle \phi_{jk}.$$

On other hand, we have

$$\left\langle e_0 \wedge \cdots \wedge e_{j-1} \wedge \tilde{f}, \bar{g}_{\lambda(0)} \wedge \cdots \wedge \bar{g}_{\lambda(j)} \right\rangle$$
$$= \sum_{k=j}^{n} \tilde{f}_k \left\langle e_0 \wedge \cdots \wedge e_{j-1} \wedge e_k, \bar{g}_{\lambda(0)} \wedge \cdots \wedge \bar{g}_{\lambda(j)} \right\rangle$$
$$= \sum_{k=j}^{n} \tilde{f}_k \psi_{jk}.$$

Therefore, we obtain the following equation:

$$\sum_{k=0}^{j} \langle \tilde{f}, \bar{g}_{\lambda(k)} \rangle \phi_{jk} = \sum_{k=j}^{n} \tilde{f}_k \psi_{jk}. \tag{A.18}$$

Further, write

$$s(0) = 0, \quad s(j) = \sum_{i=0}^{j-1} i = \frac{(j-1)j}{2} \ (j \geq 1),$$

$$\psi_0 = 1, \quad \psi_j = \prod_{i=0}^{j-1} \psi_{ii} \in \mathcal{L}_{s(j)} \ (1 \leq j \leq n),$$

$$\alpha_{00} = 1, \ \alpha_{jk} = \prod_{i=k}^{j-1} \psi_{ii} \in \mathcal{L}_{s(j)-s(k)} \ (1 \leq j \leq n, \ 0 \leq k \leq j-1),$$

$$\beta_{jk} = \phi_{jk}\alpha_{j-1,k} \in \mathcal{L}_{s(j)-s(k)} \ (1 \leq j \leq n, \ 0 \leq k \leq j-1),$$

$$\gamma_{jk} = \psi_{j-1}\psi_{jk} \in \mathcal{L}_{s(j)+1} \ (j \leq k \leq n),$$

and set

$$P_j = \mathbf{H}_{m-s(j)}^{s(j)+1} : \mathcal{L}^*(m+1) \longrightarrow \mathcal{L}^*(m-s(j)) \ (j \geq 1),$$

$$\mathcal{B}_{jk} = \mathbf{H}_{\beta_{jk}, m-s(j)} : \mathcal{L}^*(m-s(k)) \longrightarrow \mathcal{L}^*(m-s(j)),$$

$$\Gamma_{jk} = \begin{cases} \dot{\mathbf{H}}_{\gamma_{jj}, m-s(j)} : \mathcal{L}^*(m+1) \longrightarrow \mathcal{L}^*(m+1) & : \ k = j \\ \mathbf{H}_{\gamma_{jk}, m-s(j)} : \mathcal{L}^*(m+1) \longrightarrow \mathcal{L}^*(m-s(j)) & : \ j < k \leq n \end{cases}.$$

Linear mappings

$$M_0 : V(m) \longrightarrow \mathcal{L}^*(m), \quad L_0 : V(m) \longrightarrow V(m)$$

are defined respectively by

$$M_0 \left(\sum_{i=0}^{n} z_i \otimes e_i \right) = z_0 + \sum_{i=1}^{n} \mathbf{H}_{g_{\lambda(0)i}}(z_i),$$

$$L_0 \left(\sum_{i=0}^{n} z_i \otimes e_i \right) = M_0 \left(\sum_{i=0}^{n} z_i \otimes e_i \right) \otimes e_0 + \sum_{i=1}^{n} z_i \otimes e_i,$$

where $z_0 \in \mathcal{L}^*(m)$ and $z_i \in \mathcal{L}^*(m+1)$ for $i = 1, ..., n$. Obviously, L_0 is an isomorphism. Then

$$\begin{aligned} M_0 \circ \tilde{F}_m &= M_0 \left(\tilde{f}_0 G_0 \tilde{G}^{\mathrm{II}m} \otimes e_0 + \sum_{i=1}^{n} \tilde{f}_i \tilde{G}^{\mathrm{II}m+1} \otimes e_i \right) \\ &= \tilde{f}_0 G_0 \tilde{G}^{\mathrm{II}m} + \sum_{i=1}^{n} \mathbf{H}_{g_{\lambda(0)i}} \left(\tilde{f}_i \tilde{G}^{\mathrm{II}m+1} \right) \\ &= \tilde{f}_0 G_0 \tilde{G}^{\mathrm{II}m} + \sum_{i=1}^{n} \tilde{f}_i G_0 g_{\lambda(0)i} \tilde{G}^{\mathrm{II}m} \\ &= G_0 \left\langle \tilde{f}, \bar{g}_{\lambda(0)} \right\rangle \tilde{G}^{\mathrm{II}m} \end{aligned} \tag{A.19}$$

and

$$L_0 \circ \tilde{F}_m = G_0 \left\langle \tilde{f}, \bar{g}_{\lambda(0)} \right\rangle \tilde{G}^{\mathrm{II}m} \otimes e_0 + \sum_{i=1}^{n} \tilde{f}_i \tilde{G}^{\mathrm{II}m+1} \otimes e_i. \tag{A.20}$$

Now take $j \in \mathbb{Z}[1, n]$ and define linear mappings

$$M_j : V(m) \longrightarrow \mathcal{L}^*(m+1), \quad L_j : V(m) \longrightarrow V(m)$$

respectively by

$$M_j \left(\sum_{i=0}^{n} z_i \otimes e_i \right) = -\sum_{k=0}^{j-1} \mathcal{B}_{jk} \left(P_k(z_k) \right) + \sum_{k=j}^{n} \Gamma_{jk}(z_k),$$

$$L_j \left(\sum_{i=0}^{n} z_i \otimes e_i \right) = \sum_{k=0}^{j-1} z_k \otimes e_k + M_j \left(\sum_{i=0}^{n} z_i \otimes e_i \right) \otimes e_j + \sum_{k=j+1}^{n} z_k \otimes e_k,$$

where $z_0 \in \mathcal{L}^*(m)$ and $z_i \in \mathcal{L}^*(m+1)$ for $i = 1, ..., n$, and where $P_0 = id : \mathcal{L}^*(m) \longrightarrow \mathcal{L}^*(m)$. It is easy to show that L_j is an isomorphism, and, therefore, one obtains an isomorphism

$$L_\lambda = L_n \circ L_{n-1} \circ \cdots \circ L_0 : V(m) \longrightarrow V(m).$$

Denote the mapping

$$\dot{F}_0 = G_0 \left\langle \tilde{f}, \bar{g}_{\lambda(0)} \right\rangle \tilde{G}^{\mathrm{II}m} : \kappa \longrightarrow \mathcal{L}^*(m).$$

For $j \in \mathbb{Z}[1, n]$, define

$$Q_j = Q_{\gamma_{jj}, m-s(j)} : \kappa \longrightarrow \mathrm{Ker} \mathbf{H}_{m-s(j)}^{s(j)+1} \subset \mathcal{L}^*(m+1),$$

$$\dot{F}_j = G_0^{s(j)+1} \psi_j \left\langle \tilde{f}, \bar{g}_{\lambda(j)} \right\rangle \tilde{G}^{\mathrm{II}m-s(j)} + \tilde{f}_j Q_j : \kappa \longrightarrow \mathcal{L}^*(m+1).$$

Put $\vec{F}_{-1} = \tilde{F}_m$ and write

$$\vec{F}_j = \sum_{k=0}^{j} \dot{F}_k \otimes e_k + \sum_{k=j+1}^{n} \tilde{f}_k \tilde{G}^{\mathrm{II}m+1} \otimes e_k : \kappa \longrightarrow V(m)$$

for $j = 0, 1, ..., n$.

Lemma A.17. *For $j \in \mathbb{Z}[0, n]$,*

$$M_j \circ \vec{F}_{j-1} = \dot{F}_j, \quad L_j \circ \vec{F}_{j-1} = \vec{F}_j.$$

Proof. For $j = 0$, these are the equalities (A.19) and (A.20). Assume $j \in \mathbb{Z}[1, n]$ and that the statement is proved for $j - 1$. We see

$$\begin{aligned}
\mathcal{B}_{j0} \circ \dot{F}_0 &= \mathbf{H}_{\beta_{j0}, m-s(j)} \left(G_0 \left\langle \tilde{f}, \bar{g}_{\lambda(0)} \right\rangle \tilde{G}^{\mathrm{II}m} \right) \\
&= G_0 \left\langle \tilde{f}, \bar{g}_{\lambda(0)} \right\rangle G_0^{s(j)} \beta_{j0} \tilde{G}^{\mathrm{II}m-s(j)} \\
&= \left\langle \tilde{f}, \bar{g}_{\lambda(0)} \right\rangle \phi_{j0} \psi_{j-1} G_0^{s(j)+1} \tilde{G}^{\mathrm{II}m-s(j)}.
\end{aligned}$$

For $k \in \mathbb{Z}[1, j-1]$, recalling that

$$P_k = \mathbf{H}_{m-s(k)}^{s(k)+1} : \mathcal{L}^*(m+1) \longrightarrow \mathcal{L}^*(m-s(k))$$

is the projection, and since

$$\tilde{f}_k \mathcal{Q}_k = \tilde{f}_k \mathcal{Q}_{\gamma_{kk}, m-s(k)} : \kappa \longrightarrow \operatorname{Ker}\mathbf{H}_{m-s(k)}^{s(k)+1},$$

$$G_0^{s(k)+1} \psi_k \left\langle \tilde{f}, \bar{g}_{\lambda(k)} \right\rangle \tilde{G}^{\amalg m-s(k)} : \kappa \longrightarrow \mathcal{L}^*(m-s(k)),$$

and

$$\mathcal{L}^*(m+1) = \mathcal{L}^*(m-s(k)) \oplus \operatorname{Ker}\mathbf{H}_{m-s(k)}^{s(k)+1},$$

we have

$$
\begin{aligned}
P_k \circ \dot{F}_k &= \mathbf{H}_{m-s(k)}^{s(k)+1} \left(G_0^{s(k)+1} \psi_k \left\langle \tilde{f}, \bar{g}_{\lambda(k)} \right\rangle \tilde{G}^{\amalg m-s(k)} + \tilde{f}_k \mathcal{Q}_k \right) \\
&= G_0^{s(k)+1} \psi_k \left\langle \tilde{f}, \bar{g}_{\lambda(k)} \right\rangle \tilde{G}^{\amalg m-s(k)} : \kappa \longrightarrow \mathcal{L}^*(m-s(k)),
\end{aligned}
$$

and

$$
\begin{aligned}
\mathcal{B}_{jk} \circ P_k \circ \dot{F}_k &= \mathcal{B}_{jk} \left(G_0^{s(k)+1} \psi_k \left\langle \tilde{f}, \bar{g}_{\lambda(k)} \right\rangle \tilde{G}^{\amalg m-s(k)} \right) \\
&= G_0^{s(k)+1} \psi_k \left\langle \tilde{f}, \bar{g}_{\lambda(k)} \right\rangle \mathbf{H}_{\beta_{jk}, m-s(j)} \left(\tilde{G}^{\amalg m-s(k)} \right) \\
&= G_0^{s(k)+1} \psi_k \left\langle \tilde{f}, \bar{g}_{\lambda(k)} \right\rangle G_0^{s(j)-s(k)} \beta_{jk} \tilde{G}^{\amalg m-s(j)} \\
&= \left\langle \tilde{f}, \bar{g}_{\lambda(k)} \right\rangle \phi_{jk} \psi_{j-1} G_0^{s(j)+1} \tilde{G}^{\amalg m-s(j)}.
\end{aligned}
$$

Also

$$
\begin{aligned}
\Gamma_{jj} \left(\tilde{f}_j \tilde{G}^{\amalg m+1} \right) &= \tilde{f}_j \dot{\mathbf{H}}_{\gamma_{jj}, m-s(j)} \left(\tilde{G}^{\amalg m+1} \right) \\
&= \tilde{f}_j G_0^{s(j)+1} \gamma_{jj} \tilde{G}^{\amalg m-s(j)} + \tilde{f}_j \mathcal{Q}_{\gamma_{jj}, m-s(j)} \\
&= \tilde{f}_j G_0^{s(j)+1} \psi_{j-1} \psi_{jj} \tilde{G}^{\amalg m-s(j)} + \tilde{f}_j \mathcal{Q}_j.
\end{aligned}
$$

For $j+1 \leq k \leq n$, we have

$$
\begin{aligned}
\Gamma_{jk} \left(\tilde{f}_k \tilde{G}^{\amalg m+1} \right) &= \tilde{f}_k \mathbf{H}_{\gamma_{jk}, m-s(j)} \left(\tilde{G}^{\amalg m+1} \right) \\
&= \tilde{f}_k \gamma_{jk} G_0^{s(j)+1} \tilde{G}^{\amalg m-s(j)} \\
&= \tilde{f}_k \psi_{j-1} \psi_{jk} G_0^{s(j)+1} \tilde{G}^{\amalg m-s(j)}.
\end{aligned}
$$

Therefore,

$$
M_j \circ \vec{F}_{j-1} = M_j \left(\sum_{k=0}^{j-1} \dot{F}_k \otimes e_k + \sum_{k=j}^{n} \tilde{f}_k \tilde{G}^{\amalg m+1} \otimes e_k \right)
$$

$$= -\sum_{k=0}^{j-1} B_{jk} \left(P_k(\dot{F}_k) \right) + \sum_{k=j}^{n} \Gamma_{jk} \left(\tilde{f}_k \tilde{G}^{\mathrm{II}m+1} \right)$$

$$= \left(-\sum_{k=0}^{j-1} \left\langle \tilde{f}, \bar{g}_{\lambda(k)} \right\rangle \phi_{jk} + \sum_{k=j}^{n} \tilde{f}_k \psi_{jk} \right) \psi_{j-1} G_0^{s(j)+1} \tilde{G}^{\mathrm{II}m-s(j)} + \tilde{f}_j \mathcal{Q}_j$$

$$= \left\langle \tilde{f}, \bar{g}_{\lambda(j)} \right\rangle \phi_{jj} \psi_{j-1} G_0^{s(j)+1} \tilde{G}^{\mathrm{II}m-s(j)} + \tilde{f}_j \mathcal{Q}_j$$

$$= \left\langle \tilde{f}, \bar{g}_{\lambda(j)} \right\rangle \psi_{j-1,j-1} \psi_{j-1} G_0^{s(j)+1} \tilde{G}^{\mathrm{II}m-s(j)} + \tilde{f}_j \mathcal{Q}_j$$

$$= \left\langle \tilde{f}, \bar{g}_{\lambda(j)} \right\rangle \psi_j G_0^{s(j)+1} \tilde{G}^{\mathrm{II}m-s(j)} + \tilde{f}_j \mathcal{Q}_j = \dot{F}_j,$$

and, finally, we have

$$L_j \circ \vec{F}_{j-1} = \sum_{k=0}^{j-1} \dot{F}_k \otimes e_k + M_j \circ \vec{F}_{j-1} \otimes e_j + \sum_{k=j+1}^{n} \tilde{f}_k \tilde{G}^{\mathrm{II}m+1} \otimes e_k$$

$$= \sum_{k=0}^{j} \dot{F}_k \otimes e_k + \sum_{k=j+1}^{n} \tilde{f}_k \tilde{G}^{\mathrm{II}m+1} \otimes e_k = \vec{F}_j.$$

□

The lemma above means $L_\lambda \circ \tilde{F}_m = \vec{F}_n$. Hence, we obtain the equality:

$$L_\lambda \circ \tilde{F}_m = \sum_{j=0}^{n} \left\langle \tilde{f}, \bar{g}_{\lambda(j)} \right\rangle \psi_j G_0^{s(j)+1} \tilde{G}^{\mathrm{II}m-s(j)} \otimes e_j + \sum_{j=1}^{n} \tilde{f}_j \mathcal{Q}_j \otimes e_j. \tag{A.21}$$

A.5 Growth of the Steinmetz-Stoll mappings

We continue the discussions in § A.4. Now we assume that each g_j grows slower than a holomorphic curve $f : \kappa \longrightarrow \mathbb{P}(V)$. Then by (A.1), each $g_{jk} \in \tilde{\mathcal{G}}$ grows slower that f, i.e.,

$$\lim_{r \to \infty} \frac{T(r, g_{jk})}{T(r, f)} = 0, \quad k = 0, ..., n,$$

for $j \in \mathrm{Im}\lambda$ with $\lambda \in J_n(\mathcal{G})$. Therefore, all elements in $\mathcal{R}$, and hence all holomorphic curves defined over $\mathcal{R}$ grow slower than f. Then $\mathcal{R} \subset M_f(\kappa)$.

Let $\theta = (\theta_0, ..., \theta_l)$ be a base of $\mathcal{L} = \mathcal{L}_1$ with $\theta_0 = 1$ and let $\tau = (\tau_0, ..., \tau_l)$ be the dual base. Write

$$\tilde{G} = \sum_{j=0}^{l} \tilde{G}_j \tau_j.$$

Then $\tilde{G}_j = G_0 \theta_j, j = 0, ..., l$. Thus, the mapping G is defined over $\mathcal{R}$, and hence

$$\lim_{r \to \infty} \frac{T(r, G)}{T(r, f)} = 0.$$

Note that $G_0 = \tilde{G}_0 = \langle \theta_0, \tilde{G} \rangle$. Put $b = \mathbb{P}(\theta_0) \in \mathbb{P}(\mathcal{L})$. Hence the first main theorem implies

$$0 \leq N\left(r, \frac{1}{G_0}\right) = N_G(r, b) \leq T(r, G) + O(1).$$

Therefore

$$\lim_{r \to \infty} \frac{N\left(r, \frac{1}{G_0}\right)}{T(r, f)} = 0.$$

Note that $\tilde{G} \neq 0$ which implies $\tilde{G}^{\mathrm{II}m+1} \neq 0$ and $\tilde{G}^{\mathrm{II}m} \neq 0$. Thus, we have

$$N_{F_m}(r) = N\left(r, \frac{1}{\tilde{F}_m}\right) \leq N\left(r, \frac{1}{G_0}\right),$$

and hence

$$\lim_{r \to \infty} \frac{N_{F_m}(r)}{T(r, f)} = 0.$$

Let $|\ |$ be a norm defined over the base θ of $\mathcal{L}$ which induces norms on $\mathcal{L}^*$ and

$$\{\mathrm{II}_m \mathcal{L}^* \otimes V'\} \oplus \{\mathrm{II}_{m+1} \mathcal{L}^* \otimes V''\} \supset V(m),$$

respectively, and hence restricts a norm on $V(m)$. Note that

$$\begin{aligned}
|\tilde{F}_m| &\leq \max\left\{ |\tilde{f}_0 G_0 \tilde{G}^{\mathrm{II}m}|, |\tilde{G}^{\mathrm{II}m+1} \otimes (\tilde{f} - \tilde{f}_0 e_0)| \right\} \\
&\leq \max\left\{ |\tilde{f}_0||\langle 1, \tilde{G}\rangle||\tilde{G}|^m, \left(\max_{1 \leq j \leq n} |\tilde{f}_j| \right) |\tilde{G}|^{m+1} \right\} \\
&\leq |\tilde{f}||\tilde{G}|^{m+1},
\end{aligned}$$

which gives

$$\|\ \ \mu(r, \tilde{F}_m) \leq \mu(r, \tilde{f})\mu(r, \tilde{G})^{m+1}.$$

Since μ is continuous and $|\kappa|$ is dense in $\mathbb{R}[0, +\infty)$, we obtain, for all $r > 0$,

$$\mu(r, \tilde{F}_m) \leq \mu(r, \tilde{f})\mu(r, \tilde{G})^{m+1}.$$

Therefore we have

$$T(r, F_m) \leq T(r, f) + (m+1)T(r, G) + O(1) = \{1 + o(1)\}T(r, f).$$

Let $\tau_1, ..., \tau_{q(m)}$ be a base of $\mathcal{L}^*(m)$ over κ and let $\tau_{j,1}, ..., \tau_{j,q(m+1)}$ be a base of $\mathcal{L}^*(m+1)$ over κ for $j = 1, ..., n$. Then

$$e(m) = \{\tau_i \otimes e_0, \tau_{jk} \otimes e_j \mid 1 \leq j \leq n, 1 \leq i \leq q(m), 1 \leq k \leq q(m+1)\}$$

is a base of $V(m)$ over which the norm is defined. Take $c \in \bigwedge_{k(m)+1} V(m)^*$ with $|c| = 1$. Then

$$\mathbf{W}\left[e(m), \tilde{F}_m\right] = \left\langle \tilde{F}_m \wedge \tilde{F}_m' \wedge \cdots \wedge \tilde{F}_m^{(k(m))}, c \right\rangle$$

is the Wronski determinant of $\tilde{F}_m$ with respect to $e(m)$, with

$$\left| \mathbf{W}\left[e(m), \tilde{F}_m\right] \right| = \left| \tilde{F}_m \wedge \tilde{F}_m' \wedge \cdots \wedge \tilde{F}_m^{(k(m))} \right|.$$

A meromorphic function on κ is defined by

$$\mathbf{S}_m = \frac{\mathbf{W}[e(m), \tilde{F}_m]}{\mathbf{K}_m},$$

where

$$\mathbf{K}_m = \tilde{f}_0^{q(m)} (\tilde{f}_1 \cdots \tilde{f}_n)^{q(m+1)} G_0^{(m+1)(k(m)+1)}.$$

Lemma A.18. *If f is linearly non-degenerate over $\mathcal{R}$, then there exists a meromorphic function $\phi_m \in \mathcal{R}$ such that $\mathbf{K}[e(m), \tilde{F}_m] = \phi_m \mathbf{K}_m$.*

Proof. By Lemma A.11, there are entire functions B_i and B_{jk} such that

$$\tilde{G}^{\amalg m} = \sum_{i=1}^{q(m)} B_i \tau_i, \quad \tilde{G}^{\amalg m+1} = \sum_{k=1}^{q(m+1)} B_{jk} \tau_{jk}$$

and $\theta_i = G_0^{-m} B_i \in \mathcal{L}_m \subset \mathcal{R}$, $\theta_{jk} = G_0^{-m-1} B_{jk} \in \mathcal{L}_{m+1} \subset \mathcal{R}$, and such that $\theta_1, ..., \theta_{q(m)}$ is a base of $\mathcal{L}_m$ and $\theta_{j,1}, ..., \theta_{j,q(m+1)}$ is a base of $\mathcal{L}_{m+1}$. Thus

$$\tilde{F}_m = \sum_{i=1}^{q(m)} \tilde{f}_0 G_0 B_i \tau_i \otimes e_0 + \sum_{j=1}^{n} \sum_{k=1}^{q(m+1)} \tilde{f}_j B_{jk} \tau_{jk} \otimes e_j,$$

and hence

$$\mathbf{K}\left[e(m), \tilde{F}_m\right] = \left(\prod_{i=1}^{q(m)} \tilde{f}_0 G_0 B_i \right) \cdot \left(\prod_{j=1}^{n} \prod_{k=1}^{q(m+1)} \tilde{f}_j B_{jk} \right) = \phi_m \mathbf{K}_m,$$

where

$$\phi_m = \left(\prod_{i=1}^{q(m)} \theta_i \right) \left(\prod_{j=1}^{n} \prod_{k=1}^{q(m+1)} \theta_{jk} \right) \in \mathcal{R}.$$

$\square$

If f is linearly non-degenerate over $\mathcal{R}$, then (6.3) implies

$$m\left(r, \mathbf{S}\left[e(m), \tilde{F}_m\right]\right) = O(1).$$

Note that

$$\mathbf{S}_m = \phi_m \mathbf{S}\left[e(m), \tilde{F}_m\right].$$

We have

$$m(r, \mathbf{S}_m) \leq m(r, \phi_m) + m\left(r, \mathbf{S}\left[e(m), \tilde{F}_m\right]\right) = o(T(r, f)).$$

Define

$$a_i = \mathbb{P}(\epsilon_i) \in \mathbb{P}(V^*), \quad i = 0, ..., n.$$

Note that

$$N(r, \mathbf{S}_m) \leq q(m) N_f(r, a_0) + q(m+1) \sum_{i=1}^{n} N_f(r, a_i)$$
$$+ (m+1)(k(m)+1) N\left(r, \frac{1}{G_0}\right).$$

Then, for $r \to \infty$,

$$T(r, \mathbf{S}_m) \leq q(m) N_f(r, a_0) + q(m+1) \sum_{i=1}^{n} N_f(r, a_i) + o(T(r, f)).$$

Take $0 < m \in \mathbb{Z}$ with $m > s(n)$. For any $\lambda \in J_n(\mathcal{G})$, an entire function $\mathbf{K}_{m,\lambda} \not\equiv 0$ is defined by

$$\mathbf{K}_{m,\lambda} = \left(\prod_{j=0}^{n} \langle \tilde{f}, \tilde{g}_{\lambda(j)} \rangle^{q(m-s(j))}\right) \left(\prod_{j=1}^{n} \tilde{f}_j^{q(m+1)-q(m-s(j))}\right) G_0^{(m+1)(k(m)+1)}.$$

Lemma A.19. *If f is linearly non-degenerate over $\mathcal{R}$, then for any $\lambda \in J_n(\mathcal{G})$ there exist a base $e(m)$ of $V(m)$ and a meromorphic function $\phi_{m,\lambda} \in \mathcal{R}$ such that*

$$\mathbf{K}[e(m), L_\lambda \circ \tilde{F}_m] = \phi_{m,\lambda} \mathbf{K}_{m,\lambda}.$$

Proof. Let $\tau_1, ..., \tau_{q(m)}$ be a base of $\mathcal{L}^*(m)$ over κ. By Lemma A.11, there are entire functions B_i such that

$$\tilde{G}^{\text{II}m} = \sum_{i=1}^{q(m)} B_i \tau_i,$$

with $\theta_i = G_0^{-m} B_i \in \mathcal{L}_m \subset \mathcal{R}$, and such that $\theta_1, ..., \theta_{q(m)}$ is a base of $\mathcal{L}_m$. Take an integer $j \in \mathbb{Z}[1, n]$, and recall that

$$\mathcal{L}^*(m+1) = \mathcal{L}^*(m-s(j)) \oplus \text{Ker} \mathbf{H}_{m-s(j)}^{s(j)+1}.$$

Therefore there is a base $\tau_{j,1}, ..., \tau_{j,q(m+1)}$ of $\mathcal{L}^*(m+1)$ over κ such that $\tau_{j,1}, ..., \tau_{j,q_j}$ is a base of $\mathcal{L}^*(m-s(j))$, and $\tau_{j,q_j+1}, ..., \tau_{j,q(m+1)}$ is a base of $\text{Ker} \mathbf{H}_{m-s(j)}^{s(j)+1}$, where

$$q_j = q(m-s(j)).$$

By Lemma A.11, there are entire functions B_{jk} such that

$$\tilde{G}^{\text{II}m+1} = \sum_{k=1}^{q(m+1)} B_{jk} \tau_{jk},$$

with $\theta_{jk} = G_0^{-m-1} B_{jk} \in \mathcal{L}_{m+1} \subset \mathcal{R}$, and such that $\theta_{j,1}, ..., \theta_{j,q(m+1)}$ is a base of $\mathcal{L}_{m+1}$. Again by (A.16), we have

$$\tilde{G}^{\mathrm{II}m+1} = G_0^{s(j)+1} \tilde{G}^{\mathrm{II}m-s(j)} + \mathcal{Q}_{m-s(j)}^{s(j)+1}.$$

Hence

$$G_0^{s(j)+1} \tilde{G}^{\mathrm{II}m-s(j)} = \sum_{k=1}^{q_j} B_{jk} \tau_{jk},$$

and thus the functions

$$\tilde{B}_{jk} = G_0^{-s(j)-1} B_{jk}, \quad k = 1, ..., q_j$$

are entire. Now by Lemma A.11,

$$\tilde{\theta}_{jk} = G_0^{-m+s(j)} \tilde{B}_{jk} = G_0^{-m-1} B_{jk}, \quad k = 1, ..., q_j$$

constitute a base of $\mathcal{L}_{m-s(j)}$, which is a linear subspace of $\mathcal{L}_{m+1}$. Recall that

$$\mathcal{Q}_j = \mathcal{Q}_{\gamma_{jj}, m-s(j)} : \kappa \longrightarrow \mathrm{Ker}\mathbf{H}_{m-s(j)}^{s(j)+1} \subset \mathcal{L}^*(m+1).$$

We can write

$$\mathcal{Q}_j = \sum_{k=q_j+1}^{q(m+1)} C_{jk} \tau_{jk},$$

where C_{jk} are entire functions. Let $\tau_{j,1}^*, ..., \tau_{j,q(m+1)}^*$ be the dual base of $\tau_{j,1}, ..., \tau_{j,q(m+1)}$. Then

$$C_{jk} = \langle \tau_{jk}^*, \mathcal{Q}_j \rangle = G_0^{m+1} \Psi_{jk},$$

where $\Psi_{jk} \in \mathcal{L}_{m+1}$, according to Lemma A.14. Combining with (A.21), we have

$$L_\lambda \circ \tilde{F}_m = \sum_{j=0}^n \sum_{k=1}^{q_j} \left\langle \tilde{f}, \bar{g}_{\lambda(j)} \right\rangle \psi_j G_0^{m+1} \tilde{\theta}_{jk} \tau_{jk} \otimes e_j + \sum_{j=1}^n \sum_{k=q_j+1}^{q(m+1)} \tilde{f}_j G_0^{m+1} \Psi_{jk} \tau_{jk} \otimes e_j. \quad (A.22)$$

Therefore

$$\begin{aligned} \mathbf{K}\left[e(m), L_\lambda \circ \tilde{F}_m\right] &= \left(\prod_{j=0}^n \prod_{k=1}^{q_j} \left\langle \tilde{f}, \bar{g}_{\lambda(j)} \right\rangle \psi_j G_0^{m+1} \tilde{\theta}_{jk} \right) \left(\prod_{j=1}^n \prod_{k=q_j+1}^{q(m+1)} \tilde{f}_j G_0^{m+1} \Psi_{jk} \right) \\ &= \phi_{m,\lambda} \mathbf{K}_{m,\lambda}, \end{aligned}$$

where

$$\phi_{m,\lambda} = \left(\prod_{j=0}^n \prod_{k=1}^{q_j} \psi_j \tilde{\theta}_{jk} \right) \left(\prod_{j=1}^n \prod_{k=q_j+1}^{q(m+1)} \Psi_{jk} \right).$$

$\square$

Thus, for each $\lambda \in J_n(\mathcal{G})$, we obtain a meromorphic function

$$\mathbf{S}_{m,\lambda} = \frac{\mathbf{W}\left[e(m), L_\lambda \circ \tilde{F}_m\right]}{\mathbf{K}_{m,\lambda}},$$

with

$$\mathbf{S}_{m,\lambda} = \phi_{m,\lambda}\mathbf{S}\left[e(m), L_\lambda \circ \tilde{F}_m\right],$$

and hence

$$m(r, \mathbf{S}_{m,\lambda}) = o(T(r, f)).$$

Note that, for a constant $c_\lambda \in \kappa$,

$$\mathbf{W}\left[e(m), L_\lambda \circ \tilde{F}_m\right] = c_\lambda \mathbf{W}\left[e(m), \tilde{F}_m\right].$$

Then we obtain the Ru-Stoll main identity

$$\frac{\mathbf{S}_{m,\lambda}}{\mathbf{S}_m} = c_\lambda \prod_{j=0}^{n}\left(\frac{\langle \tilde{f}, \epsilon_j \rangle \langle e_0, \tilde{g}_{\lambda(j)} \rangle}{\langle \tilde{f}, \tilde{g}_{\lambda(j)} \rangle}\right)^{q(m-s(j))}$$

Hence,

$$\frac{\mu(r, \mathbf{S}_{m,\lambda})}{\mu(r, \mathbf{S}_m)} = |c_\lambda| \prod_{j=0}^{n}\left(\frac{\mu(r, f\angle a_j)\mu(r, e^0\angle g_{\lambda(j)})}{\mu(r, f\angle g_{\lambda(j)})}\right)^{q(m-s(j))},$$

where $e^0 = \mathbb{P}(e_0) \in \mathbb{P}(V)$. Define, for all $r > 0$,

$$\Xi(r) = \frac{1}{\mu(r, \mathbf{S}_m)}\prod_{j=0}^{n}\left(\frac{1}{\mu(r, f\angle a_j)}\right)^{q(m)},$$

$$\Psi_\lambda(r) = \prod_{j=0}^{n}\frac{1}{\mu(r, f\angle g_{\lambda(j)})},$$

and

$$R_\lambda(r) = \frac{\mu(r, \mathbf{S}_{m,\lambda})}{|c_\lambda|}\prod_{j=0}^{n}\frac{\mu(r, f\angle a_j)^{q(m)-q(m-s(j))}}{\mu(r, f\angle g_{\lambda(j)})^{q(m)-q(m-s(j))}\mu(r, e^0\angle g_{\lambda(j)})^{q(m-s(j))}}.$$

An easy calculation enable us to rewrite the main identity as

$$R_\lambda(r)\Xi(r) = \Psi_\lambda(r)^{q(m)}.$$

Then

$$\begin{aligned}
\log\Xi(r) &= N(r, \mathbf{S}_m) - N\left(r, \frac{1}{\mathbf{S}_m}\right) + q(m)\sum_{j=0}^{n}m_f(r, a_j) + O(1) \\
&\leq q(m)N_f(r, a_0) + q(m+1)\sum_{i=1}^{n}N_f(r, a_i) \\
&\quad + q(m)\sum_{j=0}^{n}m_f(r, a_j) + o(T(r, f)) \\
&\leq (q(m) + nq(m+1))T(r, f) + o(T(r, f)).
\end{aligned} \tag{A.23}$$

For $\lambda \in J_n(\mathcal{G})$, we also have

$$
\begin{aligned}
\log^+ R_\lambda(r) &\leq m(r, \mathbf{S}_{m,\lambda}) + \log^+ \frac{1}{|c_\lambda|} + \sum_{j=0}^n (q(m) - q(m - s(j))) m_f(r, g_{\lambda(j)}) \\
&\quad + \sum_{j=0}^n q(m - s(j)) m_{e^0 \angle g_{\lambda(j)}}(r) \\
&\leq \sum_{j=0}^n (q(m) - q(m - s(j))) T(r, f) + o(T(r, f)).
\end{aligned}
\tag{A.24}
$$

A.6 Moving targets in subgeneral position

We continue with the situations discussed in §A.4. Let u be an integer with $n \leq u \leq q$. The finite family $\mathcal{G} = \{g_j\}_{j=0}^q$ is said to be in u-subgeneral position if for any $\eta \in J_u^q$, there is a $\lambda \in J_n(\mathcal{G})$ with $\mathrm{Im}\lambda \subseteq \mathrm{Im}\eta$. Of course, n-subgeneral position is general position.

Assume that $\mathcal{G}$ is in u-subgeneral position. Obviously, if $z \in \kappa - I$, then the orbit $\mathcal{G}(z)$ is in u-subgeneral position. For $\lambda \in J_l^q$, write

$$
\mathcal{G}_\lambda(z) = \{g_{\lambda(0)}(z), ..., g_{\lambda(l)}(z)\}.
$$

Similar to Lemma 3.2 in [113], we have the following fact:

Lemma A.20. *For any $z \in \kappa - I$, $\dim E(\mathcal{G}_\lambda(z))$ is constant.*

Proof. Take $z_1, z_2 \in \kappa - I$. Set $k + 1 = \dim E(\mathcal{G}_\lambda(z_1))$ with $0 \leq k \leq n$. Then there exists $\sigma \in J_k^q$ such that $\sigma \subset \lambda$, and $\tilde{g}_{\sigma(0)}(z_1), ..., \tilde{g}_{\sigma(k)}(z_1)$ is a base of $E(\mathcal{G}_\lambda(z_1))$. Thus $\tilde{g}_\sigma(z_1) \neq 0$, that is, $\tilde{g}_\sigma \not\equiv 0$, and so $\sigma \in J_k(\mathcal{G})$. Because $z_2 \in \kappa - I$, we also have $\tilde{g}_\sigma(z_2) \neq 0$. Hence $\tilde{g}_{\sigma(0)}(z_2), ..., \tilde{g}_{\sigma(k)}(z_2)$ are linearly independent. Therefore, $k + 1 \leq \dim E(\mathcal{G}_\lambda(z_2))$. By symmetry, we obtain the equality. $\square$

Lemma A.21. *Let V be a vector space of dimension $n+1$ over κ and let $\mathcal{G} = \{g_0, g_1, ..., g_q\}$ be a family of non-Archimedean holomorphic curves $g_j : \kappa \longrightarrow \mathbb{P}(V^*)$ in u-subgeneral position with $1 \leq n \leq u \leq q$. Then there exist a function $\omega : \mathbb{Z}[0, q] \longrightarrow \mathbb{R}(0, 1]$ and a real number $\theta \geq 1$ such that for $z \in \kappa - I$, the following properties are satisfied:*
1) $0 < \omega(j)\theta \leq 1, \quad j = 0, 1, ..., q$;
2) $q - 2u + n = \theta(\sum_{j=0}^q \omega(j) - n - 1)$;
3) $1 \leq \frac{u+1}{n+1} \leq \theta \leq \frac{2u-n+1}{n+1}$;
4) $\sum_{j=0}^k \omega(\sigma(j)) \leq \dim E(\mathcal{G}_\sigma(z))$ if $\sigma \in J_k^q$ with $0 \leq k \leq u$;
5) Let $r_0, ..., r_q$ be a family of functions defined on $\kappa - I$ with $r_j \geq 1$ for all j. Then for any $\sigma \in J_k^q$ with $0 \leq k \leq u$, setting $\dim E(\mathcal{G}_\sigma(z)) = l + 1$, there exists $\lambda \in J_l(\mathcal{G})$ with $\mathrm{Im}\lambda \subset \mathrm{Im}\sigma$ such that $E(\mathcal{G}_\lambda(z)) = E(\mathcal{G}_\sigma(z))$ and

$$
\prod_{j=0}^k r_{\sigma(j)}(z)^{\omega(\sigma(j))} \leq \prod_{j=0}^l r_{\lambda(j)}(z).
$$

Proof. Fix a point $z_0 \in \kappa - I$. Then $\mathcal{G}(z_0)$ is in u-subgeneral position. We obtain the Nochka weight function ω and the Nochka constant θ for the family $\mathcal{G}(z_0)$. Define

$$\omega(j) := \omega(g_j(z_0)),$$

for the family $\mathcal{G}(z)$ if $z \in \kappa - I$. Because $\dim E(\mathcal{G}_\sigma(z))$ is constant for $z \in \kappa - I$, properties 1-4 follow from Lemma 6.23. By 5) of Lemma 6.23, for any $\sigma \in J_k^q$ with $0 \leq k \leq u$, setting $\dim E(\mathcal{G}_\sigma(z_0)) = l + 1$, then there exists $\lambda \in J_l(\mathcal{G}(z_0))$ with $\mathrm{Im}\lambda \subset \mathrm{Im}\sigma$ such that $E(\mathcal{G}_\lambda(z_0)) = E(\mathcal{G}_\sigma(z_0))$ and

$$\prod_{j=0}^{k} r_{\sigma(j)}(z)^{\omega(\sigma(j))} \leq \prod_{j=0}^{l} r_{\lambda(j)}(z), \quad z \in \kappa - I.$$

Note that $\tilde{g}_\lambda(z_0) \neq 0$. Thus $\tilde{g}_\lambda \not\equiv 0$ and $\lambda \in J_l(\mathcal{G})$. Hence $\tilde{g}_\lambda(z) \neq 0$ since $z \in \kappa - I$, and, therefore, $E(\mathcal{G}_\lambda(z)) = E(\mathcal{G}_\sigma(z))$. □

Define the *gauge* $\Gamma(\mathcal{G})$ of $\mathcal{G}$ with respect to the norm $\rho = |\cdot|$ by

$$\Gamma(\mathcal{G})(z) = \Gamma(\mathcal{G}; \rho)(z) = \inf_{\lambda \in J_n(\mathcal{G})} \{|g_{\lambda(0)}(z) \wedge \cdots \wedge g_{\lambda(n)}(z)|\}.$$

Then $0 \leq \Gamma(\mathcal{G}) \leq 1$, and $\Gamma(\mathcal{G})(z) = \Gamma(\mathcal{G}(z)) > 0$ if $z \in \kappa - I$.

Lemma A.22. *For $x \in \mathbb{P}(V)$, $0 < b \in \mathbb{R}$, define*

$$\mathcal{G}(z)(x, b) = \{j \in \mathbb{Z}[0, q] \mid |x, g_j(z)| < b\}.$$

If $z \in \kappa - I$ and $0 < b \leq \Gamma(\mathcal{G})(z)$, then $\#\mathcal{G}(z)(x, b) \leq u$.

Proof. Assume that $\#\mathcal{G}(z)(x, b) \geq u + 1$. Then $\lambda \in J_n(\mathcal{G})$ exists such that $\mathrm{Im}\lambda \subseteq \mathcal{G}(z)(x, b)$. Hence,

$$|x, g_{\lambda(j)}(z)| < b, \quad j = 0, ..., n.$$

Then Lemma 6.4 implies

$$\begin{aligned}
0 < \Gamma(\mathcal{G})(z) &\leq |g_{\lambda(0)}(z) \wedge \cdots \wedge g_{\lambda(n)}(z)| \\
&\leq \max_{0 \leq j \leq n} |x, g_{\lambda(j)}(z)| < b \leq \Gamma(\mathcal{G})(z),
\end{aligned}$$

which is impossible. □

Lemma A.23. *Take $x \in \mathbb{P}(V)$ and $z \in \kappa - I$ such that $|x, g_j(z)| > 0$ for $j = 0, ..., q$. Then*

$$\prod_{j=0}^{q} \left(\frac{1}{|x, g_j(z)|} \right)^{\omega(j)} \leq \left(\frac{1}{\Gamma(\mathcal{G})(z)} \right)^{q-u} \max_{\lambda \in J_n(\mathcal{G})} \prod_{j=0}^{n} \frac{1}{|x, g_{\lambda(j)}(z)|} \tag{A.25}$$

and

$$\prod_{j=0}^{q} \frac{1}{|x, g_j(z)|} \leq \left(\frac{1}{\Gamma(\mathcal{G})(z)} \right)^{q-u+1} \max_{\lambda \in J_{u-1}^q} \prod_{j=0}^{u-1} \frac{1}{|x, g_{\lambda(j)}(z)|}. \tag{A.26}$$

Proof. Take $b = \Gamma(\mathcal{G})(z)$. Lemma A.22 implies $\#\mathcal{G}(z)(x,b) \leq u$. Thus $\sigma \in J_u^q$ exists such that $\mathcal{G}(z)(x,b) \subset \operatorname{Im}\sigma$. Note that $E(\mathcal{G}_\sigma(z)) = V^*$. By Lemma A.21, there exists $\lambda \in J_n(\mathcal{G})$ with $\operatorname{Im}\lambda \subset \operatorname{Im}\sigma$ such that $E(\mathcal{G}_\lambda(z)) = E(\mathcal{G}_\sigma(z))$ and

$$\prod_{j=0}^{u} \left(\frac{1}{|x, g_{\sigma(j)}(z)|} \right)^{\omega(\sigma(j))} \leq \prod_{j=0}^{n} \frac{1}{|x, g_{\lambda(j)}(z)|}$$

$$\leq \max_{\lambda \in J_n(\mathcal{G})} \prod_{j=0}^{n} \frac{1}{|x, g_{\lambda(j)}(z)|}.$$

Set $C = \mathbb{Z}[0,q] - \operatorname{Im}\sigma$. Thus $|x, g_j(z)| \geq b$ for $j \in C$. Hence

$$\prod_{j \in C} \left(\frac{1}{|x, g_j(z)|} \right)^{\omega(j)} \leq \prod_{j \in C} \frac{1}{|x, g_j(z)|}$$

$$\leq \left(\frac{1}{b} \right)^{\#C} = \left(\frac{1}{\Gamma(\mathcal{G})(z)} \right)^{q-u}.$$

Hence (A.25) is proved. Similarly, we can show (A.26). □

According to Lemma 7.2 of Ru-Stoll [112], we show the following fact:

Lemma A.24. *For any $\varepsilon > 0$, there is a positive integer $m > s(n)$ such that*

$$1 \leq \frac{q(m+1)}{q(m-s(j))} < 1+\varepsilon, \quad j = 0,1,...,n,$$

$$1 \leq \frac{q(m-s(i))}{q(m-s(j))} < 1+\varepsilon, \quad 0 \leq i \leq j \leq n,$$

and

$$0 \leq \frac{q(m) - q(m-s(j))}{q(m)} < \varepsilon, \quad j = 0,1,...,n.$$

Proof. By Steinmetz's lemma (Lemma A.12), where we take $u = s(n)+1$, then $m > s(n)$ exists such that

$$1 \leq \frac{q(m+1)}{q(m-s(n))} < 1+\varepsilon.$$

Since $q(m-s(n)) \leq q(m-s(j))$ for $j = 0,...,n$, we obtain

$$1 \leq \frac{q(m+1)}{q(m-s(j))} \leq \frac{q(m+1)}{q(m-s(n))} < 1+\varepsilon, \quad j = 0,1,...,n.$$

Because

$$q(m-s(j)) \leq q(m-s(i)) \leq q(m+1), \quad 0 \leq i \leq j \leq n,$$

we have

$$1 \leq \frac{q(m-s(i))}{q(m-s(j))} \leq \frac{q(m+1)}{q(m-s(j))} < 1+\varepsilon, \quad 0 \leq i \leq j \leq n.$$

In particular,

$$\frac{q(m)}{q(m-s(j))} < 1 + \varepsilon,$$

which implies

$$0 \leq \frac{q(m) - q(m - s(j))}{q(m)} < 1 - \frac{1}{1+\varepsilon} = \frac{\varepsilon}{1+\varepsilon} < \varepsilon, \quad j = 0, 1, ..., n.$$

$\square$

For $r > 0$, the gauge function $\Gamma_{\mathcal{G}}$ is defined by

$$\Gamma_{\mathcal{G}}(r) = \inf_{\lambda \in J_n(\mathcal{G})} \{ \mu(r, g_{\lambda(0)} \wedge \cdots \wedge g_{\lambda(n)}) \}.$$

Then

$$\| \quad \Gamma_{\mathcal{G}}(r) = \Gamma(\mathcal{G})(z) \leq 1.$$

Note that

$$\prod_{\lambda \in J_n(\mathcal{G})} |g_{\lambda(0)} \wedge \cdots \wedge g_{\lambda(n)}| \leq \Gamma(\mathcal{G}).$$

By usual method, we can derive

$$\prod_{\lambda \in J_n(\mathcal{G})} \mu(r, g_{\lambda(0)} \wedge \cdots \wedge g_{\lambda(n)}) \leq \Gamma_{\mathcal{G}}(r) \quad (r > 0).$$

For $\lambda \in J_n(\mathcal{G})$, noting that the first main theorem

$$\sum_{j=0}^{n} T(r, g_{\lambda(j)}) = N_{g_{\lambda(0)} \wedge \cdots \wedge g_{\lambda(n)}}(r) + m_{g_{\lambda(0)} \wedge \cdots \wedge g_{\lambda(n)}}(r) + O(1),$$

we have

$$0 \leq -\log \Gamma_{\mathcal{G}}(r) \leq \sum_{\lambda \in J_n(\mathcal{G})} m_{g_{\lambda(0)} \wedge \cdots \wedge g_{\lambda(n)}}(r)$$

$$\leq \sum_{\lambda \in J_n(\mathcal{G})} \sum_{j=0}^{n} T(r, g_{\lambda(j)}) + O(1).$$

Theorem A.25 (Hu-Yang [64]). *Let $\mathcal{G} = \{g_j\}_{j=0}^{q}$ be a finite family of non-Archimedean holomorphic curves $g_j : \kappa \longrightarrow \mathbb{P}(V^*)$ in u-subgeneral position with $u \leq 2u - n \leq q$. Let $f : \kappa \longrightarrow \mathbb{P}(V)$ be a non-Archimedean holomorphic curve that is linearly nondegenerate over $\mathcal{R}$. Assume that g_j grows slower than f for $j = 0, ..., q$. Take any $\varepsilon > 0$. Then*

$$\sum_{j=0}^{q} m_f(r, g_j) \leq (2u - n + 1 + \varepsilon) T(r, f),$$

for sufficiently large r.

Proof. Define $b = \#J_n(\mathcal{G})$. Let ω be the Nochka weight function and θ be the Nochka constant. By Lemma A.24, $0 < m \in \mathbb{Z}$ can be chosen such that

$$1 \leq \frac{q(m+1)}{q(m)} < 1 + \frac{\varepsilon}{3nb\theta} \leq 1 + \frac{\varepsilon}{3n\theta},$$

and

$$0 \leq \frac{q(m) - q(m - s(j))}{q(m)} < \frac{\varepsilon}{3nb\theta}.$$

Then Lemma A.23 implies

$$\prod_{j=0}^{q}\left(\frac{1}{|f, g_j|}\right)^{\omega(j)} \leq \left(\frac{1}{\Gamma(\mathcal{G})}\right)^{q-u} \sum_{\lambda \in J_n(\mathcal{G})} \prod_{j=0}^{n} \frac{1}{|f, g_{\lambda(j)}|},$$

which yields

$$\prod_{j=0}^{q}\left(\frac{1}{\mu(r, f \angle g_j)}\right)^{\omega(j)} \leq \left(\frac{1}{\Gamma_{\mathcal{G}}(r)}\right)^{q-u} \sum_{\lambda \in J_n(\mathcal{G})} \prod_{j=0}^{n} \frac{1}{\mu(r, f \angle g_{\lambda(j)})}$$

$$= \left(\frac{1}{\Gamma_{\mathcal{G}}(r)}\right)^{q-u} \sum_{\lambda \in J_n(\mathcal{G})} \Psi_\lambda(r)$$

$$= \left(\frac{1}{\Gamma_{\mathcal{G}}(r)}\right)^{q-u} \sum_{\lambda \in J_n(\mathcal{G})} (R_\lambda(r)\Xi(r))^{\frac{1}{q(m)}}.$$

We obtain

$$\sum_{j=0}^{q}\omega(j)m_f(r, g_j) \leq -(q-u)\log\Gamma_{\mathcal{G}}(r) + \frac{1}{q(m)}(\log\Xi(r)$$

$$+ \sum_{\lambda \in J_n(\mathcal{G})} \log^+ R_\lambda(r)) + \frac{1}{q(m)}\log b$$

$$\leq (q-u)\sum_{\lambda \in J_n(\mathcal{G})}\sum_{j=0}^{n} T(r, g_{\lambda(j)})$$

$$+ \left(1 + n\frac{q(m+1)}{q(m)}\right)T(r, f)$$

$$+ b\sum_{j=0}^{n} \frac{q(m) - q(m - s(j))}{q(m)}T(r, f) + o(T(r, f))$$

$$\leq \left(n + 1 + \frac{2\varepsilon}{3\theta}\right)T(r, f) + o(T(r, f)).$$

By the properties of the Nochka weights,

$$\sum_{j=0}^{q} m_f(r, g_j) = \sum_{j=0}^{q} \theta\omega(j)m_f(r, g_j) + \sum_{j=0}^{q}(1 - \theta\omega(j))m_f(r, g_j)$$

$$\leq \; \theta(n+1)T(r,f) + \frac{2\varepsilon}{3}T(r,f) + o(T(r,f))$$

$$+ \left(1 + q - \theta \sum_{j=0}^{q} \omega(j) \right) T(r,f) + O(1)$$

$$\leq \; \left(1 + q - \theta(\sum_{j=0}^{q} \omega(j) - n - 1) \right) T(r,f) + \varepsilon T(r,f)$$

$$= \; (2u - n + 1 + \varepsilon)T(r,f),$$

for sufficiently large r. $\square$

Corollary A.26. *Assumptions as in Theorem A.25. Then we have*

$$\sum_{j=0}^{q} \delta_f(g_j) \leq 2u - n + 1.$$

In the proof of Theorem A.25, by using (A.26) in Lemma A.23, we have

$$\prod_{j=0}^{q} \frac{1}{|f,g_j|} \leq \left(\frac{1}{\Gamma(\mathcal{G})} \right)^{q+1-u} \max_{\lambda \in J_{u-1}^q} \prod_{i=0}^{u-1} \frac{1}{|f,g_{\lambda(i)}|}.$$

Hence

$$\sum_{j=0}^{q} m_f(r,g_j) \; \leq \; \max_{\lambda \in J_{u-1}^q} \sum_{i=0}^{u-1} m_f(r,g_{\lambda(i)}) - (q+1-u)\log\Gamma_{\mathcal{G}}(r) + O(1)$$

$$\leq \; uT(r,f) + o(T(r,f)).$$

Further, by the first main theorem, we can obtain the following result:

Theorem A.27. *Let* $\mathcal{G} = \{g_j\}_{j=0}^{q}$ *be a finite family of non-Archimedean holomorphic curves* $g_j : \kappa \longrightarrow \mathbb{P}(V^*)$ *in u-subgeneral position. Let* $f : \kappa \longrightarrow \mathbb{P}(V)$ *be a non-Archimedean holomorphic curve such that the pair* (f, g_j) *is free for* $j = 0, ..., q$. *Assume that* g_j *grows slower than* f *for* $j = 0, ..., q$. *Then*

$$\sum_{j=0}^{q} m_f(r,g_j) \leq uT(r,f) + o(T(r,f)),$$

for sufficiently large r.

Corollary A.28. *Assumptions as in Theorem A.27. Then we have*

$$\sum_{j=0}^{q} \delta_f(g_j) \leq u.$$

A.7 Moving targets in general position

We continue with the situation of §A.4. Assume that $\mathcal{G} = \{g_j\}_{j=0}^q$ is in general position and assume that f is k-flat over $\mathcal{R}$ with $0 \leq k \leq n \leq q$ such that each pair (f, g_j) is free for $j = 0, ..., q$. By Lemma A.8, there is a base $\xi = (\xi_0, ..., \xi_n)$ of $V_{\mathcal{R}}$ over $\mathcal{R}$ with dual base $\alpha = (\alpha_0, ..., \alpha_n)$ such that for the reduced representation $\tilde{f} : \kappa \longrightarrow V$,

$$\tilde{f} = \sum_{j=0}^k \langle \tilde{f}, \alpha_j \rangle \xi_j.$$

Moreover, the base can be chosen such that $\alpha_j \in V^*$ for $j = 0, ..., k$, and

$$\langle \tilde{f}, \alpha_0 \rangle, ..., \langle \tilde{f}, \alpha_k \rangle$$

are holomorphic and linearly independent over $\mathcal{R}$. We also have

$$\tilde{g}_j = \sum_{i=0}^n \langle \xi_i, \tilde{g}_j \rangle \alpha_i, \quad j = 0, ..., q.$$

We take a base $\epsilon = (\epsilon_0, ..., \epsilon_n)$ of V^* with $\epsilon_j = \alpha_j$ for $j = 0, ..., k$. Let $e = (e_0, ..., e_n)$ be the dual base of ϵ. Let W be the vector space spanned by $e_0, ..., e_k$ over κ. Then $\epsilon_0, ..., \epsilon_k$ span W^*. A non-Archimedean holomorphic curve $\hat{f} : \kappa \longrightarrow \mathbb{P}(W)$ is defined with a representation

$$\tilde{\hat{f}} = \sum_{j=0}^k \langle \tilde{f}, \epsilon_j \rangle e_j : \kappa \longrightarrow W.$$

Since $\langle \tilde{f}, \epsilon_0 \rangle, ..., \langle \tilde{f}, \epsilon_k \rangle$ are linearly independent over $\mathcal{R}$, the mapping $\hat{f}$ is linearly nondegenerate over $\mathcal{R}$. Note that

$$\tilde{f} = \sum_{j=0}^k \langle \tilde{f}, \epsilon_j \rangle \xi_j = \sum_{j=0}^n \langle \tilde{f}, \epsilon_j \rangle e_j.$$

We have

$$\left| \tilde{\hat{f}} \right| = \max_{0 \leq j \leq k} |\langle \tilde{f}, \epsilon_j \rangle| \leq \max_{0 \leq j \leq n} |\langle \tilde{f}, \epsilon_j \rangle| = |\tilde{f}|$$

$$\leq \max_{0 \leq j \leq k} |\langle \tilde{f}, \epsilon_j \rangle| |\xi_j| \leq |\tilde{\hat{f}}| \max_{0 \leq j \leq k} \max_{0 \leq i \leq n} |\xi_{ji}|,$$

where

$$\xi_j = \sum_{i=0}^n \xi_{ji} e_i, \quad \xi_{ji} \in \mathcal{R}.$$

Hence

$$\mu\left(r, \tilde{\hat{f}}\right) \leq \mu(r, \tilde{f}) \leq \mu\left(r, \tilde{\hat{f}}\right) \max_{0 \leq j \leq k} \max_{0 \leq i \leq n} \mu(r, \xi_{ji}) \quad (r > 0).$$

Note that $\tilde{f} \neq 0$. We have

$$N_{\hat{f}}(r) = N\left(r, \frac{1}{\tilde{f}}\right) \leq \sum_{j=0}^{k} N\left(r, \xi_j\right) \leq \sum_{j=0}^{k} \sum_{i=0}^{n} N(r, \xi_{ji}) = o(T(r, f)).$$

Therefore

$$T(r, \hat{f}) + O(1) \leq T(r, f) \leq T(r, \hat{f}) + o(T(r, f)). \tag{A.27}$$

Remark. If $k = 0$, then $T(r, \hat{f})$ is constant. The inequality (A.27) is impossible. Thus we must have $k \geq 1$.

Also non-Archimedean holomorphic curves $\hat{g}_j : \kappa \longrightarrow \mathbb{P}(W^*)$ are defined with representations

$$\tilde{\hat{g}}_j = \sum_{i=0}^{k} \langle \xi_i, \tilde{g}_j \rangle \epsilon_i : \kappa \longrightarrow W^*, \quad j = 0, ..., q.$$

Note that $\tilde{\hat{g}}_j \not\equiv 0$ since (f, g_j) is free. Obviously, each $\hat{g}_j$ is defined over $\mathcal{R}$, which implies

$$\lim_{r \to \infty} \frac{T(r, \hat{g}_j)}{T(r, f)} = 0, \quad j = 0, ..., q,$$

and hence

$$\lim_{r \to \infty} \frac{T(r, \hat{g}_j)}{T(r, \hat{f})} = 0, \quad j = 0, ..., q.$$

Lemma A.29. *The family* $\hat{\mathcal{G}} = \{\hat{g}_j\}_{j=0}^{q}$ *is in n-subgeneral position.*

Proof. Take $\sigma \in J_n^q$. Then $\tilde{g}_\sigma \not\equiv 0$, and hence

$$\det(\langle \xi_i, \tilde{g}_{\sigma(j)} \rangle) \not\equiv 0 \quad (0 \leq i, j \leq n).$$

Therefore, there is a $\lambda \in J_k^q$ with $\text{Im}\lambda \subseteq \text{Im}\sigma$ such that

$$\det(\langle \xi_s, \tilde{g}_{\lambda(t)} \rangle) \not\equiv 0 \quad (0 \leq s, t \leq k).$$

We have

$$\tilde{\hat{g}}_\lambda = \det(\langle \xi_s, \tilde{g}_{\lambda(t)} \rangle) \epsilon_0 \wedge \cdots \wedge \epsilon_k \not\equiv 0.$$

Hence $\lambda \in J_k(\hat{\mathcal{G}})$. Thus $\hat{\mathcal{G}}$ is in n-subgeneral position. $\qquad\square$

Theorem A.30 (Hu-Yang [64]). *Let* $\mathcal{G} = \{g_j\}_{j=0}^{q}$ *be a finite family of non-Archimedean holomorphic curves* $g_j : \kappa \longrightarrow \mathbb{P}(V^*)$ *in general position with* $q \geq n$. *Take an integer k with* $1 \leq k \leq n$. *Let* $f : \kappa \longrightarrow \mathbb{P}(V)$ *be a non-Archimedean holomorphic curve which is k-flat over* $\mathcal{R}$ *such that each pair (f, g_j) is free for $j = 0, ..., q$. Assume that g_j grows slower than f for each j. Then*

$$\sum_{j=0}^{q} \delta_f(g_j) \leq 2n - k + 1.$$

Proof. By Corollary A.26,

$$\sum_{j=0}^{q} \delta_f(\hat{g}_j) \leq 2n - k + 1.$$

Note that

$$\langle \tilde{f}, \tilde{g}_j \rangle = \sum_{i=0}^{n} \langle \tilde{f}, \alpha_i \rangle \langle \xi_i, \tilde{g}_j \rangle = \sum_{i=0}^{k} \langle \tilde{f}, \epsilon_i \rangle \langle \xi_i, \tilde{g}_j \rangle = \langle \hat{f}, \hat{\tilde{g}}_j \rangle.$$

Let $\vec{f}$ and $\vec{g}_j$ be reduced representations of $\hat{f}$ and $\hat{g}_j$, respectively. Then there are a holomorphic function $h \not\equiv 0$ and meromorphic functions $w_j \not\equiv 0$ such that $\hat{\vec{f}} = h\vec{f}$ and $\hat{\tilde{g}}_j = w_j \vec{g}_j$. We obtain

$$\langle \tilde{f}, \tilde{g}_j \rangle = h w_j \langle \vec{f}, \vec{g}_j \rangle.$$

Therefore

$$N_f(r, g_j) - N_{\hat{f}}(r, \hat{g}_j) = N\left(r, \frac{1}{h}\right) + N\left(r, \frac{1}{w_j}\right) - N(r, w_j).$$

Take a reduced representation $\tilde{\xi}_i$ of the non-Archimedean holomorphic curve

$$x_i = \mathbb{P} \circ \xi_i : \kappa \longrightarrow \mathbb{P}(V).$$

Then there is a meromorphic function $h_i \not\equiv 0$ such that $\xi_i = h_i \tilde{\xi}_i$. Clearly,

$$N\left(r, \frac{1}{h_i}\right) \leq \sum_{k=0}^{n} N\left(r, \frac{1}{\xi_{ik}}\right) = o(T(r, f)),$$

and

$$N(r, h_i) \leq \sum_{k=0}^{n} N(r, \xi_{ik}) = o(T(r, f)).$$

Note that

$$w_j \vec{g}_j = \hat{\tilde{g}}_j = \sum_{i=0}^{k} h_i \langle \tilde{\xi}_i, \tilde{g}_j \rangle \epsilon_i \not\equiv 0.$$

Take some $i \in \mathbb{Z}[0, k]$ such that the holomorphic function $\langle \tilde{\xi}_i, \tilde{g}_j \rangle \not\equiv 0$. Then

$$N\left(r, \frac{1}{w_j}\right) \leq N\left(r, \frac{1}{h_i}\right) + N_{x_i}(r, g_j) = o(T(r, f)).$$

$$N(r, w_j) \leq \sum_{i=0}^{k} N(r, h_i) = o(T(r, f)).$$

Note that

$$N\left(r, \frac{1}{h}\right) = N_{\hat{f}}(r) = o(T(r, f)).$$

We obtain

$$N_f(r, g_j) - N_{\hat{f}}(r, \hat{g}_j) = o(T(r, f)),$$

which leads easily to

$$\delta_f(g_j) = \delta_{\hat{f}}(\hat{g}_j),$$

and the theorem follows. $\qquad \square$

Bibliography

[1] Adams, W. W. & Straus, E. G., Non archimedean analytic functions taking the same values at the same points, Illinois J. Math. 15(1971), 418-424.

[2] Azarin, V. S. and Ronkin, A. L., On the inequality of B.Shiffman for meromorphic mappings, Teor. Funktsiĭ Funktsional. Anal. i prilozhen. 44(1985), 3-16; English transl. in J. Soviet Math. 48(3) (1990).

[3] Baker, I. N., The existence of fixpoints of entire functions, Math. Z. 73(1960), 280-284.

[4] Baker, I. N., On a class of meromorphic functions, Proc. Amer. Math. Soc. 17 (1966), 819-822.

[5] Beardon, A. F., Iteration of rational functions, Springer-Verlag, 1991.

[6] Berkovich, V., Spectral theory and analytic geometry over non-Archimedean fields, Amer. Math. Soc., 1990.

[7] Bézivin, J. P. & Boutabaa, A., Decomposition of p-adic meromorphic functions, Ann. Math. Blaise Pascal 2(1995), 51-60.

[8] Biancofiore, A., A defect relation for linear system on compact complex manifolds, Illinois J. Math. 28(1984), 531-546.

[9] Bosch, S., Güntzer, U. & Remmert, R., Non-Archimedean analysis, Springer-Verlag, 1984.

[10] Boutabaa, A., Theorie de Nevanlinna p-adique, Manuscripta Math. 67 (1990), 251-269.

[11] Boutabaa, A., Applications de la théorie de Nevanlinna p-adic, Collect. Math. 42(1991), 75-93.

[12] Boutabaa, A., On some p-adic functional equations, Lecture Notes in Pure and Applied Mathematics N. 192 (1997), 49-59, Marcel Dekker.

[13] Boutabaa, A., A transcendental p-adic entire solution of a linear differential equation, preprint.

[14] Boutabaa, A. & Escassut, A., Property $f^{-1}(S) = g^{-1}(S)$ for entire and meromorphic p-adic functions, preprint.

[15] Boutabaa, A. & Escassut, A., Applications of the p-adic Nevanlinna theory to functional equations, preprint.

[16] Boutabaa, A. & Escassut, A., On uniqueness of p-adic meromorphic functions, preprint.

[17] Boutabaa, A. & Escassut, A., URS and URSIM for p-adic unbounded analytic functions inside a disk, preprint.

[18] Boutabaa, A. & Escassut, A., URS and URSIM for p-adic unbounded meromorphic functions inside a disk, preprint.

[19] Boutabaa, A., Escassut, A. & Haddad, L., On uniqueness of p-adic entire functions, Indagationes Mathematicae 8(1997), 145-155.

[20] Cartan, H., Sur les zéros des combinaisons linéaires de p fonctions holomorphes données, Mathematica (cluj) 7(1933), 80-103.

[21] Chen, W., Cartan's conjecture: defect relations for meromorphic maps from parabolic manifold to projective space, University of Notre Dame Thesis, 1987.

[22] Cherry, W., A survey of Nevanlinna theory over non-Archimedean fields, Bull. Hong Kong Math. Soc. 1(2) (1997), 235-249.

[23] Cherry, W., Non-Archimedean analytic curves in Abelian varieties, Math. Ann. 300 (1994), 393-404.

[24] Cherry, W., A non-Archimedean analogue of the Kobayashi semi-distance and its non-degeneracy on Abelian varieties, Illinois Journal of Mathematics 40 (1) (1996), 123-140.

[25] Cherry, W. & Yang, C. C., Uniqueness of non-Archimedean entire functions sharing sets of values counting multiplicity, Proc. Amer. Math. Soc. 127(4)(1999), 967-971.

[26] Cherry, W. & Ye, Z., Non-Archimedean Nevanlinna theory in several variables and the Non-Archimedean Nevanlinna inverse problem, Trans. Amer. Math. Soc. 349(1997), 5043-5071.

[27] Clark, D. N., A note on the p-adic convergence of solution of linear differential equations, Proc. Amer. Math. Soc. 17 (1966), 262-269.

[28] Corrales-Rodrigáñez, C., Nevanlinna theory in the p-adic plane, Ph.D. Thesis, University of Michigan, 1986.

[29] Corrales-Rodrigáñez, C., Nevanlinna theory on the p-adic plane, Annales Polonici Mathematici LVII (1992), 135-147.

[30] Dwork, B., Gerotto, G. & Sullivan, F. J., An introduction to G-functions, Annals of Mathematics Studies, Princeton University Press, 1994.

[31] Eremenko, A. E. and Sodin, M. L., The value distribution of meromorphic functions and meromorphic curves from the point of view of potential theory, St. Petersburg Math. J. 3(1)(1992), 109-136.

[32] Escassut, A., Analytic elements in p-adic analysis, World Scientific Publishing Co. Pte. Ltd., 1995.

[33] Frank, G. & Reinders, M., A unique range set for meromorphic functions with 11 elements, Complex Variables 37(1998), 185-193.

[34] Fujimoto, H., On uniqueness of meromorphic functions sharing finite sets, preprint.

[35] Fujimoto, H., The uniqueness problem of meromorphic maps into the complex projective space, Nagoya Math. J. 58(1975), 1-23.

[36] Fujimoto, H., Value distribution theory of the Gauss map of minimal surfaces in $\mathbb{R}^m$, Aspects of Mathematics E21, Vieweg, 1993.

[37] Gackstatter, F. & Laine, I., Zur Theorie der gewöhnlichen Differentialgleichungen im Komplexen, Ann. Polon. Math. 38(1980), 259-287.

[38] Gopalakrishna, H. S. & Bhoosnurmath, S. S., Uniqueness theorems for meromorphic functions, Math. Scand. 39(1976), 125-130.

[39] Gouvêa, F. Q., p-adic numbers, Springer, 1997.

[40] Griffiths, P., Holomorphic mappings: Survey of some results and discussion of open problems, Bull. Amer. Math. Soc. 78(1972), 374-382.

[41] Gross, F., On the equation $f^n + g^n = 1$, Bull. Amer. Math. Soc. 72 (1966), 86-88, Correction: 72 (1966), p. 576.

[42] Gross, F. & Yang, C. C., On preimage and range sets of meromorphic functions, Proc. Japan Acad. 58(1982), 17-20.

[43] Gundersen, G. G., Meromorphic solutions of $f^6 + g^6 + h^6 = 1$, Analysis 18 (1998), 285-290.

[44] Hà, Huy Khóai, On p-adic meromorphic functions, Duke Math. J. 50(1983), 695-711.

[45] Hà, Huy Khóai, Heights for p-adic holomorphic functions of several variables, Max-Planck Institut Für Mathematik 89-83(1989).

[46] Hà, Huy Khóai, La hauteur des fonctions holomorphes p-adiques de plusieurs variables, C. R. Acad. Sci. Paris Sér. I Math. 312 (1991), 751-754.

[47] Hà, Huy Khóai, Théorie de Nevanlinna et problèmes Diophantiens, Vietnam J. Math. 23(1995), 57-81.

[48] Hà, Huy Khóai, A note on URS for p-adic meromorphic functions, preprint.

[49] Hà, Huy Khóai & Mai, Van Tu, p-adic Nevanlinna-Cartan theorem, International J. of Math. 6(1995), 719-731.

[50] Hà, Huy Khóai & My, Vinh Quang, On p-adic Nevanlinna theory, Lecture Notes in Math. 1351(1988), 146-158, Springer-Verlag.

[51] Hayman, W. K., Picard values of meromorphic functions and their derivatives, Ann. of Math., II. Ser. 70(1959), 9-42.

[52] Hayman, W. K., Meromorphic functions, Oxford: Clarendon Press, 1964.

[53] Hayman, W. K., Waring's problem für analytische Funktionen, Bayer. Akad. Wiss. Math.-Natur. Kl. Sitzungsber 1984, (1985), 1-13.

[54] He, Y. Z. & Xiao, X. Z., Algebroid functions and ordinary differential equations (Chinese), Science Press, Beijing, 1988.

[55] Hille, E., Analytic function theory II, Ginn and Company, 1962.

[56] Hsia, L. C., Closure of periodic points, preprint.

[57] Hu, P. C., Value distribution and admissible solutions of algebraic differential equations, J. of Shandong University (2)28(1993), 127-133.

[58] Hu, P. C., An extension of Hayman's inequality, J. of Math.(PRC) (4)10(1990), 405-412.

[59] Hu, P. C., Value distribution theory and complex dynamics, PhD thesis, The Hong Kong University of Science and Technology, 1996.

[60] Hu, P. C. & Mori, S., Elimination of defects of non-Archimedean holomorphic curves, Proc. of ISAAC'99 Congress.

[61] Hu, P. C. & Yang, C. C., Uniqueness of meromorphic functions on $\mathbb{C}^m$, Complex Variables 30(1996), 235-270.

[62] Hu, P. C. & Yang, C. C., Value distribution theory of p-adic meromorphic functions, Izvestiya Natsionalnoi Academii Nauk Armenii (National Academy of Sciences of Armenia) 32 (3) (1997), 46-67.

[63] Hu, P. C. & Yang, C. C., A unique range set of p-adic meromorphic functions with 10 elements, Acta Math. Viet. 24 (1) (1999), 95-108.

[64] Hu, P. C. & Yang, C. C., The Cartan conjecture for p-adic holomorphic curves, Izvestiya Natsionalnoi Akademii Nauk Armenii. Matematika 32 (4) (1997), 45-80.

[65] Hu, P. C. & Yang, C. C., Malmquist type theorem and factorization of meromorphic solutions of partial differential equations, Complex Variables 27(1995), 269-285.

[66] Hu, P. C. & Yang, C. C., Further results on factorization of meromorphic solutions of partial differential equations, Results in Mathematics 30(1996), 310-320.

[67] Hu, P. C. & Yang, C. C., The Second Main Theorem for algebroid functions of several complex variables, Math. Z. 220(1995), 99-126.

[68] Hu, P. C. & Yang, C. C., Differentiable and complex dynamics of several variables, Kluwer Academic Publishers, 1999.

[69] Hu, P. C. & Yang, C. C., Fatou-Julia theory in high dimensional spaces, International Workshop on The Value Distribution Theory and Its Applications, Hong Kong, 1996; Bull. of Hong Kong Math. Soc. 1 (2) (1997), 273-287.

[70] Hu, P. C. & Yang, C. C., Notes on second main theorems, Complex Variables 37 (1-4) (1998), 251-277.

[71] Hu, P. C. & Yang, C. C., A note on Shiffman's conjecture, preprint.

[72] Hu, P. C. & Yang, C. C., The "abc" conjecture over function fields, preprint.

[73] Hu, P. C. & Yang, L. Z., Admissible solutions of algebraic differential equations, J. of Shandong University (1)26(1991), 19-25.

[74] Khrennikov, A., p-adic valued distributions in mathematical physics, Math. and Its Appl. 309, Kluwer Academic Publishers, 1994.

[75] Khrennikov, A., Non-Archimedean analysis: quantum paradoxes, dynamical systems and biological models, Kluwer Academic Publishers, 1997.

[76] Koblitz, N., p-adic analysis: a short course on recent work, Cambridge Univ. Press, 1980.

[77] Koenigs, G., Recherches sur les intégrales de certaines équations fonctionnelles, Ann. Sci. École Norm. Sup. (3rd series) 1(1884), pp. Supplément 3-41.

[78] Krasner, M., Prolongement analytique uniforme et multiforme dans les corps valués complets: éléments analytique, préliminaires du théorème d'unicité, C. R. A. S. Paris, A 239 (1954), 468-470.

[79] Krasner, M., Rapport sur le prologement analytique dans les corps valués complets par la méthode des éléments analytiques quasi-connexes, Bull. Soc. Math. France, Mém. 39-40 (1974), 131-254.

[80] Laine, I., On the behaviour of the solutions of some first order differential equations, Ann. Acad. Sci. Fenn. A I 497(1971).

[81] Laine, I., Admissible solutions of some generalized algebraic differential equations, Publ. Univ. Joensun. Ser. B 10(1974).

[82] Lang, S., Fundamentals of Diophantine geometry, Berlin Heidelberg New York: Springer, 1983.

[83] Lang, S., Introduction to complex hyperbolic spaces, Springer-Verlag, 1987.

[84] Lang, S., Old and new conjectured Diophantine inequalities, Bull. Amer. Math. Soc. 23 (1990), 37-75.

[85] Langley, J. K., On differential polynomials and results of Hayman and Doeringer, Math. Z. 187(1984), 1-11.

[86] Li, P. & Yang, C. C., On the unique range sets of meromorphic functions, Proc. Amer. Math. Soc. 124 (1996), 177-185.

[87] Li, P. & Yang, C. C., Some further results on the unique range sets of meromorphic functions, Kodai Math. J. 18 (1995), 437-450.

[88] Malmquist, J., Sur les fonctions á un nombre fini de brances satisfaisant á une érentielle du premier order, Acta Math. 42(1920), 433-450.

[89] Mason, R. C., Equations over function fields, Lecture Notes in Math. 1068 (1984), 149-157, Springer.

[90] Mason, R. C., Diophantine equations over function fields, London Math. Soc. Lecture Note Series, Vol. 96, Cambridge University Press, United Kingdom, 1984.

[91] Mason, R. C., The hyperelliptic equation over function fields, Math. Proc. Cambridge Philos. Soc. 93(1983), 219-230.

[92] Milloux, H., Les fonctions méromorphes et leurs dérivées, Paris, 1940.

[93] Mokhon'ko, A. Z., Estimates for Nevanlinna characteristic of algebroid functions, I, Function Theory, Functional Analysis and Their Applications 33(1980), 29-55; II, 35(1981), 69-73 (Russian).

[94] Mori, S., Elimination of defects of meromorphic mappings of $\mathbb{C}^m$ into $\mathbb{P}^n(\mathbb{C})$, Ann. Acad. Sci. Fenn. 24(1999).

[95] Mori, S., Elimination of defects of meromorphic mappings by small deformation, Proc. of US-Japan Seminar of the Delaware Conference, Kluwer Acad. Publ. (1999).

[96] Mues, E. & Reinders, M., Meromorphic function sharing one value and unique range sets, Kodai Math. J. 18(1995), 515-522.

[97] Mues, E. & Steinmetz, N., Meromorphe funktionen, die mit ihrer ableitung werte teilen, Manuscripta Math. 29(1979), 195-206.

[98] Nevanlinna, R., Le théorème de Picard-Borel et la théorie des fonctions meromorphes, Gauthier Villars, Paris, 1929; Reprint, Chelsea, New York, 1974.

[99] Nochka, E. I., Defect relations for meromorphic curves, Izv. Akad. Nauk. Moldav. SSR Ser. Fiz.-Teklam. Mat. Nauk 1(1982), 41-47.

[100] Nochka, E. I., On a theorem from linear algebra, Izv. Akad. Nauk. Moldav. SSR Ser. Fiz.-Teklam. Mat. Nauk 3(1982), 29-33.

[101] Nochka, E. I., On the theory of meromorphic curves, Dokl. Akad. Nauk SSSR 269(1983), 377-381.

[102] Noguchi, J., Holomorphic curves in algebraic varieties, Hiroshima Math. J. 7(1977), 833-853.

[103] Ostrovskii, I., Pakovitch, F. & Zaidenberg, M., A remark on complex polynomials of least deviation, IMRN International Mathematics Research Notices, No. 14, 1996, 699-703.

[104] Pakovitch, F., Sur un problème d'unicité pour les polynômes, prépublication, Institut Fourier de Mathématiques, 324, Grenoble, 1995.

[105] Picard, E., Traité d'analyse II, Gauthier-Villars, Paris, 1925.

[106] Poole, E. G. C., Introduction to the theory of linear differential equations, Oxford University Press, Oxford, 1936.

[107] Reinders, M., Unique range sets ignoring multiplicities, Bull. Hong Kong Math. Soc. 1(1997), 339-341.

[108] Rellich, F., Über die ganzen Lösungen einer gewöhnlichen Differentialgleichungen erster Ordnung, Math. Ann. 117(1940), 587-589.

[109] Rosenbloom, P. C., The fix-points of entire functions, Medd. Lunds Univ. Mat. Sem. Tome Suppl. M. Riesz (1952), 186-192.

[110] Ru, M., The second main theorem with moving targets on parabolic manifolds, preprint.

[111] Ru, M., A note on p-adic Nevanlinna theory, preprint.

[112] Ru, M. & Stoll, W., The second main theorem for moving targets, J. of Geom. Analysis 1(1991), 99-138.

[113] Ru, M. & Stoll, W., The Cartan conjecture for moving targets, Proc. of Symposia in Pure Math. 52(1991), Part 2, 477-508.

[114] Ru, M. & Vojta, P., Schmidt's subspace theorem with moving targets, Invention Math. 127(1997), 51-65.

[115] Ru, M. & Wong, P. M., Integral points of $\mathbb{P}^n - \{2n+1$ hyperplanes in general position $\}$, Invent. Math. 106 (1991), 195-216.

[116] Rubel, L. A. & Yang, C. C., Values shared by an entire function and its derivative, Lecture Notes in Math. Vol. 599 (1977), 101-103, Springer.

[117] Schlickewei, H. P., The p-adic Thue-Siegel-Roth-Schmidt theorem, Arch. Math. 29(1977), 267-270.

[118] Schmidt, W. M., Simultaneous approximation to algebraic numbers by elements of a number field, Monatsh. Math. 79(1975), 55-66.

[119] Schmidt, W. M., Diophantine approximation, Lecture Notes in Math. 785(1980), Springer.

[120] Schmieder, G., A uniqueness theorem for complex polynomials , preprint.

[121] Shirosaki, M., Another proof of the defect relation for moving targets, Tôhoku Math. J. 43(1991), 355-360.

[122] Shiffman, B., On holomorphic curves and meromorphic maps in projective space, Indiana Univ. Math. J. 28(1979), 627-641.

[123] Shnirelman, L. G., O funkcijah v normirovannyh algebraičeski zamknutyh telah (on functions in normed algebraically closed division rings), Izvestija AN SSSR 2 (1938), 487-498.

[124] Smiley, L. M., Geometric conditions for unicity of holomorphic curves, Contemporary Math. 25(1983), 149-154.

[125] Steinmetz, N., Eigenschaften eindeutiger Lösungen gewöhnlichen Differentialgleichungen im Komplexen, Karlsurche; dissertation, 1978.

[126] Steinmetz, N., Eine Verallgemeinerung des zweiten Nevanlinnaschen Hauptsatzes, J. Reine Angew. Math. 368(1986), 134-141.

[127] Stoll, W., An extension of the theorem of Steinmetz- Nevanlinna to holomorphic curves, Math. Ann. 282(1988), 185-222.

[128] Stoll, W., Value distribution theory for meromorphic maps, Aspects of Math. E7, Vieweg, 1985.

[129] Toda, N., On the growth of meromorphic solutions of an algebraic differential equations, Proc. Japan Acad. 60 Ser. A (1984), 117-120.

[130] Toda, N., On the functional equation $\sum_{i=0}^{p} a_i f_i^{n_i} = 1$, Tohoku, Math. J. 23(1971), 289-299.

[131] Tu, Z. H., A Malmquist type theorem of some generalized higher order algebraic differential equations, Complex Variables 17(1992), 265-275.

[132] Van der Waerden, B. L., Algebra, Vol. 2, 7-th ed., Springer-Verlag, 1991.

[133] Vojta, P., Diophantine approximation and value distribution theory, Lecture Notes in Math. 1239, Springer, 1987.

[134] Vojta, P., Roth's theorem with moving targets, International Math. Research Notices, (1996), 109-114.

[135] Wittich, H., Neuere Untersuchungenrn Über eindeutige analytische Funktionen, Springer, Berlin-Göttingen-Heidelberg, 1955.

[136] Wu, H., The equidistribution theory of holomorphic curves, Princeton University Press, Princeton, 1970.

[137] Yang, C. C., A problem on polynomials, Rev. Roum. Math. Pures et Appl. 22(1977), 595-598.

[138] Yang, C. C. & Hu, P. C., A survey on p-adic Nevanlinna theory and its applications to differential equations, Taiwanese J. of Math. 3 (1) (1999), 1-34.

[139] Yang, L., Multiple values of meromorphic functions and its combinations, Acta Math (in Chinese) 14 (1964), 428-437.

[140] Yi, H. X., On a question of Gross, Sci. China, Ser. A 38(1995), 8-16.

[141] Yi, H. X., The unique range sets of entire or meromorphic functions, Complex Variables 28(1995), 13-21.

[142] Yi, H. X., Meromorphic functions that share one or two values, Complex Variables 28(1995), 1-11.

[143] Yi, H. X. & Yang, C. C., On the uniqueness theory of meromorphic functions (in Chinese), Science Press, Beijing, 1996.

[144] Yosida, K., On algebroid-solutions of ordinary differential equations, Japan J. Math. 10(1934), 119-208.

[145] Yu, K. W. & Yang, C. C., On the meromorphic solutions of $\sum_{j=1}^{p} a_j(z)f_j^{k_j}(z) = 1$, preprint.

Symbols

291

Index

293